ECONOMIC MARKET DESIGN AND PLANNING FOR ELECTRIC POWER SYSTEMS

IEEE Press
445 Hoes Lane
Piscataway, NJ 08854

A complete list of titles in the IEEE Press Series on Power Engineering appears at the end of this book.

ECONOMIC MARKET DESIGN AND PLANNING FOR ELECTRIC POWER SYSTEMS

Edited by

JAMES MOMOH
LAMINE MILI

IEEE PRESS SERIES ON POWER ENGINEERING
Mohamed E. El-Hawary, *Series Editor*

IEEE
Celebrating 125 Years
of Engineering the Future

WILEY
A JOHN WILEY & SONS, INC., PUBLICATION

demand bids and partial capacity cost compensations for day-ahead electricity auctions.

- Chapter 5, which is authored by R. Mookherjee, B. F. Hobbs, T. L. Friesz, and M. A. Rigdon, puts forward a dynamic game theoretic model of oligopolistic competition in spatially distributed electric power markets having a 24-hour planning horizon.
- Chapter 6, which is authored by G. Deltas and C. Hadjicostis, investigates the interaction between system availability/reliability, economic restructuring, and regulating constraints.
- Chapter 7, which is authored by J. A. Momoh, P. Fanara Jr., H. Kurban, and L. J. Iwarere, introduces economic, technical, modeling and performance indices for reliability measures across boundary disciplines.
- Chapter 8, which is authored by L. Mili and K. Dooley, investigates the decision making processes associated with the risk assessment and management of bulk power transmission systems under a unified methodological framework of security and survivability objectives.
- Chapter 9, which is authored by J. McCalley, R. Kumar, V. Ajjarapu, O. Volij, H. Liu, L. Jin, and W. Shang, introduces models for power transmission system enhancement by integrating economic analysis of the transmission cost to accommodate an informed business decision. Finally,
- Chapter 10, which is authored by J. Momoh, elaborates on next generation optimization for electric power systems.

We are grateful to Katherine Drew from ONR for providing financial and moral support of this initiative, Ed Zivi from ONR for providing the benchmarks, colleagues from ONR and NSF for providing a fostering environment to this work to grow and flourish. We thank former NSF Division Directors, Dr. Rajinder Khosla and Dr. Vasu Varadan, who provided seed funding for this initiative. We also thank Dr. Paul Werbos and Dr. Kishen Baheti from NSF for facilitating interdisciplinary discussions on power systems reliability and education. We are thankful to NSF-DUE program directors, Prof. Rogers from the NSF Division of Undergraduate Education and Dr. Bruce Hamilton of NSF BES Division, and.

We acknowledge graduate students from Howard University and Virginia Tech for helping us to put together this book.

CONTRIBUTORS

Venkataramana Ajjarapu
Professor
Department of Electrical and Computer Engineering
Iowa State University
Ames, IA

William E. Blankson
Congestion Analyst
American Electric Power
Columbus, OH

Ying Chen
Market Risk Analyst
Edison Mission Marketing and Trading
Boston, MA

George Deltas
Associate Professor
Department of Economics
University of Illinois, Urbana-Champaign
Urbana, IL

Kevin Dooley
Professor
Department of Supply Chain Management
W. P. Carey School of Business
Arizona State University
Tempe, AZ

Philip Fanara, Jr.
Professor of Finance
Department of Finance, International Business and Insurance, School of Business
Howard University
Washington, DC

Terry L. Friesz
Marcus Chaired Professor of Industrial Engineering
Department of Industrial and Manufacturing Engineering
Pennsylvania State University
University Park, PA

Alfredo Garcia
Associate Professor
Department of Systems and Information Engineering
University of Virginia
Charlottesville, VA

Christoforos Hadjicostis
Associate Professor
Coordinated Science Laboratory and Department of Electrical and Computer Engineering
Urbana, IL

Benjamin F. Hobbs
Theodore M. and Kay W. Schad Professor in Environmental Management
Department of Geography and Environmental Engineering
Johns Hopkins University
Baltimore, MD

L. Jide Iwarere
Associate Professor
Department of Finance, International Business and Insurance, School of Business
Howard University
Washington, DC

Licheng Jin
Network Applications Engineer
California Independent System Operator Corporation
Folsom, CA

Ratnesh Kumar
Professor
Department of Electrical and Computer Engineering
Iowa State University
Ames, IA

Haydar Kurban
Associate Professor
Department of Economics
Howard University
Washington, DC

cutting courses are Financial Engineering, Power Market and Cost Benefit Analysis and Power Environment, Advances System Theory and Computational Intelligence, Power Economics, and Computational Tools for Deregulated Power Industry.

We recommend that all multidisciplinary courses use canonical benchmark systems for verification/validation of developed theories and tools. When possible, the courses should be co–taught by professors across disciplines. To promote broader dissemination of knowledge and understanding, courses should be developed for both undergraduate and graduate students. These courses should also be made available through workshops and lectures, electronically, and should be posted on the host institution website. Furthermore, an assessment strategy should be developed and applied on an ongoing basis to ensure sustainability of the program and its impact on attracting students and improving workforce competencies in promoting or developing an efficient and reliable power systems enterprise.

1.2 POWER SYSTEM CHALLENGES

The EPNES initiative is designed to engender major advances in the integration of new concepts in control, modeling, component technology, and social and economic theories for electrical power networks' efficiency and security. It challenges educators and scientists to develop new interdisciplinary research-based curricula and pedagogy that will motivate students' learning and increase their retention across affected disciplines. As such, interdisciplinary research teams of engineers, scientists, social scientists, economists, and environmental experts are required to collaborate on the grand challenges. These challenges include but are not limited to the following categories.

A. ***Systems and Security***

- Advanced Systems Theory: Advanced theories and computer-aided modeling tools to support and validate complex modeling and simulation, advanced adaptive control theory, and intelligent-distributed learning agents with relevant controls for optimal handling of systems complexity and uncertainty.
- Robust Systems Architectures and Configurations: Advanced analytical methods and tools for optimizing and testing configurations of functional elements/architectures to include control of power electronics and systems components, complexity analysis, time-domain simulation, dynamic priority load shedding for survivability, and gaming strategies under uncertainties.
- Security and High-Confidence Systems Architecture: New techniques and innovative tools for fault-tolerant and self-healing networks, situational awareness, smart sensors, and analysis of structural changes. Applications include adaptive control algorithms, systems and component security, and damage control systems for continuity of service during major disruptions.

B. ***Economics, Efficiency and Behavior***

- Regulatory Constraints and Incentives: New research ideas that explore the influence of regulations on the economics of electric networks.

- Risk Assessment, Risk Perceptions, and Risk Management: Novel methods and applications for linking technical risk assessments, public risk perceptions, and risk management decisions.
- Public Perceptions, Consumer Behavior, and Public Information: Innovative approaches that improve public perception of electric power systems through increased publicity and education about the electric power networks.

C. ***Environmental Issues***

- Environmental Systems and Control: Innovative environmental sensing techniques for system operation and maintenance, improvements in emission control technologies, and/or network operation for minimization of environmental impact, among others. The interplay of these factors with the other topics in this solicitation is a requirement.
- Technology for Global Sustainability: Cross-disciplinary efforts that contribute to resource and environmental transitions that are needed to ensure long-term sustainability of global economic growth.

D. ***New Curricula and Pedagogy***

- New Curricula and Pedagogy: Innovative and integrated curricula and pedagogy incorporating advanced system theory, economics, and other social science perspectives, as well as environmental science, policy, and technical issues are desirable. New and innovative curricula to raise interest levels of both students and faculty, and better prepare the workforce for the power industry of the future are also desirable. Pedagogy and curricula must be developed at both the undergraduate and graduate students' level.

E. ***Benchmark Test Systems***

- Benchmark Test Systems: These are required for validation of models, advanced theories, algorithms, numerical and computational efficiency, distributed learning agents, robust situational awareness for hierarchical and/or decentralized systems, adaptive controls, self-healing networks, and continuity of service despite faults. A Navy power systems baseline ship architecture is available at the United States Naval Academy, website, http://www.usna.edu/EPNES.
- Both civil and Navy test beds will be available from the Howard University website: http://www.cesac.howard.edu/.

1.2.1 The Power System Modeling and Computational Challenge

Today, power system architectures are being made more complex as they are enhanced with new grid technology or new devices such as Flexible AC Transmission System devices (FACTS), Distributed Generation (DG), Automatic Voltage Regulator (AVR), and advanced control systems. The introduction of these systems will affect overall network performance. Performance assessments to be done can be of two types, either static and dynamic, or quasi-static dynamic behaviors under different (N-1) and (N-2) contingencies.

Several methods are commonly used for evaluating the performance of power systems under different conditions. For small and large disturbances, the methods include Lyapunov stability analysis, Power flow, Bode plots, reliability stability assessment and other frequency response techniques. These tools allow us to determine the various capabilities of the power system in an online or offline mode.

The tools will enable us to achieve better performance analyses, even taking into account other interconnecting networks on the power systems. These can include wireless communication devices, distributed generation and control devices such as generation schedulers, phase shifters, tap changing transformers, and FACTS devices. In addition to new modeling techniques that incorporate uncertainties, advanced simulation tools are needed.

1.2.2 Modeling and Computational Techniques

Develop techniques that consider all canonical devices, as well as new devices and technologies for power systems, such as FACTS and Distributed Generation, transformer taps, phase shifters with generation, load, transmission lines, DC/AC converters and their optimal location within the power system. The development of new load flow programs for DC/AC systems for ship and utility systems that take into consideration the peculiarities of both systems is desirable.

1.2.3 New Curriculum that Incorporates the Disciplines of Systems Theory, Economic and Environmental Science for the Electric Power Network

EPNES supports research that is performed in interdisciplinary groups with the objective of generating new concepts and approaches stimulated by the interaction of diverse disciplines. This will foster the development of pedagogy and education material for undergraduate and graduate level students. The initiative supports education, outreach and curriculum improvements to most effectively educate the future workforce via an interdisciplinary research approach of significant intellectual merit and broader impacts to the country as well as the global scientific community.

1.3 SOLUTION OF THE EPNES ARCHITECTURE

The explanation of the interaction of different phases of the EPNES framework is presented in terms of sustainability, survivability, efficiency and behavior. It satisfies the economic, technical and environmental constraints and other social risk factors under different contingencies. It is modeled using advanced systems concepts and accommodates new technology and testable data using the utility and military systems.

1.3.1 Modular Description of the EPNES Architecture

Module 1: High Performance Electric Power systems (HPEPs)
This is the ultimate automated power systems architecture to be built with the attributes of survivability, security, affordability, and sustainability. The tools developed in the modules below are needed to achieve the proposed HPEPs.

Module 2: Mathematical Analysis Toolkit

This module is dedicated to providing models of devices using the elements of advanced system theory and concepts, intelligent distributed learning agents and controls for optimal handling of systems complexity, robust architectures and reconfiguration, and secure, high confidence systems architecture. The toolkit will require development of new techniques and innovative tools for the optimization and testing of functional elements for electronics and systems components, complexity analysis, time domain simulation, dynamic priority load shedding for survivability, and gaming strategies under uncertainties. Additionally, for secured and high confidence systems architectures, these tools develop new techniques and analysis techniques for self-healing networks, situational awareness, smart sensors, and structural changes. This toolkit will also utilize adaptive controls, component security and damage control systems for continuity of service during major disruptions.

Module 3: Behavior and Market Model Tool

This module is to be designed based on the design parameters and cost data from the mathematical analysis tool, in order to define the economic and public perception for HPEPS. The module computes regulatory constraints and incentives that economically influence the operation of electric networks. The module provides innovative methods for linking risk assessments, public perceptions and risk management decisions. The computation of risk indices based on uncertainties and adequate pricing mechanisms is performed in this module. The computation of cost benefit analysis of different strategies is also to be included.

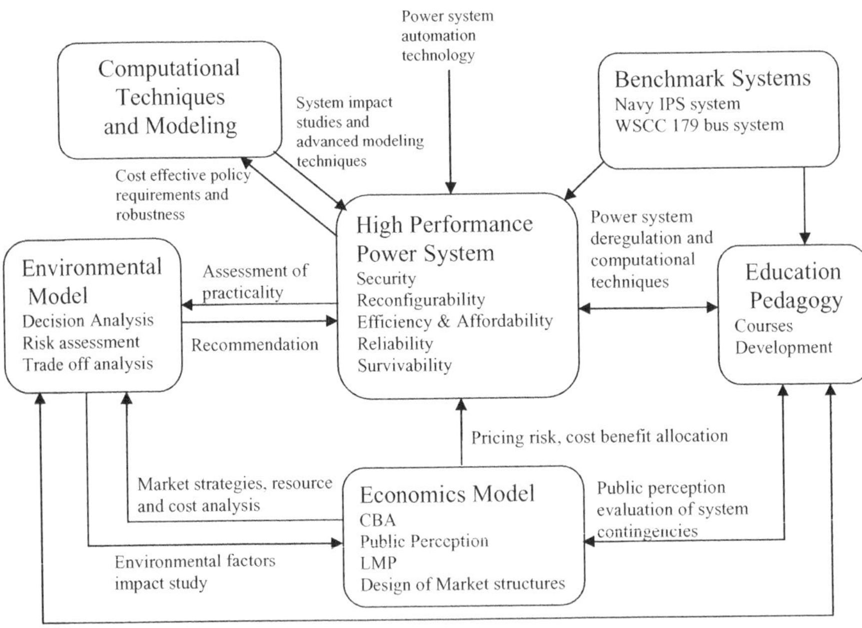

Figure 1.2. Modular representation of the EPNES framework.

Module 4: Environment Issues and Control

This module utilizes innovative environmental sensing techniques for system operation and maintenance. Improvements in emission controls techniques for minimization of environmental impact are required. To achieve this objective, several indices are needed to compute the environmental constraints that will be included in the global optimization for developing the risk assessment and cost-benefit analysis tools. The trade-off computed in this module will be used to determine new input for optimizing the HPEPS.

Module 5: Benchmark Test System

The validation of the models, advanced algorithms, numerical methods and computational efficiency will be done using the tools developed in the previous modules using the benchmark systems. Representative test beds and some useful associated models will be described in a later section of the paper. Different performance parameters or attributes of the HPEPS will be analyzed using appropriate models based on hierarchical and decentralized control systems, to ensure continuity of service and abilities in the design and operation of the proposed power system.

1.3.2 Some Expectations of Studies Using EPNES Benchmark Test Beds

Two test beds, involving civilian and military ship power systems, are proposed to support the evaluation of the performance, behavior, efficiency and security of the power systems as designed. The first is a representative civilian utility system which can be a US utility system, or the EPRI/WSCC 180 bus system. Also, the US Navy benchmark Integrated Power System (IPS) system designed by Professor Edwin Zivi of the US Navy Academy is a representative Navy testbed example. Both systems consist of generator models, transmission networks and interties, various types of loads and controls and new technology such as FACTs, AC/DC transmission, distributed generation and other control devices. To ensure that all of the elements of EPNES are considered by the researchers, including the issues of environmental constraints (such as emission from generators, plants or other devices), public perception, and pricing and cost parameters for economic and end-risk assessment.

Stemming from studies done on the benchmark systems, we plan to assess the security and reliability of the systems in different scenarios. For the economics studies, we plan to assess the cost benefit analysis acquisition tradeoff (cost versus security) and also determine the optimum market structures that will enhance the efficiency of the power system production and delivery. We plan to evaluate the risk assessment and public perception of different operational planning scenarios, given the environmental constraints. The 'why' and 'how' of the analysis of multiple objectives and constraints will be analyzed/visualized using the advanced optimization techniques. We also expect that researchers will take advantage of distributed controls and hierarchical structures to handle the challenges of designing the best automation scheme for future power systems that will adapt itself to different situations, reconfigure itself, sustain faults and still remain reliable and affordable.

1.4 IMPLEMENTATION STRATEGIES FOR EPNES

1.4.1 Performance Measures

To design reliable and secure power systems of the future, a multi-function performance metric is needed. In EPNES, we want the development of tools for measuring reliability, stability and security, affordability, sustainability and behavior of the power system under duress while taking into account environmental issues, public perception, and social impacts. Below is a summary of some of the key objectives in the EPNES framework.

1.4.2 Definition of Objectives

1. Survivability, in general terms, can be defined as the ability of a system, subsystem, or hardware component to withstand the effects of harsh disturbances, adverse environmental conditions, and/or structurally damaging natural or man-made effects. The goal of enhancing survivability is to reduce technical and human risks, while maintaining primary operational coordination, communication, and control functions during contingencies, as well as maintaining system structural integrity for autonomous healing with minimum disruptions. Thus, enhancing survivability is an indirect approach to improved risk levels for operation of the network under anomalies of loadings, man-made attacks, outages, cascading ruptures, effects of nature, and other source of disturbances.
2. Affordability is the process of minimizing system costs subject to the cost constraints associated with all needed components and services of associated resources. In the framework of this work, the costs associated with a high performance power system include installation of infrastructure, fuel and energy requirements, damage control in post fault scenarios, as well as the costs associated with implementing new or old control measures. Affordability is used to meet a setpoint performance requirement at a sufficient level of quality service (an aspect of public perception) and response of a service in need, when needed and regardless of the price (demand-supply balance). Who is willing to pay? To answer this question, research is needed to model and evaluate public perception and social impacts of decisions.
3. Efficiency of electric power networks has technical and market-driven economic components. This includes the cost of ancillary services that are required to sustain the operation of the power network. Efficiency is often seen as a performance measure of cost minimization subject to the constraints of fuel prices, value-added bidding strategies for competing resources, and effective use of resources in normal operation as well as during system faulted conditions. The cost minimization process should be extended to include the constraints on the environment in the economic model of the network.

4. Sustainability is an index that provides insight as to how well the system can maintain a relatively safe and economical margin of reliability, grid/network integrity, and system capability to function under conditions of shock, isolation, or heavy loading. In the short term, robust power network controls should provide suitable levels of stability and reliability to prevent localized brown-outs/black-outs, cascading failures, or system-wide interruption of service. This is true in the long term but requires emphasis on economic and environmental constraints in a competing market of scarce resources.

1.4.3 Selected Objective Functions and Pictorial Illustrations

This section broadly specifies the nature of the objective functions for survivability, affordability, efficiency, and sustainability of the electric power network. Accurate models for the various performance indices as well as market dynamics are needed. Overall, these objectives and several others will form the backbone of a comprehensive computational tool that will be used to solve the new breed of electric power networks operating under various conditions. The mathematical models for the selected objectives are summarized below.

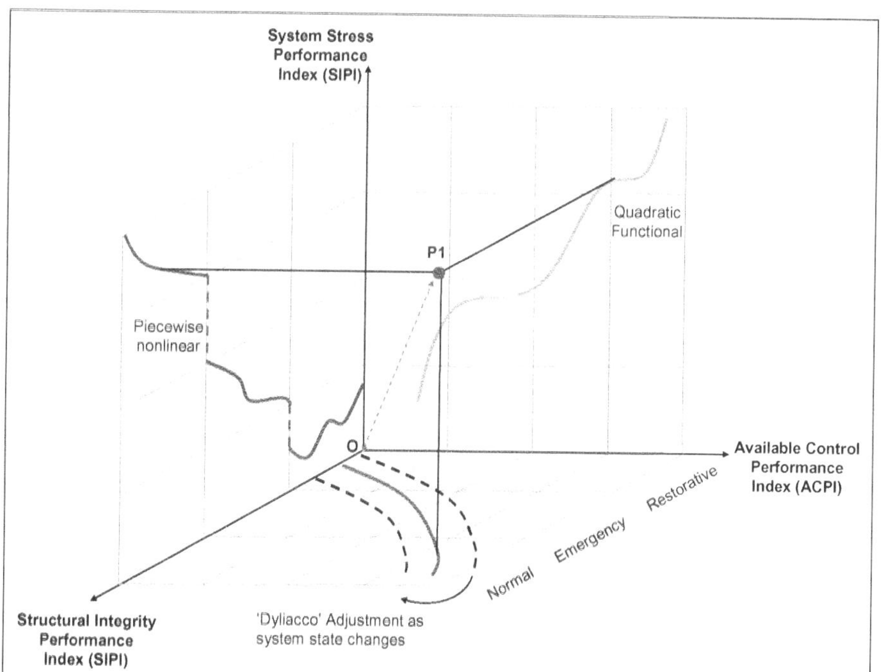

Figure 1.3. Sketch of the survivability objective function.

1.4.3.1 Survivability Objective This objective characterizes the ability of the system or sub-system to be operated with minimum disruption using available controls to maintain structural integrity of the stressed network. The objective function (depicted in Figure 1.3) may be stated as:

$$\text{Minimize} \qquad F_{SV} = \sum_{t=0}^{T}\sum_{i=1}^{NS}\{\omega^T(k_{SS,i}SSPI_i(t)+k_{SI,i}SIPI_i(t)+k_{AS,i}ASPI_i(t))\}$$

where:

$SSPI_i(t)$: System Stress Performance Index
$SIPI_i(t)$: Structural Integrity Performance Index
$ACPI_i(t)$: Available Control performance Index
ω^T : Weightings or correction vector for the respective indices
$k_{j,i}$: Normalizing or model approximation for $j \in \{SS, SI, AC\}$
$t \in \{0, T\}$: Time frame
$i \in \{1, NS\}$: Set of subsystems in the network

1.4.3.2 Affordability Objective This objective attempts to minimize the cost of operating the network subject to the budgetary considerations. The objective function (depicted in Figure 1.4) may be stated as:

$$\text{Minimize} \qquad F_{AF} = \left|\sum_{t=0}^{T}\left\{\sum_{i=1}^{NS} a_i^T(C_{CM,i}(t)+C_{FS,i}(t)+C_{TI,i}(t)) - \mu^T(t)\hat{F}_{SB}(t)\right\}\right|$$

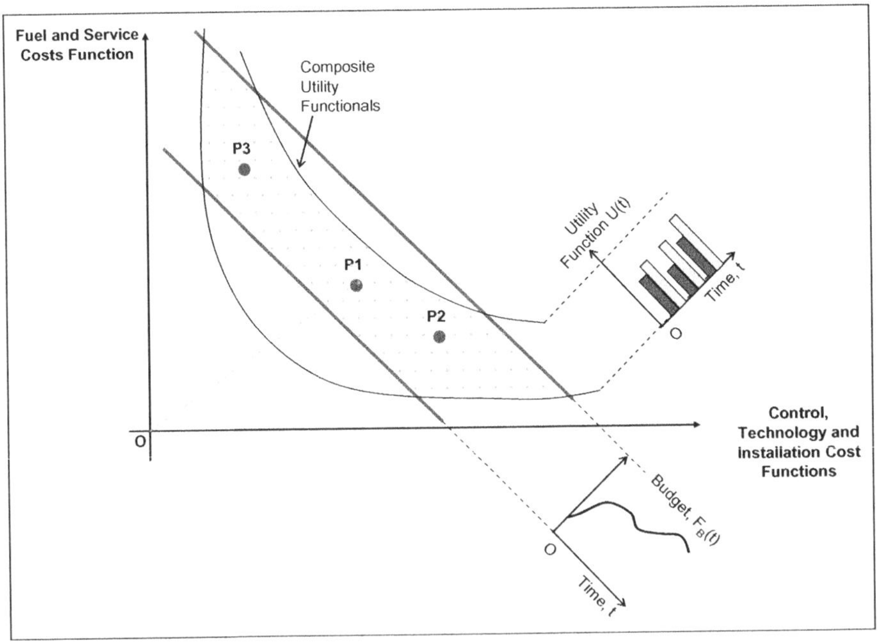

Figure 1.4. Sketch of the affordability objective function.

where:

$C_{CM,i}$: Control and Maintenance costs
$C_{FS,i}$: Fuel and Service costs
$C_{TI,i}$: New Technology and Installation costs
a_i^T : vector of weights and correction multipliers
μ^T : Willingness-to-Pay Penalty functions
$i \in \{1, NS\}$: Set of subsystems in the network
$t \in \{0, T\}$: Time frame

1.4.3.3 *Efficiency Objective* This objective characterizes the cost-effective usage of energy, control, and ancillary support services in the electric power networks and as such, it has technical and market-driven economic components. The objective function (depicted in Figure 1.5) may be stated as:

$$\text{Minimize} \qquad F_{AE} = \sum_{t=0}^{T}\sum_{i=1}^{NS}\left\{\frac{\Delta_i^T}{\omega_{C_i}}(C_{AS,i}+C_{FC,i}+C_{AC,i})(t)-\left(\Delta_i^T\lambda_i^T f_{\substack{budget\\ constraint}}\right)(t)\right\}$$

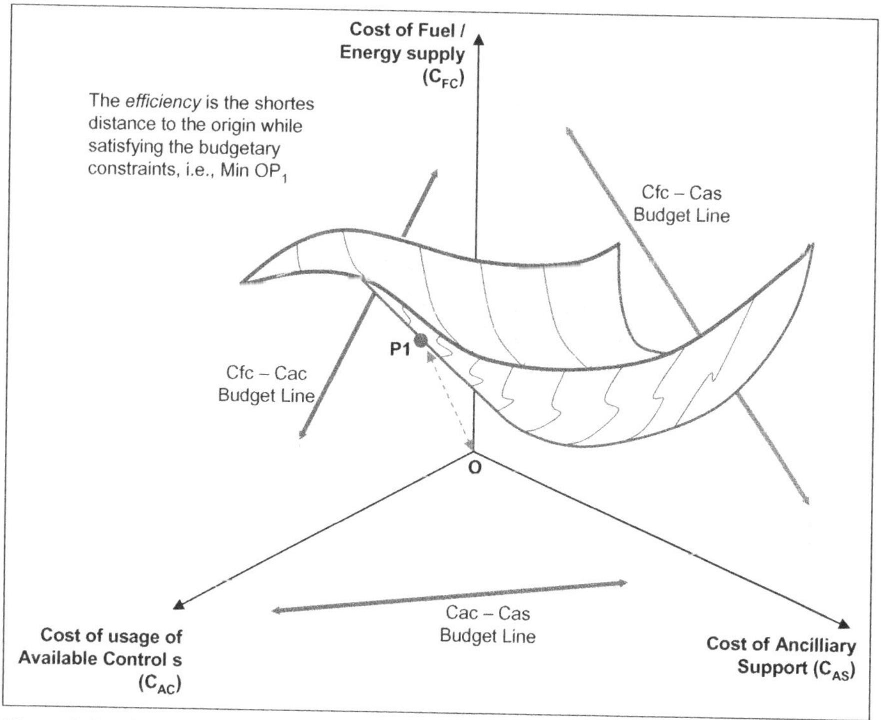

Figure 1.5. Sketch of the efficiency objective function.

where:

$C_{AS,i}$: Cost of Ancillary Service support
$C_{FC,i}$: Cost of Fuel / Energy
$C_{AC,i}$: Cost of Usage of Available Control options
Δ_i^T : Past and Present time span, $[t, t\text{-}1]$
ω_{C_i} : Scaling multipliers
λ_i^T : Penalty functions
$f_{\substack{budget\\constraint}}$: Budgetary constraints
$i \in \{1, NS\}$: Set of subsystems in the network
$t \in \{0, T\}$: Time frame

1.4.3.3 *Sustainability Objective* Sustainability, loosely stated as 'minimizing intervention,' is an objective that measures network capability relative to safe and economical margins of reliability, grid/network integrity, and system capability to function under conditions of shock, isolation, or heavy loading. The objective function (depicted in Figure 1.6) may be stated as:

$$\text{Minimize } F_{SU} = \sum_{t=0}^{T}\sum_{i=1}^{NS}\{k_1[\beta_{rel,i}(1-I_{rel,i}(t))+\beta_{sta,i}(1-I_{sta,i}(t))]\} + \sum_{t=0}^{T}\sum_{i=1}^{NS}k_2\left(CBS_{oper,i}+\mu_i^T h_{econ}(t)\right)$$

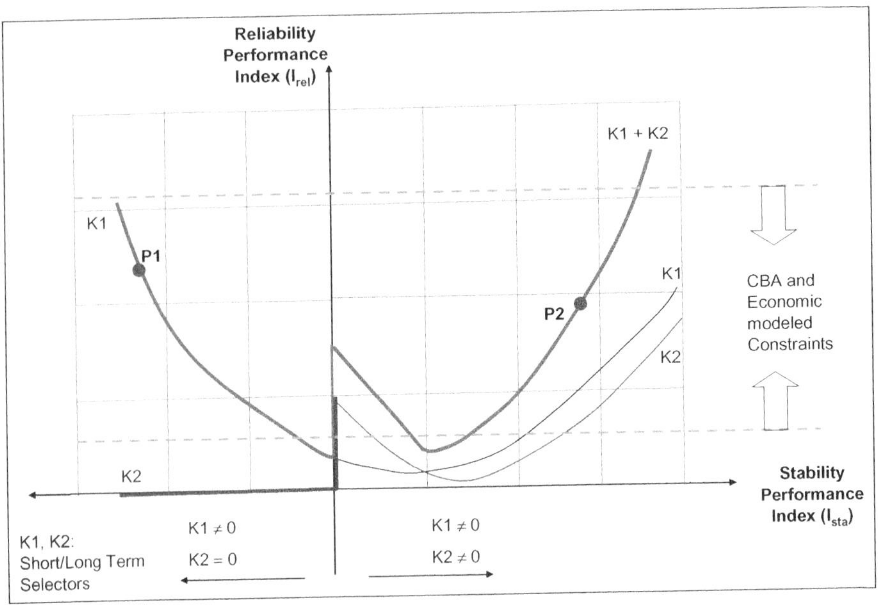

Figure 1.6. Sketch of the sustainability objective function.

where:

I_{rel} : Reliability index vector of the network
I_{sta} : Stability index vector of the network
β_{rel}, β_{sta} : Scaling multipliers for the index vectors
CBA_{oper} : Functional of Cost-Benefit for the operation of the network
$h_{econ}(t)$: Economic constraints (hard and soft)
μ_i^T : Penalty on the economic constraints
k_1, k_2 : Term selectors
$k \in \{0, 1\}$: Long term, short term values of k
$i \in \{1, NS\}$: Set of subsystems in the network
$t \in \{0, T\}$: Time frame

Finally, in an attempt to evaluate the constrained multi-objective functions, analytical hierarchical process and Pareto-optimal analysis could be used to assign priority and ranking to control options used in the general formulation of the optimal power flow problem. The next section of the chapter highlights topical areas of research towards this goal.

1.5 TEST BEDS FOR EPNES

1.5.1 Power System Model for the Navy

To build a High Performance Electric Power System (HPEPS) model for the U.S. Navy ship system, a detailed physical model and mathematical model of each component of the ship system is needed. For an integrated power system, at minimum, the generator model, the AC/DC converter, DC/AC inverter and various ship service loads need to be modeled. Because the Navy ship power system is an Integrated Power System (IPS), an AC/DC power flow program needs to be specially designed for the performance evaluation and security assessment of the naval ship system. Accurate contingency evaluation of the Naval Integrated Power System should be based on a comprehensive system model of the naval ship system.

Figure 1.7 is the AC generation and propulsion test-bed. It comprises the following elements:

- The prime mover and governor is a 150 Hp four-quadrant dynamometer system
- The synchronous machine (SM) is a Leroy Somer two bearing Alternator part number LSA432L7. It is rated for 59 kW (continuous duty) with an output line-to-line voltage of 520–590 V_{rms}. The machine is equipped with a brushless excitation system and a voltage regulator.
- The propulsion load consists of the propulsion power converter, induction motor, and load emulator:
  - A rectified, DC-link, inverter propulsion power converter

  - The propulsion motor is a 460 V_{rms} L-L, 37 kW, 1800 rpm, Baldor model number ZDM4115T-AM1 Induction Machine (IM).
  - The load emulator is a 37 kW four-quadrant dynamometer.
- The 15 KW ship service power supply (PS) consists of 480 V 3-phase AC diode rectifier bridge feeding a buck converter to produce 500 V DC. These converters provide the logical interconnection of the AC and the DC test-beds. In the future, an alternative, thyristor-based active rectifier converter may be available.
- A future pulsed load
- The harmonic filter (HF) is a wye-connected LC arrangement. The effective capacitance is 50 mF (which is implemented with two 660 V_{rms} 25 mF capacitors in series) and the design value of inductance is 5.6 mH (rated for a 40 A peak, without saturating).

Figure 1.7 also shows the DC zonal ship service distribution test-bed. It is composed of the following elements:

- Each 15 kW ship service power supply consists of a 480 V 3-phase AC diode rectifier bridge feeding a buck converter to produce 500 V DC. These converters provide the logical interconnection of the AC and DC testbeds. In the future, an alternative, thyristor-based active rectifier converter may be available.
- The 5 kW ship service converter modules convert 500 Vdc distribution power to intra-zone distribution of approximately 400 Vdc.
- The 5 kW ship service inverter modules convert the intra-zone 400 V dc to three phase 230 V AC powers.
- The Motor controller (MC) is a three-phase inverter rated at 5 kW.
- The constant power load (CPL) is a buck converter rated at 5 kW.

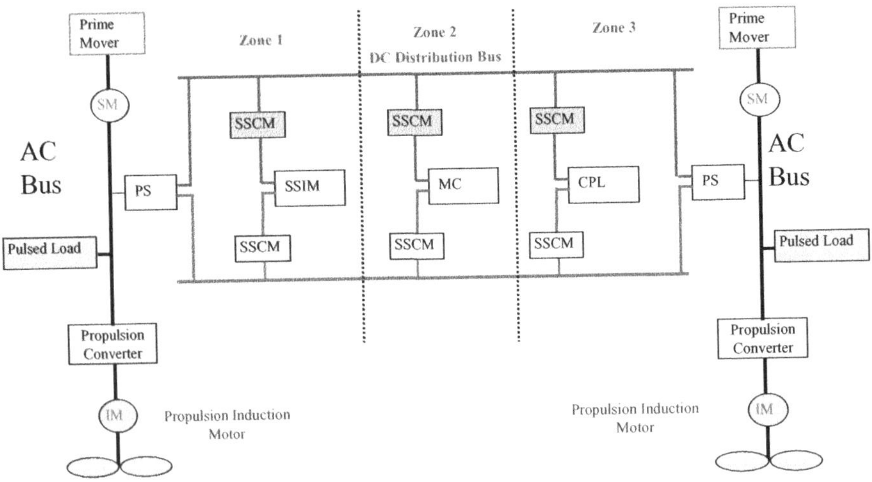

Figure 1.7. Navy Power System Topology.

1.5.2 Civil Testbed—179-Bus WSCC Benchmark Power System

The WSCC benchmark system contains 179 buses, 205 transmission lines, 58 generators, and 104 *equivalenced* loads on the high voltage transmission circuits. The system is operated at 230-, 345-, and 500-kV. Figure 1.8 shows a HV single line diagram of this system.

Also, embedded in this system are several control devices/options that include ULTC transformers, fixed series compensators, switchable series compensators, static tap changers/phase regulators, generation control, and 3-winding transformers. At 100 MVA System base, the total generation is 681.79 + j156.34 p.u. and the total load is 674.10 + j165.79 p.u.

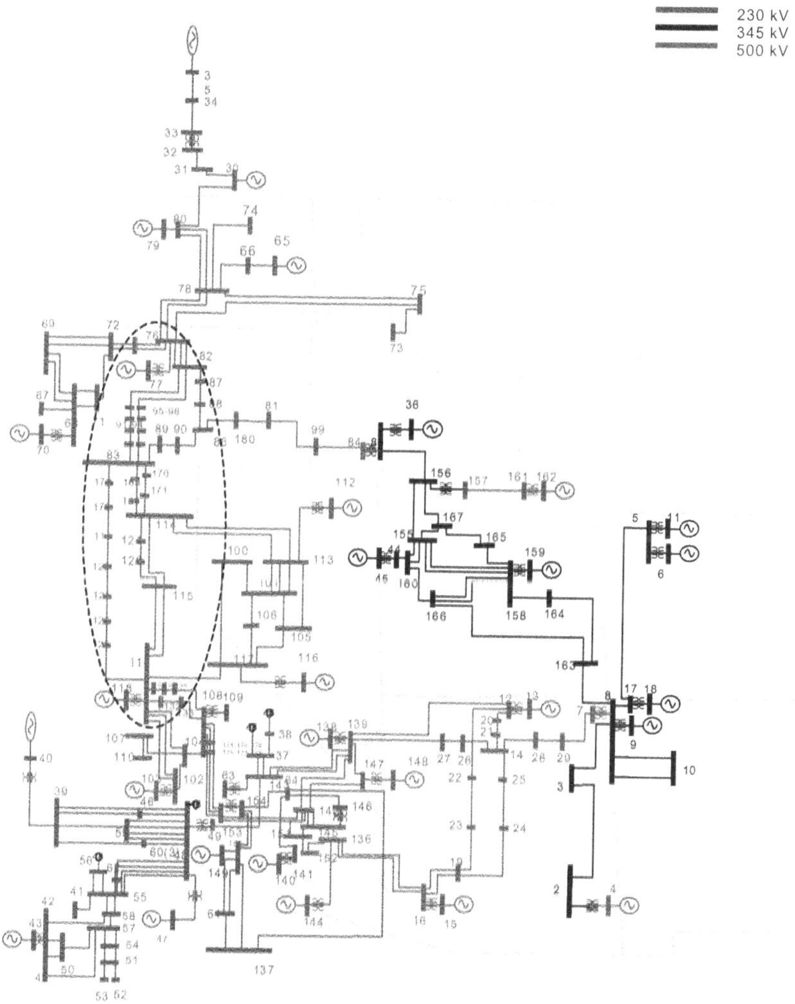

Figure 1.8. One-line diagram of the 179-Bus reduced WSCC electric power system.

1.6 EXAMPLES OF FUNDED RESEARCH WORK IN RESPONSE TO THE EPNES SOLICITATION

1.6.1 Funded Research by Topical Areas/Groups under the EPNES Award

The awarded research topical areas are grouped in four areas consisting of:

(1) Group A: system theory, security technology/communications, micro-electro-mechanical systems (MEMS);

(2) Group B: economic market efficiency;

(3) Group C: interdisciplinary research in systems, economics, and environment;

(4) Group D: interdisciplinary education. The titled of the awards for each of these groups are listed below. The four joint NSF/ONR awards are marked with an asterisk, *.

Group A: Systems Theory, Security, Technology / Communications, Micro Electro Mechanical Systems (MEMS)

- University integrated Micro-Electro-Mechanical Systems (MEMS) and advance technology for the next generation / power distribution;
- *Dynamic models in fault tolerant operation and control of energy processing systems;
- Unified power and communication infrastructure for high security electricity supply;
- Intelligent power router for distributed coordination in electric energy processing networks;
- *High confidence control of the power networks using dynamic incentive mechanism;
- Planning reconfigurable power systems control for transmission enhancement with cost recovery systems.

Group B: Economic Market Efficiency

- Forward contracts, multi-settlement equilibrium and risk management in competitive electricity markets;
- Dynamic game theoretic models of electric power markets and their vulnerability;
- Security of supply and strategic learning in restructured power markets;
- Robustness, efficiency and security of electric power grid in a market environment;
- *Dynamic transmission provision and pricing for electric power systems;
- Pricing transmission congestion to alleviate stability constraints in bulk power planning.

Group C: Interdisciplinary Research in Systems, Economics, and Environment

- Designing an efficient and secure power system using an interdisciplinary research and education approach;
- *Integrating electrical, economics, and environmental factors into flexible power system engineering;
- Modeling the interconnection between technical, social, economics, and environmental components of large scale electric power systems;
- A holistic approach to the design and management of a secure and efficient distributed generation power system;
- Power security enhancement via equilibrium modeling and environmental assessment (Collaborative effort among three universities);
- Decentralized resources and decision making.

Group D: Interdisciplinary Education Component of EPNES Initiative

- Development of an undergraduate engineering course in market engineering with application to electricity markets.
- Educational component: Modeling the interaction between the technical, social, economic and environmental components of large scale electric power systems.
- A technological tool and case studies for education in the design and management of a secure and efficient distributed generation power system.

1.6.2 EPNES Award Distribution

To date, a total of 17 awards, valuing more than U.S. 19 million, were granted to the winning proposals from 21 universities under the EPNES initiative, supporting the research activities of faculty and students. The topical areas and involved schools are listed in the previous section of this paper. Figure 1.9 shows the distribution

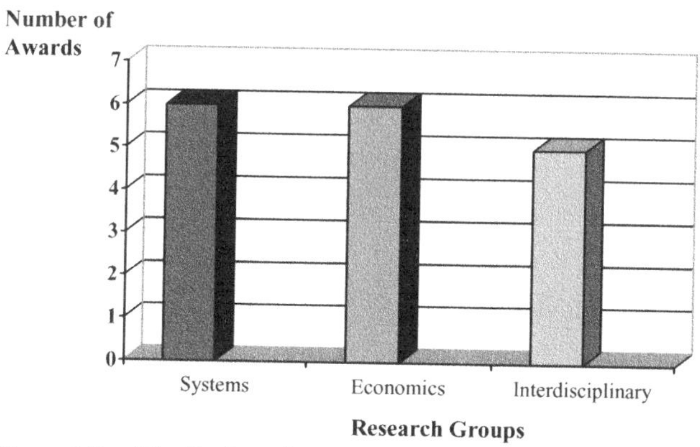

Figure 1.9. Distribution of EPNES awards among interdisciplinary research groups.

among the Systems, Economics, and Interdisciplinary groups. These three groups are spanned by the requirements of Education and Benchmark Systems.

1.7 FUTURE DIRECTIONS OF EPNES

1. Promote the implementation of the current EPNES goals by researchers for adoption in the private sector and the Navy. The underlying objective of EPNES is to unify cross-disciplinary research in systems theory, economics principles, and environmental science for the electric power system of the future.
2. Continue to involve industry and government agencies as partners. For example, utilize EPNES as a vehicle for collaboration with U.S. Department of Energy in addressing future needs of the industry such as blackouts, intelligent networks, and power network efficiency.
3. Include more mathematics and system engineering concepts in the scope of EPNES. This includes development of an initiative that is geared to include applied mathematics, systems theory, and security in addressing the needs of the power networks.
4. Extend the economic foundations from markets to cost-benefit analysis and pricing mechanisms for the new age high-performance power networks, both terrestrial and naval.
5. Continue to support reform in power systems with better education pedagogy and more adequate curricula in the colleges and universities. Enforce 'learning and research' via collaboration for increased activities that cut across engineering, science, mathematics, environmental, and social science disciplines. Promote and distribute the new education programs throughout the universities and colleges.
6. Use EPNES as a benchmark for proposal requirements of other NSF initiatives. Subsequent proposals submitted by Principal Investigators to an NSF multidisciplinary announcement should not be limited to the component level of problem-solving but should reflect a broader and more comprehensive interdisciplinary thinking, together with a plan for real-time implementation of the research by the private sector. Future initiatives will be structured toward the areas of Human Social Dynamics (HSD), Critical Cyber Infrastructure (CCI), and Information Technology Research (ITR).

1.8 CONCLUSIONS

In this vision of the Electric Power Networks Security and Efficiency (EPNES) initiative, we have described the framework of interdisciplinary research work and the underlying needs that drove the initiative. EPNES has many challenging research and education tasks to be finished, which will require state-of-the-art knowledge and technologies to solve. However, the research results of the EPNES project will be significant and useful for the improvement of both terrestrial and naval power system performance in terms of survivability, sustainability, efficiency and security as well as environment.

The funded research under the EPNES collaboration illustrates the breadth of the initiative and we believe that the research results will enhance power system security reliability, and affordability, help efforts for environment protection, and maintain high system sustainability. The results of EPNES will have significant impact to the education of students in multiple fields of engineering, science, and economics.

ACKNOWLEDGMENTS

On behalf of the National Science Foundation (NSF) and the Office of Naval Research (ONR), the author would like to acknowledge the participation of all the Principal Investigators who submitted winning proposals from various educational institutions. They have effectively risen to the challenges of the EPNES initiative.

The author acknowledges the support of the Office of Naval Research in the definition, execution, and partial funding of the EPNES collaboration, in particular, Katherine Drew of the Engineering and Physical Sciences Department.

In addition, the author acknowledges Prof. Edwin Zivi of the United States Naval Academy for his formulation and definition of the Naval benchmark testbed system.

The author also would like to extend his gratitude to the supporting management teams and staff of NSF.

BIBLIOGRAPHY

[1] Program Solicitation for NSF/ONR Partnership in Electric Power Networks Efficiency and Security (EPNES), NSF-02-041, National Science Foundation, http://www.cesac.howard.edu/NSF_proposals/nsf02041.htm.

[2] ONR/NSF EPNES Control Challenge Problem Website, United States Naval Academy, http://www.usna.edu/EPNES.

[3] Center for Energy Systems and Control (CESaC), Howard University, http://www.cesac.howard.edu/.

[4] Power System Data Information, Arizona State University, http://www.public.asu.edu/~huini/WsccDataFiles.htm.

CHAPTER 2

MODELING ELECTRICITY MARKETS: A BRIEF INTRODUCTION

Alfredo Garcia[1], *Lamine Mili*[2], *and James Momoh*[3]
[1]University of Virginia, [2]Virginia Tech, [3]Howard University

EDITORS' SUMMARY: This chapter provides a comprehensive overview of the economic structure of present and future electricity markets from the combined perspectives of economics and electrical engineering. It describes the basic structure of an electricity market and defines concepts such as consumer surplus, congestion rents, and market power. Furthermore, it outlines the mechanisms resulting in strategic bidding by generators and provides definitions and applications of the different equilibrium models to effectively analyze associated outcomes (prices and quantities). Examples from different equilibrium models (e.g. Cournot, auction-based) are presented. LMP calculations are then described via examples and economic dispatch formulation. Finally, their possible extension in stochastic and dynamic markets is highlighted via adaptive dynamic programming.

2.1 INTRODUCTION

Electricity markets have emerged all around the world since the early 1990s. In general, they tend to be characterized by an oligopoly of generators, very little demand-side elasticity in the short term, and complex administered market mechanisms. The market mechanisms are designed to facilitate both financial trading and physical (real-time) system balancing. After many decades of treating generation, transmission, distribution, and retail of electricity as a vertically integrated regulated monopoly, many economists raised doubts about the appropriateness of this particular organizational structure for the electric power industry. In highly industrialized economies, the main motivation for these claims was inspired by technological

Economic Market Design and Planning for Electric Power Systems, Edited by James Momoh and Lamine Mili

breakthroughs that resulted in more efficient and less capital-intensive combined-cycle natural gas fueled power plants. This new feature led economists to argue that the extent of economies of scale did not justify endowing a regulated utility with a legal monopoly in generation. Instead, opening generation to competition, they argued, would induce more efficient decisions for new investments and/or maintenance of installed capacity. In developing economies with strained public finances, the state's involvement in the provision of electricity was thought to create perverse incentives for investments (e.g. through corrupt procurement) and politically-motivated pricing policies that included subsidies and induced welfare losses.

Restructuring the electricity industry typically consists of a series of reforms. Vertical disintegration of generation, transmission, distribution, and retail businesses is accompanied by the introduction of a spot market for generation. Typically, transmission and distribution remain regulated activities and rules governing open access to the transmission and/or distribution systems are implemented in order to facilitate entry by new power generators and/or retailers.

Up until now, all experiences with restructured electricity markets show that electricity trading may give rise to highly volatile prices. This issue is intrinsic to electricity as a flow commodity, which cannot be economically stored. To accommodate for real-time balancing, day-ahead price formation is complemented with successive transactions or settlements for required adjustments on real-time operations. Since electrical energy is not economically storable, restructured electricity markets are more complex than the traditional commodity markets. Hence, existing economic models of price formation in commodity markets are not applicable. Moreover, the high levels of industry concentration make the occurrence of strategic behavior almost inevitable. In light of these features, theoretical economic analyses have tended to be based upon highly stylized models. Power engineers have sometimes criticized these economic models, because they fail to take into account non-trivial features such as loop-flow and reactive power. Nonetheless, these simplified models have been very useful for guiding regulatory policy-making. In this chapter, we provide a brief introduction to the economic modeling of electricity markets. Our intention is to provide non-economists with a quick overview of the existing models.

2.2 THE BASIC STRUCTURE OF A MARKET FOR ELECTRICITY

A market can be roughly defined as an environment that allows potential buyers, sellers and retailers of a given economic product to engage in trade. Consider for instance, the famous Fulton "fresh" fish market in Manhattan. Every day, producers (i.e., fishers) make available their recent catch directly or indirectly through retailers (i.e., firms that specialize in dealing with potential customers and storing recently caught fish in industrial scale refrigerators). Potential customers stroll around this market place evaluating and comparing the different offers posted. Consider further a specific homogeneous product, say tuna. Through bargaining and comparing posted offers, a "clearing" price for tuna slowly but surely emerges as the trading day passes. This "clearing" price has the following dual property: any producer

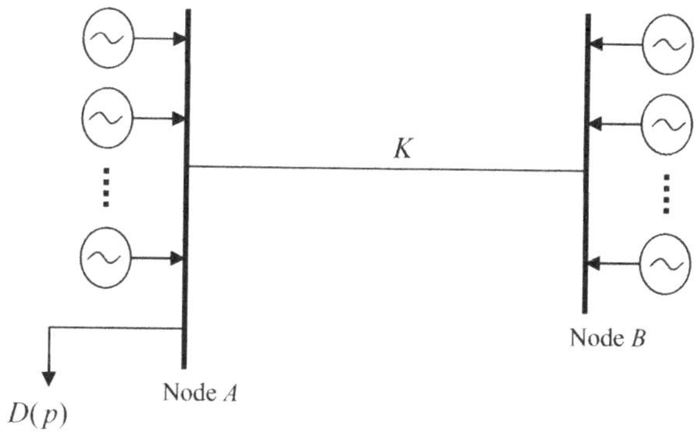

Figure 2.1. One-line diagram of a 2-bus system.

requesting a higher ask price will not be able to sell its catch, while any customer trying to "lowball" sellers will simply not be able to buy anything. Could the trade of electricity be undertaken in a similar fashion?

Consider the following example represented in Figure 2.1. This electricity market features for a given hour within a day, with generation capacity at both nodes A and B, and cheaper generation located at node B (i.e. *aggregate* costs C_A and C_B at node A and B satisfy $C'_A \geq C'_B$, and C''_A, $C''_B > 0$) and all demand located at node A and equal to $D(p)$.

The most immediate difference to the simple fish market example relates to the spatial configuration of the market: there is no single marketplace as the transmission line connects buyers and sellers, serving as a platform for trade. Assuming the transmission line connecting the markets has "infinite" capacity, we are interested in characterizing the market dispatch (q_A^*, q_B^*) and clearing prices (p_A^*, p_B^*). We further assume for simplicity that producers are "price takers:" either they take the going price if it exceeds their marginal costs (i.e., they effectively sell their capacity at the given price) or they abstain from selling. Hence:

$$p_A^* = C'_A(q_A^*) \text{ and } p_B^* = C'_B(q_B^*) \tag{2.1}$$

It follows that $p_A^* = p_B^*$. To see this, assume $p_A^* > p_B^*$. This implies that an "arbitrage" opportunity exists: any trader could buy electricity at node B at price p_B^* and sell it at node A at a price p_A^*. On the other hand, $p_A^* < p_B^* = C'_B(q_B^*)$ is a contradiction to node B producers' price taking behavior for they would be selling at prices below their marginal cost. Hence, $p_A^* = p_B^* = p^*$ and $D(p^*) = q_A^* + q_B^*$.

2.2.1 Consumer Surplus

Let us assume that the demand at node A is of the form $D(p^*) = D - \alpha p$. Let us denote by $\overline{p}$, the price level at which demand is zero, i.e., $0 = D(\overline{p}) = D - \alpha\overline{p}$. Or equivalently, $\overline{p} = \frac{D}{\alpha}$. In other words, customers would not demand any electricity if the market price exceeds $\overline{p}$. Let $P(Q)$ denote the inverse demand function. In this example:

$$P(Q) = \overline{p} - \frac{Q}{\alpha} \tag{2.2}$$

Here a simple interpretation is useful: for an aggregate level of consumption Q, the "marginal" customer has a willingness to pay a price equal to $P(Q)$. Thus, along the market dispatch this "marginal" customer with an infinitesimal consumption of dQ experiences a net surplus of $(P(Q) - p^*)dQ$. The (gross) aggregate consumer surplus is $\int_0^{D(p^*)} P(Q)dQ$. A widely-used measure of social welfare consists of adding up net consumer surplus and producers' profits. This measure is equivalent to subtracting production costs from (gross) aggregate consumer surplus. We now argue that the market dispatch described above maximizes social welfare. To see this, consider an infinitesimal increase in market dispatch, say dq_A^* and dq_B^*. This brings about an increase in (gross) consumer surplus by the amount $p^*(dq_A^* + dq_B^*)$ and a cost increase of $C'_A dq_A^* + C'_B dq_B^*$. Thus, at the market dispatch the net effect is null:

$$(p^* - C'_A)dq_A^* + (p^* - C'_B)dq_B^* = 0 \tag{2.3}$$

2.2.2 Congestion Rents

Suppose the line has a capacity of $K < q_B^*$. Now the "no arbitrage" condition fails as there is a limit to how much production from the cheapest generator can discipline prices at node A. Along the constrained market dispatch $(\tilde{q}_A, \tilde{q}_B)$ and $(\tilde{p}_A, \tilde{p}_B)$, it must hold that cheap generation uses up transmission capacity, i.e. $\tilde{q}_B = K$ and $\tilde{p}_A > p^* > \tilde{p}_B$, since more expensive generators at node A must cover residual demand $D(\tilde{p}_A) - K$. This disparity in clearing prices induces a "congestion rent:"

$$K(\tilde{p}_A - \tilde{p}_B) \tag{2.4}$$

This amount is also known as "merchandizing surplus" to emphasize the fact that whoever owns the line can buy at low prices and sell at much higher prices to earn an intermediation rent.

2.2.3 Market Power

Let us now assume that all the power plants at node A are owned by a single firm, while producers at node B continue to behave as "price takers." This implies $\hat{p}_B = C'_B(K) = \tilde{p}_B$. Nonetheless, the one producer at node A maximizes profit by solving for the optimal price $\hat{p}_A$ where:

$$\hat{p}_A \in \arg\max_p [p(D(p) - K) - C_A(D(p) - K)]. \tag{2.5}$$

It follows that $\hat{p}_A > \tilde{p}_A > p^*$ and $\hat{q}_A < \tilde{q}_A \leq q_A^*$. Consequently, the congestion rent is increased:

$$K(\hat{p}_A - \tilde{p}_B) > K(\tilde{p}_A - \tilde{p}_B). \tag{2.6}$$

That is, the generator located at node A, has a "captive load" or a "residual monopoly" and would therefore bid its capacity well above marginal cost. The ability to price above marginal cost is sometimes referred to as "market power" and constitutes evidence of the market not being perfectly competitive since in a perfectly competitive market, producers behave as "price takers."

2.2.4 Architecture of Electricity Markets

The above example would suggest that various market architectures ranging from highly decentralized to highly centralized trading structures would be equally effective in implementing electricity trades. Nonetheless in this example we have abstracted away from two important features: *forward contracts* and *power flow*.

First, it is typically not the case that all electricity is traded "at one spot" as in the above example. Sometimes producers and buyers enter into contractual arrangements known as "forward contracts" well in advance of the actual time at which the electricity is produced and consumed. A forward contract is an agreement between two parties to buy or sell electricity at a pre-agreed price and a future point in time. Therefore, the trade date and delivery date are separated. The forward price of such a contract is commonly compared with the "spot price," i.e., the price of electricity that is traded "on the spot." The difference between the spot and the forward price is the forward premium.

Second, the nature of power flows which are basically absent in the two-node example have strong implications for possible market architectures. To illustrate let us consider the three-node network in Figure 2.2. Flows on the network are governed by Kirchhoff's laws and not by contracts. Hence, a trade between generator 1 and load 2 for example, will cause flows on all three lines (1-2, 1-3, and 2-3). Therefore, in a highly decentralized architecture, generator 1 and load 2 would have to acquire "rights" for the use of these lines. In a complex network with a large number of bilateral trades taking place simultaneously, it is very difficult to determine the specific nature of the usage "right" that will be required by a particular trade. Hence, these high transaction costs imply that bilateral trading through "physical" rights is not a feasible market architecture. In an alternative, more centralized scheme, firms and retailers buy "financial" rights over the transmission network usage and inform an independent system operator (ISO) of the technical features of their trade. This

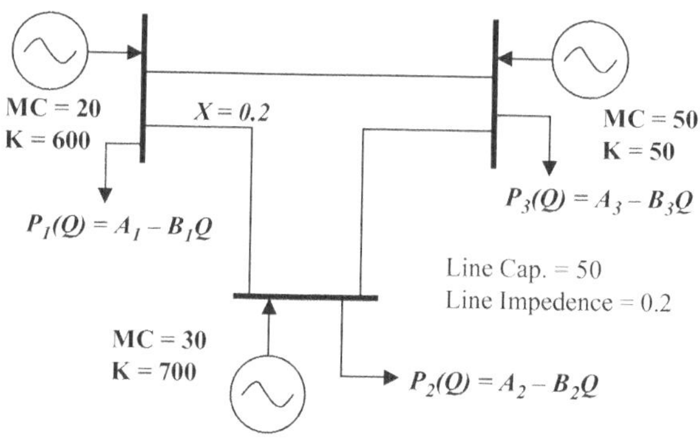

Figure 2.2. One-line diagram of a three player, 3-bus system.

agency is in turn in charge of finding the cheapest way to support all submitted or scheduled trades. Financial rights provide holders with a hedge over the potential congestion costs that may arise when all the requested trades can not be implemented as desired.

In yet another market architecture, with a higher degree of centralization, firms submit bids to a "market maker" which in turn computes the price that maximizes *estimated* social welfare. We emphasize the word *estimated* because the bids may not be truthful: producers may bid above marginal costs for their capacity and retailers may bid well below their willingness to pay for demand; thus the estimated aggregate welfare may not be equal to the true welfare. As discussed in Section 2.3, only firms with market power are able to bid above their true marginal costs. Given the high levels of ownership concentration and the existence of locational market power, appropriate models of electricity markets must somehow capture this trait. In the next section, we provide a brief overview of the modeling of strategic behavior in electricity markets.

2.3 MODELING STRATEGIC BEHAVIOR

2.3.1 Brief Literature Review

Strategic bidding by generators in restructured electricity markets has now been studied by the researchers for almost two decades. Given high levels of concentration, these markets have evidenced a great deal of strategic behavior. This makes game theory an ideal modeling tool for their study. The literature on game-theoretic models of oligopolistic competition in electricity markets can be classified into three groups: Bertrand (price-based competition), Cournot (quantity-based competition), and supply function models.

Price-based models has been used in Von der Fehr et al. [45] and [18], Garcia et al. [22] and [23], Hobbs et al. [29] and [33]. These models do not account for transmission constraints though. A price-based model has also been used to model only the transmission part of the market where the transmission owners are assumed to be price takers [37]. Typically, price-based models predict fiercer competition among firms.

In Cournot analysis, firms are assumed to bid quantities in the market leading to a market price that "clears" the market. A Cournot-Nash equilibrium is a vector of quantities such that no player has an incentive to deviate unilaterally from it. Electricity markets have been modeled using Cournot analysis under both constrained and unconstrained transmission networks. Firms owning generation plants are assumed to bid their output quantity and an independent "market maker" clears the market so as to equate supply and demand via some process (e.g., typically by solving a constrained optimal dispatch problem). Cournot models are popular due to their analytical tractability. For an excellent review of the literature on the use of the Cournot approach, see Day and Hobbs [16], Hobbs [30], and Neuhoff et al. [38]. Nearly all aspects of electricity markets i.e., pricing, market power analysis, transmission investment analysis, market coupling and other policy related questions have been studied using Cournot models. For example, Cournot model has been used for modeling equilibrium prices [2], for analyzing market power ([6], [10], [34], and [39]) and transmission capacity issues [7], for co-optimization of ancillary services [12], and for modeling non-constant marginal costs [15].

Supply function models were proposed recently by Klemperer and Meyer [36]. Green and Newbery [25] used it in the context of electricity markets. These models were extended in [4], [24] and Day et al. [16] proposed a Conjectured Supply function approach. In a Supply Function model the players are assumed to bid supply curves rather than only price or only capacity and the supply function equilibrium is reached when no player can profit by unilaterally deviating from the equilibrium play. These models are more realistic but their analysis complicated and few theoretical results have been seen in the literature.

2.3.2 Price-Based Models

Let us start with a simple illustrative example with two generating firms having constant marginal cost $c > 0$ and capacities K_i and K_j respectively. The firms must supply an inelastic demand for electricity denoted by D. The spot market for electricity operates as follows: firms submit price bids $(b_i, b_j) \in [c, \bar{p}]$ to the ISO who solves for the economic dispatch of resources. Given bids (b_i, b_j) the fraction of total demand $D(b_i, b_j)$ that is to be supplied by generator $i \neq j \in (1, 2)$ is:

$$D_i(b_i, b_j) = \begin{cases} \min\{K_i, D\} & b_i < b_j \\ \frac{1}{2}\cdot\min\{K_i, D\} + \frac{1}{2}\cdot\max\{0, D - K_j\} & b_i = b_j \\ \max\{0, D - K_j\} & b_i > b_j \end{cases} \tag{2.7}$$

The spot price $p(b_i, b_j)$ is set to equal to the price submitted by the *marginal* firm and bids are constrained by a price cap $\bar{p} > c$ stipulated by the regulatory commission. Thus, the firms' profits are $\Pi_i(b_i, b_j) = (p(b_i, b_j) - c)D_i(b_i, b_j)$. A bidding equilibrium (also referred to as a Bertrand-Nash equilibrium) is a combination (b_i^*, b_j^*) such that for any $b \in [c, \bar{p}]$, it holds that:

$$\Pi_i(b_i^*, b_j^*) \geq \Pi_i(b, b_j^*), \text{ for } i \in \{1, 2\} \tag{2.8}$$

An alternative definition of equilibrium involves the use of a "best reply" function. That is, for each firm, given an opponents' decision we compute the best pricing decision or "reply." In other words, given b_j we solve for $BR_i^*(b_j)$ where:

$$BR_i^*(b_j) \in \arg\max_{b \in [c, \bar{p}]} \Pi_i(b, b_j^*) \tag{2.9}$$

Note that a two-tuple (b_i^*, b_j^*) is a bidding equilibrium if $BR_i^*(b_j) = b_i^*$ and $BR_j^*(b_i) = b_j^*$.

1. Numerical Illustration

Let us consider the case in which the two firms have constant marginal cost c = \$20/*MWh* and capacities K_1 = 200 MW and K_2 = 200 MW respectively, and demand D = 150 MW. That is, $K_i > D$ for $i \in (1, 2)$. Note that whenever $b_j \leq c$, it is not rational to undercut the opponent and $b_i^*(b_j) = c$. If $b_j \in [c, \bar{p}]$ then there exist no optimal solution to problem (2), in a strict sense. This impasse is typically addressed by introducing the notion of "slightly undercutting" the opponent, i.e., firm i's "best" course of action is to bid $BR_i^*(b_j) = b_j - \varepsilon$ for some "small" $\varepsilon > 0$. We conclude that the only two-tuple of prices (b_1^*, b_2^*) that satisfy (1) is (c, c). This result is known as the "Bertrand paradox:" with only two firms, competition is so fierce that the only bidding equilibrium is the perfectly competitive outcome, i.e., firms bid marginal costs.

Suppose now that K_i = 100 MW for $i \in \{1, 2\}$. In this case, the asymmetric two-tuples $(c, \bar{p})$ and $(\bar{p}, c)$ are in equilibrium and the spot price is set at $\bar{p}$. These equilibria are somewhat difficult to rationalize without to recurring to exogenous arguments. The asymmetry of equilibrium payoffs makes one wonders why the marginal firm has settled for such a role. These considerations lead to search for a symmetric equilibrium. For example, both firms bid at \$200/MWh. This is however *not* an equilibrium. Both firms bidding \$200/MWh yields the following payoff:

$$\frac{1}{2}100(200-20)+\frac{1}{2}50(200-20)=\$13{,}500 \tag{2.10}$$

By undercutting (say bidding \$199/MWh) a firm would guarantee a payoff of:

$$100(200-20)=\$18{,}000 \tag{2.11}$$

Suppose firms were to choose their bids according to independent samples from the uniform distribution on $[c, \bar{p}]$, namely:

$$\Pr(b_i \le x)=\frac{x-c}{\bar{p}-c} \tag{2.12}$$

To check that this is indeed an equilibrium we write the expected profit for firm i, should he/she bid $b \in [c, \bar{p}]$:

$$\begin{aligned} E[\Pi_i(b)] &= \Pr(b_j < b)(b-c)(D-K_j)+\int_b^{\bar{p}} \frac{b_j-c}{\bar{p}-c} K_i db_j \\ &= \frac{b-20}{180}(b-20)50+\frac{100}{180}\left[\left(\frac{200^2}{2}-4000\right)-\left(\frac{b^2}{2}-20b\right)\right] \\ &= \$9{,}000 \end{aligned} \tag{2.13}$$

Note that $\dfrac{dE[\Pi_i(b)]}{db}=0$. Thus, firm 1 is indifferent between choosing any bid b in the interval $[c, \bar{p}]$. In other words, randomizing its bid choice according to the uniform distribution is the optimal course of action for the firm.

2. Tacit Collusion

Competition may be weakened when a number of firms engage in what economists refer to as "tacit collusion:" while the verb colluding refers to explicit collaboration amongst competitors to jointly exercise market power, the qualifier "tacit" specifically points to coordinated behavior amongst competitors that emerges endogenously without any explicit agreement. To illustrate this phenomena in electricity markets, consider the previous example (i.e., c = \$20/MWh, $K_1 = K_2 = 200\,\text{MW}$) within a different context: there are more generating firms participating in the market but also a higher level of demand, with an overall level of excess capacity in the market of only *50* MW. In this situation, when both firms *1* and *2* bid marginal cost they are effectively allowing some other firm in the market to set the price, say at a level $\tilde{p}$ = \$120/MWh. In a highly simplified game, the firms must decide whether to bid $\bar{p}$ and effectively set the price at the cap with a low level of dispatch or bid marginal cost and potentially allow some other firm to set the price. The normal form for the resulting 2 × 2 simultaneous game is depicted in Table 2.1.

Note "*both firms bid marginal cost*" is a Nash Equilibrium. However, let us consider the strategy combination according to which play begins in a *phase 1* (in which the two firms "take turns" in setting the spot price equal to $\bar{p}$, and transitions to *phase 2* (in which firms bid marginal cost forever). Assuming *firm 1* is to start bidding marginal cost (and consequently, *firm 2* sets the price at $\bar{p}$) its discounted payoff should *phase 1* hold indefinitely is:

$$18000 + 9000\beta + 18000\beta^2 + 9000\beta^3 \ldots = \frac{18000 + 9000\beta}{1-\beta} \tag{2.14}$$

If *firm 1* deviates, say at even period *2*, play follows *phase 2* after the second period and her total discounted payment will be:

$$18000 + 10000\beta + 10000\beta^2 + \ldots = 18000 + \frac{10000\beta}{1-\beta} \tag{2.15}$$

Note that deviating is not profitable whenever $\beta > \frac{1}{8}$. Moreover, deviations at later stages of the game are also not profitable if $\beta > \frac{1}{8}$. To see this, suppose for example, that *firm 1* is to deviate at *period 4*, then its payoff is:

$$\begin{aligned} &18000 + 9000\beta + 18000\beta^2 + 10000\beta^3 + 10000\beta^4 + \ldots \\ &= 18000 + 9000\beta + \left(18000 + \frac{10000\beta}{1-\beta}\right)\beta^2 \end{aligned} \tag{2.16}$$

which is again less than the discounted payoff should *phase 1* hold indefinitely whenever $\beta > \frac{1}{8}$.

TABLE 2.1. Normal form for the 2 × 2 Simultaneous Game.

	$p = \bar{p}$	$p = c$
$p = \bar{p}$	(13500; 13500)	(9000; 18000)
$p = c$	(18000; 9000)	$(100(\bar{p} - 20);\ 100(\bar{p} - 20))$

2.3.3 Quantity-Based Models

Rather than competing in prices, firms compete by deciding how much output to make available to the market. For illustration let us suppose as before there are only two firms in the market, say *1* and *2*. Assuming a marginal cost c, the profit function for firm i, given production levels (q_i, q_j):

$$\Pi_i(q_i, q_j) = P(q_i + q_j)q_i - cq_i \tag{2.17}$$

where $P(\cdot)$ is the inverse demand function (i.e., $P(q_i + q_j)$ is the price at which a total of $Q = q_i + q_j$ would be sold or "cleared" in the market). In a Cournot-Nash equilibrium (q_i^*, q_j^*) no firm can benefit (strictly) from deviating or changing unilaterally its production decision. Formally, for all feasible production levels q_i for $i \in \{1, 2\}$:

$$\Pi_i(q_i^*, q_j^*) \geq \Pi_i(q_i, q_j^*) \tag{2.18}$$

1. Application (of Cournot model) to Electricity Markets

We assume the existence of a "fringe" of perfectly competitive producers which is modeled via a supply function $S(p)$. For a given level of (inelastic) demand for electricity, the oligopolists face a residual demand equal to $Q = D(p) - S(p)$. Like a monopolist, the dominant firms (i.e., oligopolists) face a downward sloping demand curve. However, unlike the monopolist, the dominant firms must take into account the "competitive fringe" firms in making its output decisions. Given production levels (q_i, q_j), the inverse demand function evaluated at $Q = q_i + q_j$ is the solution p to the following equation:

$$Q = D - S(p) \tag{2.19}$$

For instance, when $S(p) = \alpha p$, residual demand is of the form $D(p) = D - \alpha p$. Let us denote by $\bar{p}$, the price level at which residual demand is zero, i.e., $0 = D(\bar{p}) = D - \alpha p$. Or equivalently, $\bar{p} = \frac{D}{\alpha}$. Thus:

$$P(Q) = \bar{p} - \frac{Q}{\alpha} \tag{2.20}$$

Given q_j we compute the "best reply" to this output by solving:

$$\max_q \Pi_i(q, q_j) \tag{2.21}$$

The necessary first order condition for optimality (which is also sufficient in this case) yields:

$$\frac{\partial P(q + q_j)}{\partial q} q + P(q + q_j) = c \tag{2.22}$$

Or equivalently, marginal revenue equals marginal cost. This is equivalent to:

$$-\frac{1}{\alpha} q + \bar{p} - \frac{(q + q_j)}{\alpha} = c \tag{2.23}$$

The best reply is therefore given by:

$$BR_i^*(q_j) = \frac{1}{2}[\alpha(\bar{p} - c) - q_j] \tag{2.24}$$

As before, a Cournot-Nash equilibrium is a fixed-point of the best reply map, i.e., (q_i^*, q_j^*) is such that $q_i^* = BR_i^*(q_j)$ and $q_j^* = BR_j^*(q_i)$. This leads to the linear system of equations expressed as:

$$\begin{aligned} q_i^* &= \frac{1}{2}[\alpha(\bar{p} - c) - q_j^*] \\ q_j^* &= \frac{1}{2}[\alpha(\bar{p} - c) - q_i^*] \end{aligned} \tag{2.25}$$

Solving we obtain:

$$q_i^* = q_j^* = \frac{1}{3}[\alpha(\bar{p} - c)] \tag{2.26}$$

The associated market price is given by:

$$P(q_i^* + q_j^*) = \bar{p} - \frac{1}{\alpha}\frac{2}{3}\alpha(\bar{p} - c) = \frac{1}{3}\bar{p} + \frac{2}{3}c \tag{2.27}$$

and the firm's equilibrium profit is:

$$\Pi_i^* = \frac{1}{9}\alpha(\bar{p} - c)^2 \tag{2.28}$$

Note that there is no "paradox" in the Cournot model: the market price is well above marginal cost. Furthermore, the higher the slope of the competitive fringe supply curve, the higher the equilibrium profits. This suggests an interpretation for the Cournot-Nash equilibrium outcome in terms of capacity withholding by firms in an effort to have more expensive fringe suppliers set the spot price. Can they withhold even more capacity and increase profits? For instance, let $q^m = \frac{1}{2}\alpha(\bar{p} - c)$ and consider the two-tuple $(\frac{1}{2}q^m, \frac{1}{2}q^m)$. Note that:

$$\Pi_i^*\left(\frac{1}{2}q^m, \frac{1}{2}q^m\right) = \left(\frac{1}{2}\bar{p} + \frac{1}{2}c - c\right)\frac{1}{4}\alpha(\bar{p} - c) = \frac{1}{8}\alpha(\bar{p} - c)^2 > \Pi^* \tag{2.29}$$

However, the outcome $(\frac{1}{2}q^m, \frac{1}{2}q^m)$ is not an equilibrium, for the best reply is given by:

$$q_i^*\left(\frac{1}{2}q^m\right) = \frac{3}{8}\alpha(\bar{p} - c) \tag{2.30}$$

In other words, if both firms withhold too much capacity, there is an incentive to take up the leftover slack by increasing output.

2. Incorporating Forward Contracts

Allaz and Vila [1] have shown how the existence of long-term contracts changes the incentive structure of the Cournot oligopoly model. If an oligopolist has already committed a substantial portion of its capacity at a predetermined price, he/she will have no incentive to manipulate the market. For instance, if firm a has sold x units in a forward contract at a price $\tilde{p} > c$, its profit function given production levels (q_i, q_j) is given by:

$$\Pi_i(q_i, q_j) = P(q_i + q_j)(q_i - x) - cq_i + \tilde{p}x \tag{2.31}$$

Notice that if $q_i = x$, then the firms profit reduces to:

$$\Pi_i(q_i, q_j) = (\tilde{p} - c)x \tag{2.32}$$

In other words, the spot market has no effect whatsoever on firm i's profit. Now let us assume both firms are contracted at the same level x at the same price, we look for the ensuing Cournot equilibrium. It is worth emphasizing here that the best production decision is not affected by the contract price. First order condition is now expressed as:

$$-\frac{1}{\alpha}(q_i - x) + P(q_i + q_j) = c \tag{2.33}$$

The Cournot-Nash equilibrium is now given by:

$$q_i^* = q_j^* = \frac{1}{3}\alpha(\overline{p} - c) + \frac{x}{3} \tag{2.34}$$

The resulting market price is given by:

$$P(q_i^* + q_j^*) = \frac{1}{3}\overline{p} + \frac{2}{3}\left(c - \frac{x}{\alpha}\right) \tag{2.35}$$

Note that for high levels of contracting (e.g., $x = \frac{D - \varepsilon}{2}$) the resulting market price reduces to:

$$P(q_i^* + q_j^*) = \frac{1}{3}\overline{p} + \frac{2}{3}\left(c - \frac{D - \varepsilon}{2\alpha}\right) = \frac{2}{3}\left(c + \frac{\varepsilon}{\alpha}\right) \tag{2.36}$$

This shows that for high levels of contracting the resulting market price may well be below marginal cost.

2.4 THE LOCATIONAL MARGINAL PRICING SYSTEM OF PJM

2.4.1 Introduction

PJM has adopted the spot pricing system advocated by Schweppe et al. [43] in the early 1980s. The main advantage of this pricing system is that it accounts for the cost

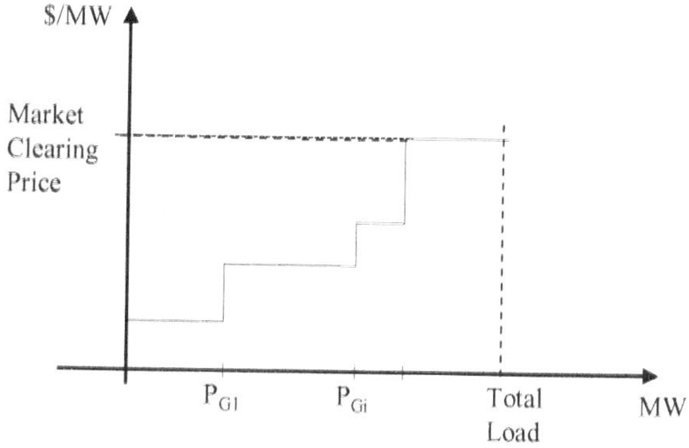

Figure 2.3. Determination of the market clearing price.

of transferring power from one location to another one when the network is constrained. This is achieved via a collection of Locational Marginal Prices (LMPs) calculated at every bus of the grid, hence their name [9, 26, 40]. As depicted in Figure 2.3, when none of the transmission lines or transformers is overloaded, the LMPs are all equal to the market clearing price defined as the highest marginal generator cost of meeting the load. On the other hand, when there exist transmission lines or transformers whose power flows exceed their thermal or stability limits, creating congestion in the transmission network, the LMPs become different across the network. They may even take negative values at some buses. Their variability results from the re-dispatch at least cost of the generating units, which is executed to alleviate the transmission congestion. These differences result from the redispatch, at least cost of the generating units, which is executed to alleviate the transmission congestion. Specifically, an LMP at a given bus is defined as the cost of an incremental change in generation of the marginal units for supplying a load increase of 1 MW at that bus. This will be explained in detail through an example in the sequel.

The LMPs are calculated every five minutes at each of the 1750 buses of the PJM system by using a constrained economic dispatch that finds the least cost generation subject to line and transformer capacity limits [13, 50]. This calculation is based on a linearized power system model of the PJM transmission network along with its neighboring systems, where the losses are neglected [13].

Based on the LMPs, the PJM-ISO also calculates zonal prices for three hubs defined as a weighted average of their associated LMPs. The hubs include the 111-bus Western hub, the 277-bus Eastern hub, and the interface hub [26]. Recently, The PJM-ISO has implemented a two-pass settlement where both the energy providers and the load entities can send bids to the auction market [36]. At the settlement, only those load entities that have accepted to pay the clearing price are served, thereby providing a certain degree of elasticity in the load demand.

2.4.2 Congestion Charges and Financial Transmission Rights

The PJM-ISO charges a congestion price to every Load Serving Entity (LSE) transferring electric energy through the PJM power network. Specifically, the congestion price charged to a LSE is equal to the difference of the LMPs from the generator bus to the load bus defining the contract path specified by the LSE to the ISO at the settlement, multiplied by the amount of power transferred to the load [40].

To hedge potential congestion charges, the PJM-ISO has put in place a forward secondary market where market participants can buy by Financial Transmission Rights (FTRs) [13, 40]. The latter provide a protection against congestion charges on a specific path, and hence constitute a hedging mechanism to manage basic risk. An FTR is a benefit when it is in the same direction as the congested flow and it is a liability when it is in the opposite direction. Specifically, an FTR credit for a power flow over a congested path, which is defined by a source and a sink, is equal to the difference between the LMPs from sink to source times the amount of power that is hedged through that path. This hedged power exceeds neither the maximum generation capacity of the FTR's owner nor the capacity limit of the congested path. In addition, the FTRs entitle their holder to receive financial credits only if the hedged power is in the same direction as the congested flow. Note that FTRs are independent from the actual energy delivery.

Because, the PJM-ISO is a non-profit organization [3, 5, 8, 11, 13, 17, 19, 21, 27, 28, 35, 41, 49, 50], all the revenues are made equal to the expenses on a monthly basis according the following rulings. When the congestion charges collected by the ISO are smaller than the FTR allocations, then all the FTR credits are reduced proportionately to the hedged amounts of energy and the deficiency is made up at the end of the month using excess congestion credits. On the other hand, when the congestion charges are larger than the FTR allocations, the excess monies accumulated are distributed monthly either to the FTR owners, proportionately to the amounts of energy that are hedged, or to cover the FTR target allocation deficiencies suffered by the ISO.

How to obtain FTRs? There are several ways for a market participant to obtain an FTR. It can be bought either via PJM e-Capacity as a Network Service from a set of generation buses to a set of aggregated load buses or via OASIS as a Firm Point-to-Point transaction [11]. It can also be purchased via the secondary market as a bilateral trading or via the centralized market, which is an FTR auction market.

2.4.3 Example of a 3-Bus System [40]

1. Market Clearing Price

Let us explain how the LMPs are calculated on an example of a 3-bus system, which is displayed in Figure 2.4. This system has two generating units attached to buses 1 and 3 with a capacity of 1000 MW and 500 MW and termed *unit 1* and *unit 3*, respectively. These units serve two loads; one load of 100 MW is connected to *node 2* and one load of 700 MW is attached to *node 3*. If *unit 1* and *unit 3* are bidding their marginal costs, which are assumed to be equal to 2 \$/MWh and 10 \$/MWh, respectively, then *unit 1* will serve the entire load while *unit 3* will not be dispatched. In other words, *unit 1* will supply 800 MW whereas *unit 3* will produce 0 MW. In this case, the market clearing price will be equal to 2 \$/MWh.

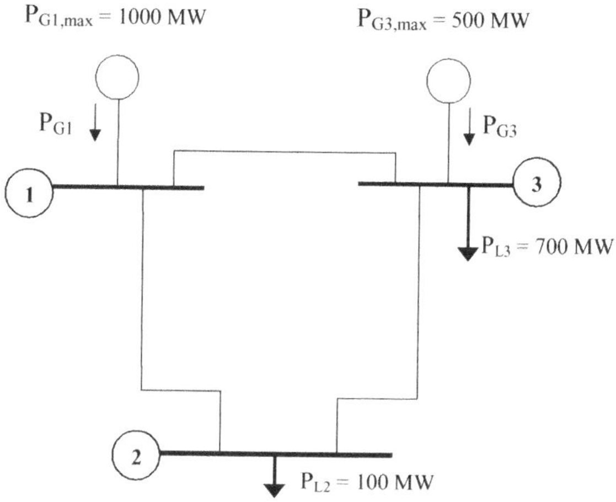

Figure 2.4. One-line diagram of a 3-bus system.

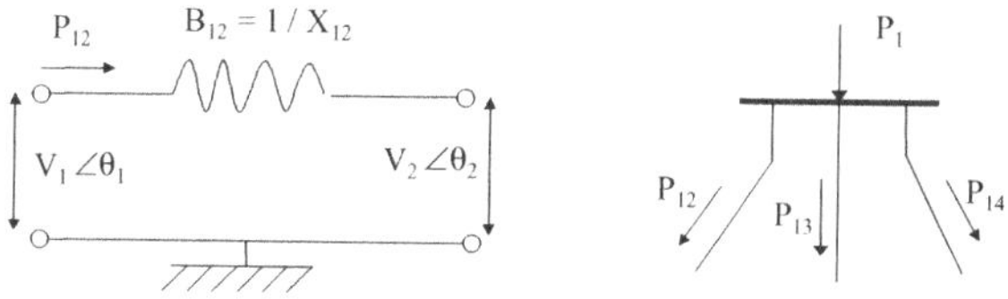

$P_{12} = B_{12}(\theta_1 - \theta_2)$ $\qquad$ $P_1 = P_{12} + P_{13} + P_{14}$

Figure 2.5. Transmission line modeling.

2. LMP Calculation Under No Congestion

Let us now assess the LMPs associated with this load profile. To this end, we need to calculate the power flows through the lines of the network to check whether there is any congestion in the system. The usual power system model being used for this calculation is the DC model. This is based on the assumption that the transmission lines are three-phase balanced and their resistances and shunt capacitances are negligible. As displayed in Figure 2.5, this allows us to represent a line as a series reactance, X_i, or equivalently as a series susceptance, $B_i = 1/X_i$. Also, it is assumed that the nodal voltage magnitudes, V_i, are close to their nominal values and that the nodal voltage phase angles, θ_i, are within a range of few degrees.

The DC model is based on the per unit system where the powers, the voltages, reactances, and susceptances are divided by their respective base values. It leads to the following equation for the power flow originating from node i and pointing toward node i of a line i-j:

$$P_{ij} = \frac{1}{X_{ij}}(\theta_i - \theta_j) = B_{ij}(\theta_i - \theta_j) \tag{2.37}$$

Now, by taking the voltage at *node 3* as a reference for the voltage phase angles, we put $\theta_3 = 0$ and write the power flows P_{12}, P_{13}, P_{23}, through lines 1-2, 1-3, and 2-3, respectively, as:

$$P_{12} = B_{12}(\theta_1 - \theta_2) \tag{2.38}$$

$$P_{13} = B_{13}\theta_1 \tag{2.39}$$

$$P_{23} = B_{23}\theta_2 \tag{2.40}$$

Using Kirchhoff's current law, which states that the sum of the powers flowing into a node is equal to the sum of the powers flowing out of that node, the power injections at *nodes 1*, *2*, and *3*, are expressed as:

$$P_1 = P_{12} + P_{13}, \tag{2.41}$$

$$P_2 = P_{21} + P_{23}, \tag{2.42}$$

$$P_3 = P_{31} + P_{32} \tag{2.43}$$

Here, the usual generator sign convention has been used. It requires putting a positive sign for a power injection that flows toward a node and a negative sign, otherwise. For simplicity, let us suppose that all three lines of the system have equal susceptances of 100 p.u. and that the base power is 100 MVA. This yields:

$$P_1 = 800\text{ MW}/100\text{ MVA} = 8\text{ pu} = 100\,(2\,\theta_1 - \theta_2), \tag{2.44}$$

$$P_2 = -100\text{ MW}/100\text{ MVA} = -1\text{ pu} = 100\,(-\theta_1 + 2\,\theta_2) \tag{2.45}$$

Solving Eqs. (2.44) and (2.45) for θ_1 and θ_2, we get:

$$\theta_1 = 0.05\text{ rads and } \theta_2 = 0.02\text{ rads} \tag{2.46}$$

It follows that the power flows through lines 1-2, 2-3, and 1-3 amount to:

$$P_{12} = 100\,(\theta_1 - \theta_2) = 3\text{ pu, that is, } P_{12} = 300\text{ MW}, \tag{2.47}$$

$$P_{23} = 100\,\theta_2 = 2\text{ pu, that is, } P_{23} = 200\text{ MW}, \tag{2.48}$$

and:

$$P_{13} = 100\,\theta_1 = 5\text{ pu, that is, } P_{13} = 500\text{ MW} \tag{2.49}$$

Under the assumption that the lines have enough capacity to carry the power flows given by (2.47), (2.48), and (2.49), implying that there is no congestion in the network, the LMPs of the three buses will settle at the market clearing price of 2 \$/MWh as depicted in Figure 2.6.

3. LMP Calculation Under Congestion

Now, let us assume that line 2-3 has a maximum capacity of 100 MW. Since, it is carrying a power of 200 MW, we conclude that congestion has occurred. Consequently, a redispatch at least cost needs to be carried out to alleviate it. Because this redispatch aims at decreasing the power flow, P_{23}, to 100 MW, we get:

$$P_{23} = 1\text{ pu} = 100\,\theta_2 \tag{2.50}$$

The other equality constraint that needs to be satisfied is the power injection at bus 2, which is equal to:

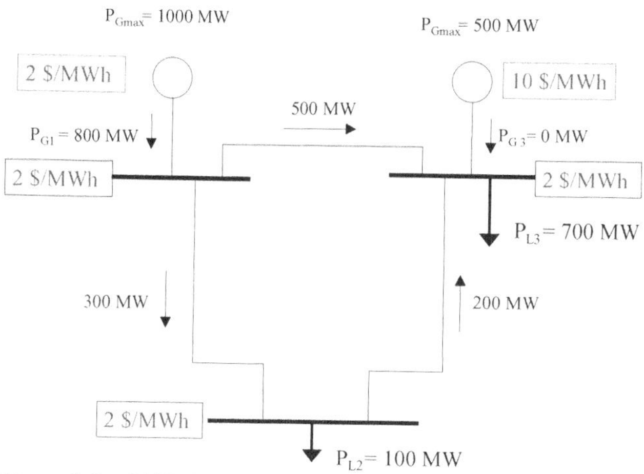

Figure 2.6. LMP determination under no congestion in the system [40].

$$P_2 = -1 \text{ pu} = 100\,(-\theta_1 + 2\,\theta_2), \tag{2.51}$$

Solving Equations (2.50) and (2.51) for θ_1 and θ_2, we obtain:

$$\theta_1 = 0.03 \text{ rads and } \theta_2 = 0.01 \text{ rads.} \tag{2.52}$$

It follows that the power generated by *unit 1* and *3* are equal respectively to:

$$P_{G1} = 100\,(2\,\theta_1 - \theta_2) = 5 \text{ pu, or } P_{G1} = 500 \text{ MW}, \tag{2.53}$$

and:

$$P_{G3} = 800 \text{ MW} - 500 \text{ MW} = 300 \text{ MW} \tag{2.54}$$

In other words, *unit 1* has to decrease its generation to 500 MW while *unit 3* is dispatched to 300 MW. This leads to the following power flows on *line 1-2* and *1-3*:

$$P_{12} = 100\,(\theta_1 - \theta_2) = 2 \text{ pu, that is, } P_{12} = 200 \text{ MW}, \tag{2.55}$$

and

$$P_{13} = 100\,\theta_1 = 3 \text{ pu, that is, } P_{13} = 300 \text{ MW}. \tag{2.56}$$

The marginal units being *unit 1* and *unit 3*, the LMPs at *buses 1* and *3* are equal to 2 $/MWh and 10 $/MWh, respectively. What about the LMP at *bus 2*? To assess its value, we need to find the incremental powers generated by *units 1* and *3* that serve an incremental load of 1 pu at *bus 2* subject to no changes in the power flow through line 2-3. As seen in Figure 2.7, we have

$$\Delta P_{23} = \Delta\theta_2 = 0 \text{ pu}, \tag{2.57}$$

and

$$\Delta P_2 = -1 \text{ pu} = 100\,(-\Delta\theta_1 + 2\Delta\theta_2), \tag{2.58}$$

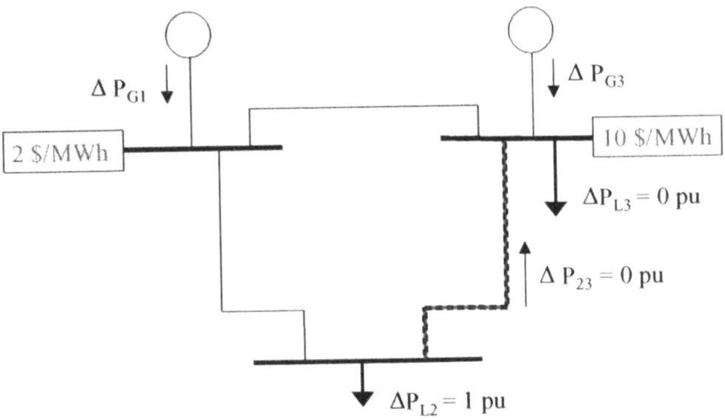

Figure 2.7. Incremental power constraints for LMP calculation [40].

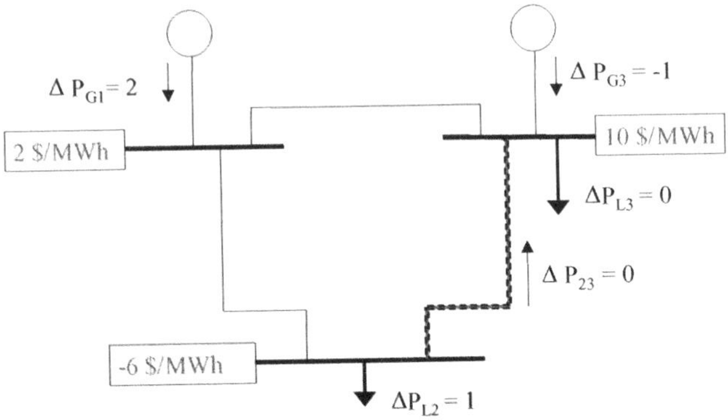

Figure 2.8. LMPs on the 3 bus-system under congestion of *line 2-3* [40].

Solving for $\Delta\theta_1$ and $\Delta\theta_2$, we get:

$$\Delta\theta_1 = 0.01 \text{ rads and } \Delta\theta_2 = 0 \text{ rads} \tag{2.59}$$

Therefore, the following power generations at *bus 1* and *3* are obtained:

$$\Delta P_1 = 100\,(2\,\Delta\theta_1 - \Delta\theta_2) = 2 \text{ pu}, \tag{2.60}$$

$$\Delta P_3 = 100\,(-\Delta\theta_1) = -1 \text{ pu}. \tag{2.61}$$

Consequently, the LMP at *bus 2* amounts to:

$$\text{LMP2} = \frac{(2)\times(2\,\$/\text{MWh}) + (-1)\times(10\,\$/\text{MWh})}{(2)+(-1)} = -6\,\$/\text{MWh}. \tag{2.62}$$

The LMPs so obtained are depicted in Figures 2.8 and 2.9. The negative value of LMP2 indicates that the load on *bus 2* will receive \$600 for a consumption of 100 MWh. On the other hand, since the LMP on *bus 3* is equal to 10 \$/MWh, the load of 700 MW on that bus will be charged \$7000. Therefore, the latter receives a clear signal to relocate itself close to *bus 2*, if it could.

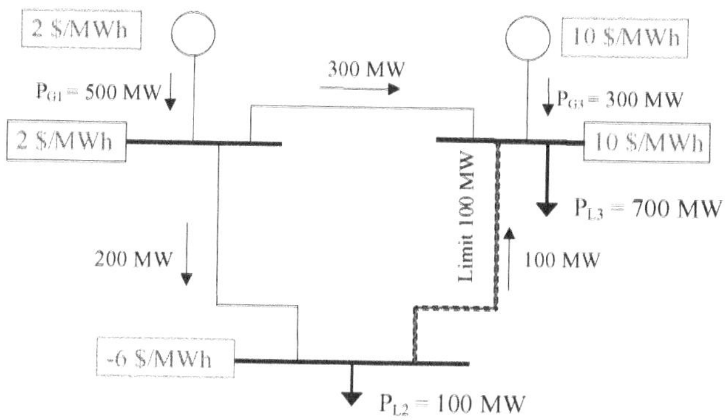

Figure 2.9. LMPs and power flows in MW under congestion on line 2-3 [40].

4. Congestion Charges and FTRs

What about the congestion charges that will be paid by each of the load serving entities (LSEs) owning respectively *units 1* and *3*, termed LSE1 and LSE3? They are calculated as follows:

$$100\ \text{MW} \times [(\$10) - (-\$6)] = \$1{,}600 \tag{2.63}$$

If LSE1 holds an FTR of 100 MW from *bus 2* to *bus 3*, then it will get a credit of:

$$100\ \text{MW} \times [(\$10) - (-\$6)] = \$1{,}600 \tag{2.64}$$

which fully compensates the congestion charge. On the other hand, if LSE3 holds an FTR of 100 MW from bus 3 to bus 2, then it will get a negative credit of:

$$100\ \text{MW} \times [(-\$6) - (\$10)] = -\$1{,}600 \tag{2.65}$$

implying that LSE3 has to pay \$1,600. In this case, the FTR does not compensate the congestion charge.

2.5 LMP CALCULATION USING ADAPTIVE DYNAMIC PROGRAMMING

2.5.1 Overview of the Static LMP Problem

The current LMP calculation is based on classical optimization previously discussed in this chapter. The determination of LMP or spot prices is obtained from optimal power flow solutions. The general formulation can be summarized as minimizing a welfare cost function subject to power balance, network, network security, and power market constraints. Specifically, let C_S and C_D denote the cost vectors of the supply bid vector P_S and the demand bid vector P_D, respectively. Then this formulation can be written as follows [51]:

$$\text{Minimize} \quad f(P_s, P_D) = C_s^T P_s - C_D^T P_D \tag{2.66}$$

subject to:

$$g(x, u, P_s, P_D) = 0 \quad \text{(Power balance)} \tag{2.67}$$

$$Q_G^{\min} \leq Q_G \leq Q_G^{\max} \quad \text{(Gen Q-Limits)} \tag{2.68}$$

$$V^{\min} \leq V \leq V^{\max} \quad \text{(Bus voltage limits)} \tag{2.69}$$

$$|P_{ij}(x)| \leq P_{ij}^{\max} \quad \text{(Thermal limits)} \tag{2.70}$$

$$\lambda_c \leq \lambda_{c_o} \quad \text{(Stability loading at 'critical points')} \tag{2.71}$$

$$0 \leq P_S \leq P_S^{\max} \quad \text{(Supply bids)} \tag{2.72}$$

$$0 \leq P_D \leq P_D^{\max} \quad \text{(Demand bids)} \tag{2.73}$$

This problem may be solved using linear or nonlinear programming methods that are make use of specialized techniques pertaining to classical optimization. These include interior point methods, Lagrangian or Newtonian approaches, and barrier penalty functions [52]. In a more general setting, we form the Lagrangian function given by:

$$L(x, u) = f(x, u) + \sum_{i \in p} \lambda_i g_i(x, u) + \sum_{j \in m} \lambda_j h_j(x, u) \tag{2.74}$$

where λ_i and λ_j are Lagrange multipliers of the equality and inequality constraints in a typical Optimal Power Flow (OPF) calculation. When applied to the latter problem, the Kuhn-Tucker necessary optimality conditions lead to:

$$\frac{\partial L}{\partial P_{S_i}} = C_{S_i} - \lambda_{S_i} + \mu P_{S_i}^{\max}, \tag{2.75}$$

$$\frac{\partial L}{\partial P_{D_i}} = C_{D_i} - \lambda_{D_i} + \mu P_{D_i}^{\max} \tag{2.76}$$

where parameter μ is now a barrier penalty function [51] and the aggregate locational marginal prices for all ith nodes in the system are $\text{LMP}_i = \lambda_i$ for $\forall i \in N_{buses}$. These values represent the shadow price or marginal costs for each market participant located at the ith node in the power system. The calculation of these LMP_i requires deterministic economic data, such as bid and cost schedules and load forecasts, together with the conventional data used in a typical security constrained optimal power flow.

By using the criteria for LMP calculation stated in a more general form, the foregoing optimization problem gives rise to lambda parameters that can be grouped into the components of energy, congestion, and losses. Their summation represents the nodal price at the reference or slack bus, which represents the marginal cost that accounts for the distribution of transmission losses, and the marginal cost of transmission congestion relative to the power injections.

2.5.2 LMP in Stochastic and Dynamic Market with Uncertainty

Currently, both day-ahead and hour-ahead markets are performed based on *ad hoc* and separated forecasts of energy needs, system congestions, and system contingencies, among others. Obviously, a better approach would be to perform all these forecasts in an integrated and unified manner. This methodology would allow market operators to achieve an optimal investment and operational decision making under uncertain dynamic conditions. Discrepancies between the predicted and observed market characteristics will reveal potential anomalies and gaming opportunities that may prevent a reliable and efficient operation of power systems. In short, it will make the market more transparent and more efficient.

To achieve the general objective of cost minimization subject to system operational and reliability constraints, we propose the development of adaptive dynamic stochastic optimization schemes that will include a prediction and a correction step of the system state and market characteristics. As shown in [53, 54], a good candidate utilizes adaptive dynamic programming based on back-propagation neural networks. Simply put, the proposed methodology consists of three components. The first component consists in a dynamic state estimation under contingencies, which incorporates load and state prediction and correction. The second component consists in the action network that is capable to adapt itself to any changes in the system state with respect to the one predicted by the dynamic state estimation. As for the third component, it consists in the critic network whose main goal is to evaluate the performance of the prediction-correction scheme carried out by the other two components. A block diagram of the adaptive dynamic programming process is displayed in Figure 2.10.

Let $R(t)$ denote the observed vector of the state vector $X(t)$ of a system to be controlled via the action vector $u(t)$ and let $v(t)$ denote the observation noise. The objective is to maximize a scalar-valued performance index $J(R(t), u(t), v(t))$ over the long run through $u(t)$. This performance index is directly related to the utility function $U(R(t), u(t))$, which is defined by the designer as follows:

$$J(R(t),u(t),v(t)) = U(R(t),u(t),v(t)) + \langle J(R(t+1),u(t+1),v(t+1))\rangle + U_0 \tag{2.77}$$

where $\langle \cdot \rangle$ stands for the expectation operator and U_0 is an intercept or bias term that prevents the system state to become unbounded. Then LMP can be defined as a function of the derivative of the performance index with respect to $R(t)$, which is given by

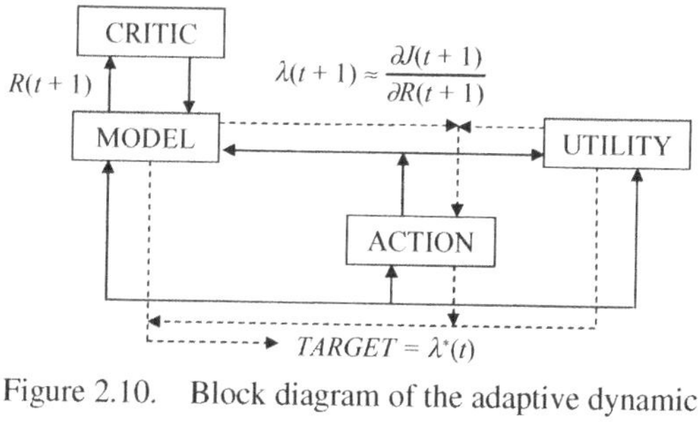

Figure 2.10. Block diagram of the adaptive dynamic programming process.

$$\lambda = \frac{\partial J(R(t), u(t), v(t))}{\partial R(t)} \tag{2.78}$$

This LMP can be regarded as a generalization of the conventional LMP in the stochastic and dynamic sense.

2.6 CONCLUSIONS

Ranked among the largest and most dynamic economic structures in the world, energy markets are of paramount importance to the sustenance of proving a basic commodity—electricity. Over the decades, small-scale energy markets have evolved in size, operational complexity, and regulatory practices into a more competitive environment in which power system deregulation replaces the conventional, vertically-integrated monopolies. This trend has given rise to various market designs to generate, transmit, and deliver electric energy in a growing number of countries, worldwide. In the U.S., this has led to the development of several market models developed by PJM-ISO, California-ISO, New-York-ISO, to cite a few.

In all of these models, the LMP signals are calculated from the results of static state estimation techniques and separated forecasts of energy consumptions and contingency analyses. A unified approach based on adaptive dynamic programming is needed to account for the stochastic and dynamic characteristics of the market. This approach has the potential to lead to the development of more transparent and more efficient electricity markets.

BIBLIOGRAPHY

[1] B. Allaz, and J. Vila, "Cournot Competition, Forwards Markets and Efficiency," *Journal of Economic Theory*, Vol. 59, pp. 1–16, 1993.

[2] B. L. Andersson and L. Bergman, "Market Structure and the Price of Electricity: An Ex-ante Analysis of Deregulated Swedish Markets," *The Energy Journal*, Vol. 16, No. 2, pp. 97–109, 1995.

[3] S. Awerbuch and A. Preston (Eds.). *The Virtual Utility: Accounting, Technology & Competitive Aspects of the Emerging Industry*. Kluwer Academic Pub., 1997.

[4] R. Baldick, R. Grant, and E. Kahn, "Theory and Application of Linear Supply Function Equilibrium in Electricity Markets," *Journal of Regulatory Economics*, Vol. 25, No. 2, pp. 143–167, 2004.

[5] B. R. Barkovich and D. V. Hawk, "Charting a New Course in California," *IEEE Spectrum*, pp. 26–31, July 1996.

[6] S. Borenstein and J. Bushnell, "An Empirical Analysis of the Potential for Market Power in California's Electricity Industry," *Journal of Industrial Economics*, Vol. 47, No. 3, pp. 285–323, 1999.

[7] S. Borenstien, J. Bushnell, and S. Stoft, "The Competitive Effects of Transmission Capacity in a Deregulated Electricity Industry," *RAND Journal of Economics*, Vol. 31, No. 2, pp. 294–325, 2000.

[8] S. Borenstein and J. Bushnell, "Electricity Restructuring: Deregulation or Reregulation," *Regulation*, Vol. 23, No. 2, pp. 46–52, 2000.

[9] E. A. Bretz, "PJM Interconnection: Model of a Smooth Operator," *IEEE Spectrum*, pp. 50–55, June 2000.

[10] J. B. Cardell, C. C. Hitt, and W. W. Hogan, Market Power and Strategic Interaction in Electricity Networks. *Resource and Energy Economics*, Vol. 19, No. 1–2, pp. 109–137, 1997.

[11] G. Cauley, P. Hirsch, A. Vojdani, T. Saxton, and F. Cleveland, "Information Network Supports Open Access," *IEEE Computer Applications in Power*, pp. 12–19, July 1996.

[12] D. Chattopadhyay, "Multicommodity Spatial Cournot Model for Generator Bidding Analysis," *IEEE Trans. on Power Systems*, Vol. 19, No. 1, pp. 267–275, 2004.

[13] R. D. Christie and B. F. Wollenberg, "Transmission Management in the Deregulated Environment," *Proceedings of the IEEE*, Vol. 88, No. 2, pp. 170–195, Feb. 2000.

[14] J. Contreras, M. Klusch, and J. B. Krawczyk, "Numerical Solutions to Nash—Cournot Equilibria in Coupled Constraint Electricity Markets," *IEEE Trans. on Power Systems*, Vol. 19, No. 1, pp. 195–206, 2004.

[15] L. B. Cunningham, R. Baldick, and M. L. Baughman, "An Empirical Study of Applied Game Theory: Transmission Constrained Cournot Behavior," *IEEE Trans. on Power Systems*, Vol. 17, No. 1, pp. 166–172, 2004.

[16] C. J. Day, B. F. Hobbs, and J. Pang, "Oligopolistic Competition in Power Networks: A Conjectured Supply Function Approach," *IEEE Trans. on Power Systems*, Vol. 17, No. 3, pp. 597–607, 2002.

[17] M. Einhorn, and R. Siddiqi (Eds.). *Electricity Transmission Pricing and Technology*. Kluwer Academic Publishers, 1996.

[18] N. Fabra, N. Von der Fehr, and D. Harbord, "Designing Electricity Auctions," *RAND Journal of Economics*, 2006.

[19] FERC Order 888, 889, 2000. http://www.ferc.gov

[20] D. Fudenberg and D. Levine, *The Theory of Learning in Games*. MIT Press, Cambridge, MA, 1998.

[21] D. Gabel and D. F. Weiman (Eds.). *Opening Networks to Competition: The Regulation and Pricing of Access*. Kluwer Academic Publishers, 1998.

[22] A. Garcia, E. Campos-Nãnez, and J. D. Reitzes, "Dynamic Pricing and Learning in Electricity Markets," *Operations Research*, Vol. 53, No. 2, pp. 231–241, 2005.

[23] A. Garcia, J. D. Reitzes, and E. Stacchetti, "Strategic Pricing When Electricity is Storable," *Journal of Regulatory Economics*, Vol. 20, No. 3, pp. 223–247, 2001.

[24] R. Green, "Increasing Competition in the British Electricity Spot Market," *Journal of Industrial Economics*, Vol. 44, No. 2, pp. 205–216, 1996.

[25] R. Green and D. Newbery, "Competition in British Electricity Spot Market," *Journal of Political Economy*, Vol. 100, No. 5, pp. 929–953, 1992.

[26] P. G. Harris, "Impacts of Deregulation on the Electric Power Industry," *IEEE Power Engineering Review*, pp. 4–6, Oct. 2000.

[27] S. Hunt and G. Shuttleworth, *Competition and Choice in Electricity*. John Wiley, 1996.

[28] S. Hunt and S. Shuttleworth, "Unlocking the Grid," *IEEE Spectrum*, pp. 20–25, July 1996.

[29] B. F. Hobbs, "Network Models of Spatial Oligopoly with an Application to Deregulation of Electricity Generation," *Operations Research*, Vol. 34, No. 3, pp. 395–409, 1986.

[30] B. F. Hobbs, "Linear Complementarity Models of Nash—Cournot Competition in Bilateral and POOLCO Power Markets," *IEEE Trans. on Power Systems*, Vol. 16, No. 2, pp. 194–202, 2001.

[31] B. F. Hobbs, C. B. Metzler, and J-S. Pang, "Strategic Gaming Analysis for Electric Power Systems: An MPEC Approach," *IEEE Trans. on Power Systems*, Vol. 15, No. 2, pp. 638–645, 2000.

[32] B. F. Hobbs, F. A. M. Rijkers, and M. Boots, The More Cooperation, the More Competition? A Cournot Analysis of the Benefits of Electric Market Coupling. *The Energy Journal*, 26(4), 69–97, 2005.

[33] B. F. Hobbs and R. E. Schuler, Assessment of the Deregulation of Electric Power Generation Using Network Models of Imperfect Spatial Competition. *Papers Regional Science Assoc.* 57, 75–89, 1985.

[34] W. W. Hogan, A Market Power Model with Strategic Interaction in Electricity Networks. *The Energy Journal*, 18(4), 107–135, 1997.

[35] M. Ilic, F. Galiana, and L. Fink, Editors. *Power Systems Restructuring: Engineering and Economics*. Kluwer Academic Publishers, 1998.

[36] P. Klemperer and M. Meyer, "Supply Function Equilibria in Oligopoly Under Uncertainty," *Econometrica*, Vol. 57, pp. 1243–1277, 1989.

[37] C. Metzler, B. F. Hobbs, and J.-Y. Pang, "Nash-Cournot Equilibria in Power Markets on a Linearized DC Network with Arbitrage: Formulations and Properties," *Networks & Spatial Economics*, Vol. 3, No. 2, pp. 123–150, 2003.

[38] K. Neuho, J. Barquin, M. G. Boots, A. Ehrenmann, B. F. Hobbs, F. A. M. Rijkers, "Network-Constrained Cournot Models of Liberalized Electricity Markets: The Devil is in the Details," *Energy Economics*, Vol. 27, pp. 495–525, 2005.

[39] S. S. Oren, "Economic Inefficiency of Passive Transmission Rights in Congested Electricity Systems with Competitive Generation," *The Energy Journal*, Vol. 18, No. 4, pp. 63–83, 1997.

[40] PJM Documents and Training Materials. http:Materials. http://www.pjm.com.

[41] F. A. Rahimi and A. Vojdani, "Meet the Emerging Transmission Market Segments," *IEEE Power Engineering Review*, pp. 26–32, Jan. 1999.

[42] A. Rudkevich, M. Duckworth, and R. Rosen, "Modeling Electricity Pricing in a Deregulated Generation Industry, *The Energy Journal*, Vol. 19, No. 3, pp. 19–48, 1998.

[43] F. C. Schweppe, M. C. Caramanis, R. D. Tabors, and R. E. Bohn, *Spot Pricing of Electricity*. Kluwer Academic Pub., 1988.

[44] T. Li and M. Shahidehpour, "Strategic Bidding of Transmission Constrained GENCOs with Incomplete Information," *IEEE Trans. on Power Systems*, Vol. 20, No. 1, pp. 437–447, 2005.

[45] N. Von der Fehr and D. Harbord, "Spot Market Competition in the U.K. Electricity Industry," *Economic Journal*, Vol. 103, pp. 531–546, May 1993.

[46] J.-Y. Wei and Y. Smeers, "Spatial Oligopolistic Electricity Models with Cournot Generators and Regulated Transmission Prices," *Operations Research*, Vol. 47, No. 1, pp. 102–112, 1999.

[47] B. Willems, "Modeling Cournot Competition in an Electricity Market with Transmission Constraints," *The Energy Journal*, Vol. 23, No. 3, pp. 95–125, 2003.

[48] A. Wood and B. Wollenberg, *Power Generation, Operation and Control*. 2nd Ed., John Wiley, 1996.

[49] G. Zaccour (Ed.). *Deregulation of Electric Utilities*. Kluwer Academic Pub., 1998.

[50] A. Zobian and M. D. Ilic, "Unbundling of Transmission and Ancillary Services, Part I: Technical Issues and Part II: Cost-Based Pricing Framework," *IEEE Trans. on Power Systems*, Vol. 12, No. 2, pp. 539–558, May 1997.

[51] C. A. Canizares, H. Chen, and W. Rosehart, "Pricing System Security in Electricity Markets," *Proceedings, Bulk Power System Dynamics and Control—V*, Onomichi, Japan, August 2001.

[52] J. A. Momoh, "*Electric Power Systems Applications of Optimization*", Marcel Dekker Inc., New York, First Edition 2001.

[53] S. Jennie, A. G. Barto, W. B. Powell, and D. Wunsch (Eds.). *Handbook of Learning and Approximate Dynamic Programming*. John Wiley & Sons, Inc., pp. 561–598, 2004.

[54] D. V. Prokhorov and D. C. Wunsch, "Adaptive Critic Designs," *IEEE Trans. Neural Network*, Vol. 8, pp. 997–1007, Sept. 1997.

CHAPTER 3

ALTERNATIVE ECONOMIC CRITERIA AND PROACTIVE PLANNING FOR TRANSMISSION INVESTMENT IN DEREGULATED POWER SYSTEMS

Enzo E. Sauma[1] *and Shmuel S. Oren*[2]

[1]Pontificia Universidad Católica de Chile at Santiago

[2]University of California at Berkeley

EDITOR'S SUMMARY: This chapter advocates the use of a multistage game model for transmission expansion as a new planning paradigm that incorporates the effects of strategic interaction between generation and transmission investments and the impact of transmission on spot energy prices. The paper also examines the policy implication of different conflicting incentives for generation and transmission investments. To this end, the authors formulate transmission planning as an optimization problem under alternative conflicting objectives. The inter-relationship between generation and transmission investment as it affects social value of transmission capacity is investigated. A simple illustrative example is provided to investigate the policy implications of divergent expansion plans resulting from the planner's level of anticipation of strategic responses and co-optimization of generation and transmission investment. First, it is found that the transmission expansion plans may be very sensitive to supply and demand parameters and hence will be affected by the assumption regarding generation investment and costs. Secondly, it is shown that the transmission investment has an important distributional impact, inducing acute conflicts of interests among market participants. To overcome this problem, a three-stage game theoretic model for transmission investment is proposed to foster proactive transmission

Economic Market Design and Planning for Electric Power Systems, Edited by James Momoh and Lamine Mili

expansion. A comparison between proactive and reactive network planners is made. It is stated that unlike the former, the reactive network planner does not account for the ability of generation investment to respond strategically in response to transmission expansion.

3.1 INTRODUCTION

Transmission investment in vertically integrated power industries were traditionally motivated by reliability considerations as well as by the economic objective of connecting load areas to remote, cheap generation resources. This was done within the framework of an integrated resource planning paradigm in order to minimize investment in transmission generation and energy cost while meeting forecasted demand and reliability criteria. The cost of such investments, once approved by the regulator, plus an adequate return on investment, has been incorporated into customers' rate base. Vertical unbundling of the electricity industry and the reliance on market mechanisms for pricing and return on investments have increased the burden of economic justification for investment in the electricity infrastructure. The role of regional assessment of transmission expansion needs and approval of proposed projects has shifted in many places from the integrated utility to a regional transmission organization (RTO), which is under the jurisdiction of the Federal Energy Regulatory Commission (FERC), while the funding of such projects through the regulated rates is still under the jurisdiction of state regulators.

In evaluating the economic implications of transmission expansions the RTO and state regulators must take into consideration that, in a market-based system, such expansions may create winners and losers, even when the project as a whole is socially justifiable on the grounds of reliability improvements and energy cost savings. Furthermore, in the new environment, transmission expansion may also be justified as a means for facilitating free trade and as a market mitigation approach to reducing locational market power.

From an economic theory perspective, the proper criterion for investment in the transmission infrastructure is the maximization of social welfare, which is composed of consumers' and producers' surplus, which also accounts for investment cost and may account for reliability by including the social cost of unreliability in this objective function. When demand is treated as inelastic, social welfare maximization is equivalent to total cost minimization including energy cost, investment cost and cost of lost load or another measure of unreliability cost. The validity of this economic objective is premised on the availability of *adequate and costless* (without transaction costs) transfer mechanisms among market participants, which assures that increases in social welfare will result in Pareto improvements (making all participants better off or neutral).

However, this principle is not always true in deregulated electric systems, where transfers are not always feasible and even when attempted are subject to many imperfections. In the U.S. electric system, which was originally designed to serve a vertically integrated market, there are misalignments between payments and rewards associated with use and investments in transmission. In fact, while payments for transmission investments and for its use are made locally (at state level), the economic impacts from these transmission investments extend beyond state boundaries so that the planning and approval process for such investment falls under FERC jurisdiction. As a result of such jurisdictional conflict, adequate side payments among market participants are not always physically or politically feasible (for instance, this would be the case of a network expansion that benefits a particular generator or load in another state, so that the cost of the expansion is not paid for by those who truly benefit from it).[1] Consequently, the maximization of social welfare may not translate to Pareto efficiency and other optimizing objectives should be considered. Unfortunately, alternative objectives may produce conflicting results with regard to the desirability of transmission investments.

One potential solution to the aforementioned jurisdictional conflict is the so-called "participant funding," which was proposed by FERC in its 2002 Notice of Proposed Rulemaking (NOPR) on Standard Market Design (FERC 2002, 98–115). Roughly, participant funding is a mechanism whereby one or more parties seeking the expansion of a transmission network (who will economically benefit from its use) assume funding responsibility. This scheme would assign the cost of a network expansion to the beneficiaries from the expansion, thus eliminating (or, at least, mitigating) the above-mentioned side-payments' problem. This policy is based on the rationale that, although most network expansions are used by and benefit all users, some few network expansions will only benefit an identifiable customer or group of customers (such as a generator building to export power or a load building to reduce congestion).

Although participant funding would potentially encourage greater regional cooperation to get needed facilities sited and built, this approach has some caveats in practice. The main shortcomings of participant funding are:

- The benefits from network upgrades are difficult to quantify and to allocate among market participants (and, thus, it could be difficult to identify and avoid detrimental expansions that benefit some participants, either at the expense of others or by decreasing social welfare).
- Mitigation of network bottlenecks is likely to require a program of system-wide upgrades, from which almost all market participants are likely to benefit, but for which the cumulative benefits can be difficult to capture through participant funding.
- After some period of time (but less than the economic life of the upgrade), if the benefits begin to accrue to a broader group of customers, then some form of crediting mechanism should be established to reimburse the original funding participants. However, this would basically be a reallocation of sunk costs.

[1]For example, it is really hard to convince people in Idaho that they should pay for a transmission line connecting Idaho and California to carry their cheap power to Californians. On the contrary, they would probably be worried about both a likely increase in their electricity prices and a potential reduction in the reliability of their own system because of the increased risk of cascading failures (due to the expansion).

- Participant funding could lead to a sort of "incremental expansions" over time. Because transmission investments tend to be lumpy, these incremental expansions may be inefficient in the long run and more costly to consumers.
- Providing some form of physical (capacity-reservation) rights in exchange for participant-funded investments could allow the exercise of market power by the withholding of the new capacity and, thereby, create new transmission bottlenecks.
- An extensive reliance on participant funding and incentive rates for transmission could lead to accelerated depreciation lives for ratemaking purposes, which will increase the risk profile for this portion of the industry.

Most of the works found in the literature about transmission planning in deregulated electric systems consider single-objective optimization problems (maximization-of-social-welfare in most of the cases), while literature that considers multiple optimizing objectives is scarce. London Economics International LLC (2002) developed a methodology to evaluate specific transmission proposals using an objective function for transmission appraisal that allows the user to vary the weights applied to producer and consumer surpluses. However, London Economics' study has no view on what might constitute appropriate weights nor on how changes in the weights affect the proposed methodology. Sun and Yu (2000) propose a "multiple-objective" optimization model for transmission expansion decisions in a competitive environment. To solve this model, however, the authors convert it into a single-objective optimization model by using fuzzy set theory. Styczynski (1999) uses a multiple-objective optimization algorithm to clarify some issues related to the transmission planning in a deregulated environment. The fact that most of this work is directly applied to the European distribution expansion problem, which is nearly optimally solved, makes uncertain the real value of this model in practice. Shrestha and Fonseka (2004) utilize a trade-off between the change in the congestion cost and the investment cost associated with a transmission expansion in order to determine the optimal expansion decision. Unfortunately, this work is not very useful in practice because of some excessively simplistic assumptions made in their decision model (e.g., ignoring the exercise of market power by generation firms).

Although some authors have used multiple optimizing objectives for transmission planning, none of them has analyzed the conflicts among these different objectives and their policy implications. This chapter attempts to show that different desired optimizing objectives can result in divergent optimal expansions of a transmission network and that this fact entails some very important policy implications, which should be considered by any decision maker concerned with transmission expansion.

The rest of the chapter is organized as follows. In Section 3.2, we present a simple radial-network example that illustrates how different optimizing objectives can result in divergent optimal expansion plans of a network. Section 3.3 explains the policy implications of the conflicts among these different optimizing objectives. In Section 3.4, we suggest a three-period model of transmission investments to evaluate transmission expansion projects. This model takes into account the policy implications of the conflicting incentives for transmission investment and explicitly considers the interrelationship between generation and transmission investments in oligopolistic power systems. In Section 3.5, we illustrate the results of our three-period model with a numerical example. Section 3.6 concludes the chapter and describes future work.

3.2 CONFLICTING OPTIMIZATION OBJECTIVES FOR NETWORK EXPANSIONS

3.2.1 A Radial-Network Example

For any given network, the network planner would ideally like to find and implement the transmission expansion that maximizes social welfare, minimizes the local market power of the agents participating in the system, maximizes consumer surplus and maximizes producer surplus. Unfortunately, these objectives may produce conflicting results with regard to the desirability of various transmission expansion plans. In this section, we illustrate, through a simple example, the divergent optimal transmission expansions based on different objective functions, and the difficulty of finding a unique network expansion policy.

We shall use a simple two-node network example, as shown in Figure 3.1, which is sufficient to highlight the potential incompatibilities among the planning objectives and their policy implications. This example is chosen for simplicity reasons and does not necessarily represent the behavior of a real system.

As a general framework of the example presented here, we assume that the transmission system uses nodal pricing, transmission losses are negligible, consumer surplus is the correct measure of consumer welfare (e.g., consumers have quasi-linear utility), generators cannot purchase transmission rights (and, thus, their bidding strategy is independent of the congestion rent), and the Lerner index (defined as the fractional price markup, i.e. [price—marginal cost] / price) is the proper measure of local market power.

Consider a network composed of two unconnected nodes where electricity demand is served by local generators. Assume Node 1 is served by a monopoly

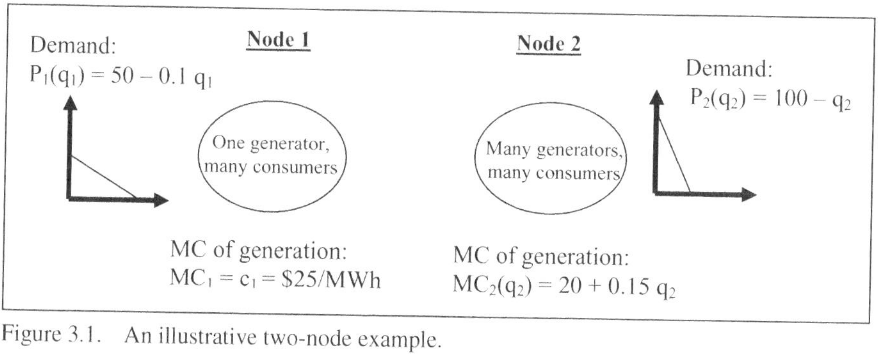

Figure 3.1. An illustrative two-node example.

producer, while Node 2 is served by a competitive fringe.[2] For simplicity, suppose that the generation capacity at each node is unlimited. We also assume both that the marginal cost of generation at Node 1 is constant (this is not a critical assumption, but it simplifies the calculations) and equal to c_1 = \$25/MWh, and that the marginal cost of generation at node 2 is linear in quantity and given by $MC_2(q_2) = 20 + 0.15 \cdot q_2$. Moreover, we assume linear demand functions. In particular, the demand for electricity at Node 1 is given by $P_1(q_1) = 50 - 0.1 \cdot q_1$, while the demand for electricity at Node 2 is given by $P_2(q_2) = 100 - q_2$.

We analyze the optimal expansion of the described network under each of the following optimizing objectives: (1) maximization of social welfare, (2) minimization of local market power, (3) maximization of consumer surplus, and (4) maximization of producer surplus.[3] We limit the analysis to only two possible network expansion options: doing nothing (that is, keeping each node as self-sufficient); or building a transmission line with "adequate" capacity (that is, building a line with high-enough capacity so that the probability of congestion is very small). For the particular cases we present here, we can easily verify that the optimal expansion under each of the four considered optimizing objectives is truly either doing nothing or building a transmission line with adequate capacity. In the general case, we can justify this simplification based on the lumpiness of transmission investments.

Under the scenario in which each node satisfy its demand for electricity with local generators (self-sufficient-node scenario), the generation firm located at Node 1 behaves as a monopolist (that is, it chooses a quantity such that its marginal cost of supply equals its marginal revenue) while the generation firms located at Node 2 behave as competitive firms (that is, they take the electricity price as given by the market-clearing rule: demand equals marginal cost of supply).

Accordingly, under the self-sufficient-node scenario (SSNS), the generation firm at Node 1 optimally produces $q_1^{(SSNS)}$ = 125 MWh and charges $P_1^{(SSNS)}$ = \$37.5/MWh. With this electricity quantity and price, the producer surplus at Node 1 (which, in this example, is equivalent to the monopolist's profit) is $PS_1^{(SSNS)}$ = \$1,563/h and the consumer surplus at this node equals $CS_1^{(SSNS)}$ = \$781/h. The Lerner index at Node 1 is $L_1^{(SSNS)} = 0.33$.[4] On the other hand, under the SSNS, the generation firms located at Node 2 optimally produce an aggregate amount equal to

[2]The fact that the generation firm located at Node 1 can exercise local market power is a crucial assumption for the purpose of this example. Without considering local market power, the results we show in this section are no longer valid. However, this supposition is fairly realistic. In fact, perfectly competitive markets are not very common in the power generation business. In our example, the perfect-competition assumption at Node 2 is only made for simplicity and it can be eliminated without changing any of the qualitative results presented in this section.

[3]In this section, we show that, for given demand functions, the optimal expansions under the four considered optimizing objectives vary depending on the cost structures of generators. To do this, we analyze the optimal expansion of the two-node network when changing the marginal cost of generation at Node 1 (i.e., when we change c_1) while keeping unaltered the cost structure of the generators at Node 2.

[4]Under monopoly, if the marginal cost of production is constant and equal to c and the demand is linear, given by $P(q) = a - b \cdot q$, where $a > c$, then the monopolist will optimally produce $q^{(M)} = (a - c) / (2b)$ and charge a price $P^{(M)} = (a + c) / 2$, making a profit of $\Pi^{(M)} = (a - c)^2 / (4b)$. Under these assumptions, the consumer surplus is equal to $CS^{(M)} = (a - c)^2 / (8b)$, and the Lerner index at the monopolist's node is equal to $L^{(M)} = (P^{(M)} - c) / P^{(M)} = (a - c) / (a + c)$.

$q_2^{(SSNS)} = 69.6$ MWh, and the market-clearing price is $P_2^{(SSNS)} = \$30.4$/MWh. With this electricity quantity and price, the producer surplus at Node 2 is $PS_2^{(SSNS)} = \$363$/h and the consumer surplus at this node is $CS_2^{(SSNS)} = \$2{,}420$/h.[5] From the previous results, we can compute the total producer surplus, the total consumer surplus, and the social welfare under the SSNS. The numerical results are given by: $PS^{(SSNS)} = PS_1^{(SSNS)} + PS_2^{(SSNS)} = \$1{,}926$/h; $CS^{(SSNS)} = CS_1^{(SSNS)} + CS_2^{(SSNS)} = \$3{,}201$/h; and $W^{(SSNS)} = PS^{(SSNS)} + CS^{(SSNS)} = \$5{,}127$/h; respectively.

Now, we consider the scenario in which there is adequate (ideally unlimited) transmission capacity between the two nodes (nonbinding-transmission-capacity scenario). Under this scenario, the generation firms face an aggregate demand given by:

$$P(Tq) = \begin{cases} 100 - Tq, & \text{if } Tq < 50 \\ 54.5 - 0.09 \cdot Tq, & \text{if } Tq \geq 50' \end{cases} \tag{3.1}$$

in which Tq is the total quantity of electricity produced. That is, $Tq = q_1 + q_2$, in which q_1 is the amount of electricity produced by the firm located at Node 1 and q_2 is the aggregate amount of electricity produced by the firms located at Node 2.

Under the nonbinding-transmission-capacity scenario (NBTCS), the two nodes may be treated as a single market in which the generator at Node 1 and the competitive fringe at Node 2 jointly serve the aggregate demand of both nodes at a single market clearing price. We assume that the monopolist at Node 1 behaves as a Cournot oligopolist interacting with the competitive fringe. That is, under the NBTCS, we assume both that the monopolist at Node 1 chooses a quantity such that its marginal cost of supply equals its marginal revenue, taking the output levels of the other generation firms as fixed, and that the generation firms at Node 2 still take the electricity price as given by the market-clearing rule.

Thus, according to the Cournot assumption, under the NBTCS, the monopolist at Node 1 optimally produces $q_1^{(NBTCS)} = 112$ MWh while the competitive fringe at Node 2 optimally produces $q_2^{(NBTCS)} = 101.2$ MWh (these output levels imply that there is a net transmission flow of 36 MWh from Node 2 to Node 1). In this case, the market-clearing price (which is the price charged by all firms to consumers) is $P^{(NBTCS)} = \$35.2$/MWh. With these new electricity quantities and prices, the producer surplus at Node 1 is equal to $PS_1^{(NBTCS)} = \$1{,}139$/h and the producer surplus at Node 2 is equal to $PS_2^{(NBTCS)} = \$768$/h.[6] As well, the consumer surpluses are

[5] Under perfect competition, if the marginal cost of supply is linear, given by $MC(q) = c + d \cdot q$, and the inverse demand function is given by $P(q) = a - b \cdot q$, where $a > c$, then the market will optimally produce a quantity $q^{(PC)} = (a - c) / (b + d)$ and the market-clearing price will be $P^{(PC)} = (a \cdot d + b \cdot c) / (b + d)$. Under these assumptions, the producer surplus is equal to $PS^{(PC)} = (d \cdot (a - c)^2) / (2 \cdot (b + d)^2)$ and the consumer surplus is $CS^{(PC)} = (b \cdot (a - c)^2) / (2 \cdot (b + d)^2)$.

[6] Under the NBTCS, assuming generators behave as Cournot firms, if the marginal costs of supply at Node 1 and Node 2 are $MC_1(q_1) = c_1$ and $MC_2(q_2) = c_2 + d_2 \cdot q_2$ respectively, and the aggregate demand is linear, given by $P(Tq) = A - B \cdot Tq$, where $A > c_1$ and $A > c_2$, then the optimal output levels solve the following two equations:

$$A - 2 \cdot B \cdot q_1 - B \cdot q_2 = c_1 \quad (\text{or } \mathrm{MR}_1 = \mathrm{MC}_1) \text{ and}$$
$$A - B \cdot (q_1 + q_2) = c_2 + d_2 \cdot q_2 \quad (\text{or } P^{(NBTCS)} = \mathrm{MC}_2)$$

$CS_1^{(NBTCS)}$ = \$1,099/h for Node 1's consumers and $CS_2^{(NBTCS)}$ = \$2,101/h for Node 2's consumers. The new Lerner index at Node 1 is $L_1^{(NBTCS)} = 0.29$.

From the above results, we can compute the total producer surplus, the total consumer surplus, and the social welfare under the NBTCS. However, these calculations require knowing who is responsible for the transmission investment costs. Without loss of generality, we assume that an independent entity (other than the existing generation firms and consumers) incurs in the transmission investment costs. Consequently, under the NBTCS, total producer surplus (not accounting for transmission investment cost) is $PS^{(NBTCS)} = PS_1^{(NBTCS)} + PS_2^{(NBTCS)}$ = \$1,907/h; total consumer surplus is $CS^{(NBTCS)} = CS_1^{(NBTCS)} + CS_2^{(NBTCS)}$ = \$3,200/h; and social welfare is $W^{(NBTCS)} = PS^{(NBTCS)} + CS^{(NBTCS)}$ − investment costs = \$5,107/h − investment costs.

Comparing both the SSNS and the NBTCS, we can observe that the expansion that minimizes local market power is building a transmission line with "adequate" capacity (at least theoretically, with capacity greater than 36 MWh) since $L^{(NBTCS)} < L^{(SSNS)}$. However, the expansion that maximizes social welfare would keep each node as self-sufficient ($W^{(NBTCS)} < W^{(SSNS)}$, even if the investment costs were negligible). Moreover, both the expansion that maximizes total consumer surplus and the expansion that maximizes total producer surplus are keeping each node as self-sufficient (i.e., $CS^{(NBTCS)} < CS^{(SSNS)}$ and $PS^{(NBTCS)} < PS^{(SSNS)}$). This means that, in this particular case, while the construction of a non-binding-capacity transmission line linking both nodes minimizes the local market power of generation firms, this network expansion decreases social welfare, total consumer surplus, and total producer surplus. Figures 3.2, 3.3 and 3.4 illustrate these findings.

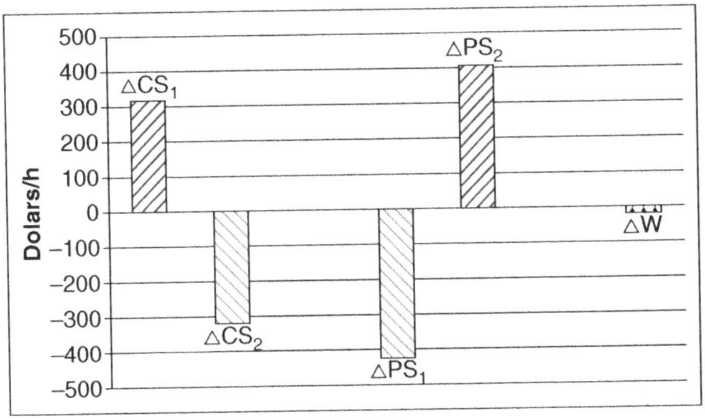

Figure 3.2. Effects on consumers and producers of building a non-binding-capacity line between both nodes, assuming that the investment cost is negligible.

The solution to this system of equations is: $q_1^{(NBTCS)} = (B \cdot (c_2 - c_1) + d_2 \cdot (A - c_1)) / (B \cdot (B + 2 \cdot d_2))$ and $q_2^{(NBTCS)} = (A - 2 \cdot c_2 + c_1) / (B + 2 \cdot d_2)$. Under these assumptions, the market-clearing price is $P^{(NBTCS)} = (d_2 \cdot (A + c_1) + c_2 \cdot B) / (B + 2 \cdot d_2)$. According to this market-clearing price and the optimal output levels, the producer surplus at Node 1 is $PS_1^{(NBTCS)} = (B \cdot (c_2 - c_1) + d_2 \cdot (A - c_1))^2 / (B \cdot (B + 2d_2)^2)$, and the producer surplus at Node 2 is $PS_2^{(NBTCS)} = (d_2 \cdot (A - 2 \cdot c_2 + c_1)^2) / (2 \cdot (B + 2d_2)^2)$.

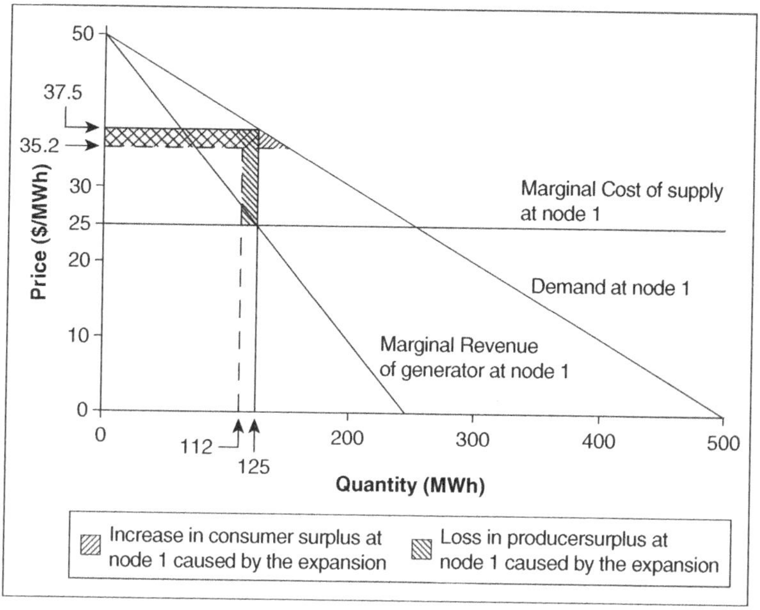

Figure 3.3. Equilibrium at Node 1 under both the SSNS and the NBTCS.

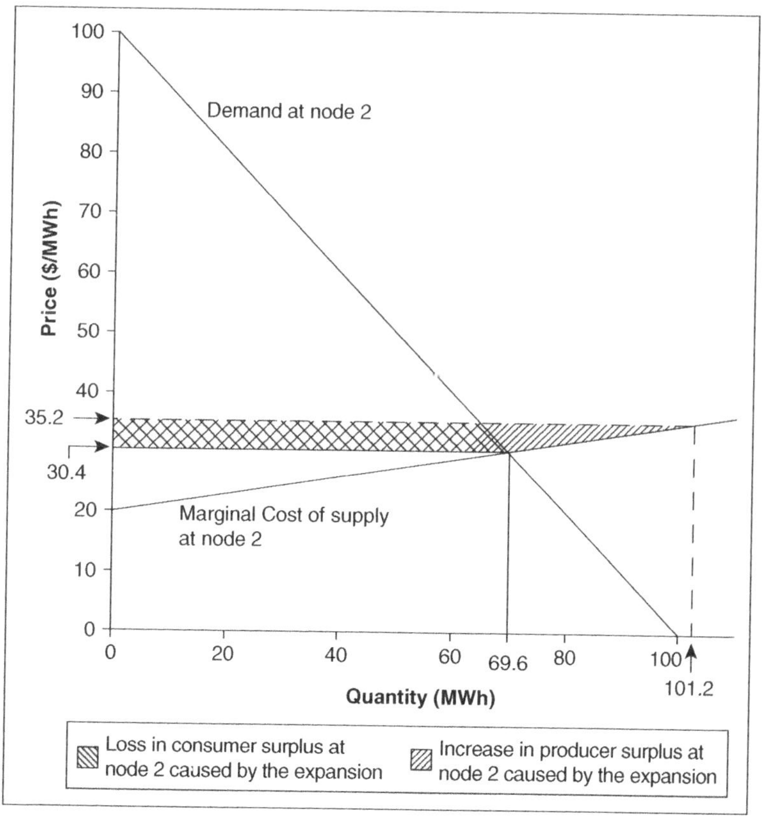

Figure 3.4. Equilibrium at Node 2 under both the SSNS and the NBTCS.

Figure 3.2 demonstrates that, in this particular case, the construction of the non-binding-capacity transmission line reduces social welfare even if the investment costs were negligible. Furthermore, this figure leads to an interesting observation: if the consumers at Node 1 (and/or the producers at Node 2) had enough political power, then they could encourage the construction of a non-binding-capacity transmission line linking both nodes even though it would decrease social welfare. That is, in this case, the "winners" from the transmission investment (consumers at Node 1 and generation firms at Node 2) can be expected to expend up to the amount of rents that they stand to win to obtain approval of this expansion project although it reduces social welfare.

It is interesting to note that, in this example, building the transmission line between the two nodes will result in flow from the expensive generation node to the cheap node, so that the transmission line cannot realize the potential *gains from trade* between the two nodes. On the contrary such flow decreases social welfare due to the exporting of power from an expensive-generation area into a cheap-generation area. This phenomenon is due to the exercise of market power by the generator at Node 1, who finds it advantageous to let the competitive fringe increase its production by exporting power to the cheap node, in order to sustain a higher market price. In economic trade theory, *gains from trade* is defined as the improvement in consumer incomes and producer revenues that arise from the increased exchange of goods or services among the trading areas (countries in international trade studies). It is well understood that, in absence of local market power (e.g., excluding all monopoly rents), the trade between areas must increase the total utility of all the areas combined. That is, *gains from trade* must be a non-negative quantity (Sheffrin, 2005). This rationale underlines common wisdom that prevailed in a regulated environment justifying the construction of transmission between cheap and expensive generation nodes on the grounds of reducing energy cost to consumers. However, as our example demonstrates, such rationale may no longer hold in a market-based environment where market power is present. Moreover, if we excluded monopoly rents from our social welfare calculations, then we would obtain zero gain from trade, in agreement with the *gains from trade* economic principle. However, even in that case, our example would still help us to illustrate that transmission expansions have distributional impacts, which create conflicts of interests among market participants.

Figure 3.3 and Figure 3.4 assists us to explain the results obtained in our particular example. These two figures show the price-quantity equilibria at each node under the two considered scenarios. In these figures, the solid lines represent the equilibria under the SSNS, while the dotted lines correspond to the equilibria under the NBTCS.

One way to explain the results obtained in the example presented in this section is through the distinction between two different effects due to the construction of the non-binding-capacity transmission line, as suggested by Leautier (2001). On one hand, competition among generation firms increases. This effect "forces" the firm located at Node 1 to decrease its retail price with respect to the SSNS. On the other hand, the transmission expansion causes a substitution (in production) of some low-cost power by more expensive power as result of the exercise of local market power.

The construction of the non-binding-capacity transmission line allows market participants to sell/buy power demanded/produced far away. This characteristic encourages competition among generation firms. In our example, the introduction of competition entails a decrease in the retail price at Node 1 with respect to the SSNS. As shown in Figure 3.3, this price reduction causes an increase in the Node 1's consumer surplus (because the demand at Node 1 increases) and a reduction in the profit of the monopolist at Node 1 with respect to the SSNS.

Moreover, because of the ability to exercise local market power, the monopolist at Node 1 can reduce its output (although the demand at Node 1 increases with respect to the SSNS) and keep a retail price higher than the SSNS market-clearing price at Node 2 in order to maximize its profit under the NBTCS. As this happens, the Node 2's firms increase their output levels (increasing both the generation marginal cost and the retail price at Node 2 with respect to the SSNS equilibrium) up to the point in which the retail prices at both nodes are equal (assuming the transmission constraint is not binding) and the total demand is met, NBTCS equilibrium. As shown in Figure 3.4, at this new equilibrium, the producer surplus at Node 2 increases while the consumer surplus at Node 2 decreases with respect to the SSNS. In other words, because the power generation at Node 1 is cheaper than the one at Node 2 for the relevant output levels, the exercise of local market power by the Node 1's firm causes a substitution of some of the low-cost power generated at Node 1 by more expensive power produced at Node 2 to meet demand. This out-of-merit generation, caused by the transmission expansion, reduces social welfare with respect to the SSNS.

In summary, while the first effect (competition effect) is social-welfare improving, the second effect (substitution effect) is social-welfare decreasing in the case of the example presented in this section. Furthermore, the substitution effect dominates in this particular example. Two facts contribute to the explanation of the dominance of the substitution effect: the generation marginal cost at Node 1 is much lower than the one at Node 2 (for the relevant output levels), although the pre-expansion price at Node 1 is higher than the equilibrium price at Node 2; and the demand and supply elasticities at Node 2 are higher than those at Node 1.

The analysis shown in this section makes it evident that the transmission expansion plan that minimizes local market power of generation firms may differ from the expansion plan that maximizes social welfare, consumer surplus, or total producer surplus, when the effect of the expansion on market prices is taken into consideration. Likewise, the transmission expansion plan that maximizes total producer surplus may differ from the expansion plan that maximizes social welfare and consumer surplus, while the transmission expansion plan that maximizes total consumer surplus may differ from the expansion plan that maximizes social welfare. These conclusions can all be drawn based on the simple two node example given above (see the Appendix for detailed calculations).

Finally, it is worth mentioning that our Cournot assumption is not essential in order to derive the qualitative results and conclusions presented here. The different optimization objectives we have considered may result in divergent optimal transmission expansion plans even when we model the competitive interaction of the generation firms as Bertrand competition.

3.2.2 Sensitivity Analysis in the Radial-Network Example

It is interesting to study the behavior of our two-node network under perturbation of some supply and/or demand parameters. Next, we present a sensitivity analysis of the optimal network expansion decision with respect to the marginal cost of supply at Node 1, c_1.

Figure 3.5 shows the changes in the optimal network expansion plan, under each of the four optimization objectives we have considered, as we vary the marginal cost of generation at Node 1 (keeping all other parameters unaltered and assuming that investment costs are negligible).

We note that none of the optimizing objectives leads to a consistent optimal expansion for all values of the parameter c_1. Moreover, this figure demonstrates that only for values of c_1 between \$5/MWh and \$12.4/MWh the four optimization objectives lead to the same optimal expansion plan. For c_1 higher than \$5/MWh, the competition among generation firms intensifies under the NBTCS, forcing the monopolist at Node 1 to reduce its retail price (i.e., $P_1^{(NBTCS)} < P_1^{(SSNS)}$), thus decreasing the monopolist's local market power. Moreover, for c_1 lower than \$12.4/MWh, under the SSNS, the monopolist at Node 1 sets a retail price lower than the equilibrium price at Node 2 (i.e., $P_1^{(SSNS)} < P_2^{(SSNS)}$). Thus, under the NBTCS, there is a net transmission flow from Node 1 to Node 2 that improves producer surplus, consumer surplus, and social welfare with respect to the SSNS.

Another interesting observation from Figure 3.5 is that the optimal network expansion plan, under most of the optimization objectives, is highly sensitive to the marginal cost of generation at Node 1 when this parameter has values between \$25/MWh and \$27/MWh.

We also performed a sensitivity analysis of the optimal network expansion plan with respect to some demand parameters. Modifying some of the demand func-

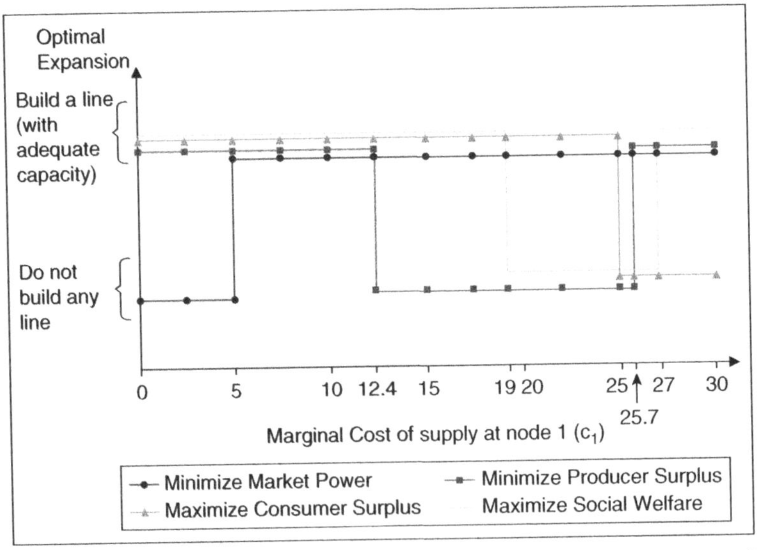

Figure 3.5. Sensitivity to the marginal cost of supply at Node 1 in the two-node network.

tion parameters, while keeping all supply parameters unaltered, leads to qualitative results that are similar to those observed when we vary the supply cost at Node 1. Such analysis shows that the optimal expansion plan under each of the four optimization objectives is highly sensitive to the demand structure.

3.3 POLICY IMPLICATIONS

The results discussed in the previous section have two important policy implications.

First, we observed that the optimal expansion of a network depends on the optimizing objective utilized and can be highly sensitive to supply and demand parameters. Even when the optimizing objective is clearly determined, the optimal network expansion plan changes depending on the cost structure of the generation firms. However, generation costs are typically uncertain and depend on factors such as the available generation capacity or the generation technology used, which in turn affect the optimal network investment plan. It follows that the interrelationship between generation and transmission investments should be considered when evaluating any transmission expansion project. Accounting for such interactions has been part of the integrated resource planning paradigm that prevailed under the regulated vertically integrated electricity industry, but is no longer feasible in the restructured industry. In Section 3.4 below, we describe a new planning paradigm that offers a way of accounting for generators response to transmission investment in an unbundled electricity industry with a competitive generation sector.

Second, our analysis shows that transmission investments have important distributional impact. While some transmission investments can greatly benefit some market participant, they may harm some other constituents. Consequently, policy makers looking after socially efficient network expansions should be aware of the distributional impact of merchant investments. Moreover, the dynamic nature of power systems entails changes over time of not only demand and supply structures, but also the mix of market participants, which adds complexity to the valuation of merchant transmission expansion projects. Even when a merchant investment appears to be beneficial under the current market structure, the investment could become socially inefficient when future generation and transmission plans and/or demand forecasts are considered.

3.4 PROACTIVE TRANSMISSION PLANNING

In this section we introduce a three-period model as a new planning paradigm that takes into consideration the policy implications reviewed in the previous section. The basic idea behind this model is that the interrelationship between the generation and the transmission investments affects the social value of the transmission capacity, so that transmission planning must take into consideration its effect both on generation investment and on the resulting market equilibrium, while recognizing that investment decisions in generation will respond to the transmission expansion plan in anticipation of the subsequent market equilibrium conditions.

3.4.1 Model Assumptions

The model does not assume any particular network structure, so that it can be applied to any network topology. Moreover, we assume that all nodes are both demand nodes and generation nodes and that all generation capacity at a node is owned by a single firm. We allow generation firms to exercise local market power and assume that their interaction can be characterized through Cournot competition, i.e., firms chose their production quantities so as to maximize their profit with respect to the residual demand function while taking the production quantities of other firms and the dispatch decisions of the system operator as given. Furthermore, the model allows many lines to be simultaneously congested as well as probabilistic contingencies describing demand shocks, generation outages and transmission line outages.

The model consists of three periods, as displayed in Figure 3.6. We assume that, at each period, players making decisions observe all previous-periods actions and form rational expectations regarding the outcome of the current and subsequent periods. That is, we define the transmission investment model as a "complete- and perfect-information" game[7] and the equilibrium as "sub game perfect."

The last period (period 3) represents the energy market operation. That is, in this period, we compute the equilibrium quantities and prices of electricity over given generation and transmission capacities determined in the previous periods. We model the energy market equilibrium in the topology of the transmission network through a DC approximation of Kirchhoff's laws. Specifically, flows on lines can be calculated by using the power transfer distribution factor (PTDF) matrix, whose elements give the proportion of flow on a particular line resulting from an injection of one unit of power at a particular node and a corresponding withdrawal at an arbitrary (but fixed) slack bus. Different PTDF matrices corresponding to different transmission contingencies, with corresponding state probabilities, characterize uncertainty regarding the realized network topology in the energy market equilibrium. We assume that generation and transmission capacities as well as demand shocks are subject to random fluctuations that are realized in Period 3 prior to the production and redispatch decisions by the generators and the system operator. We further assume that the probabilities of all such credible contingencies are public knowledge.

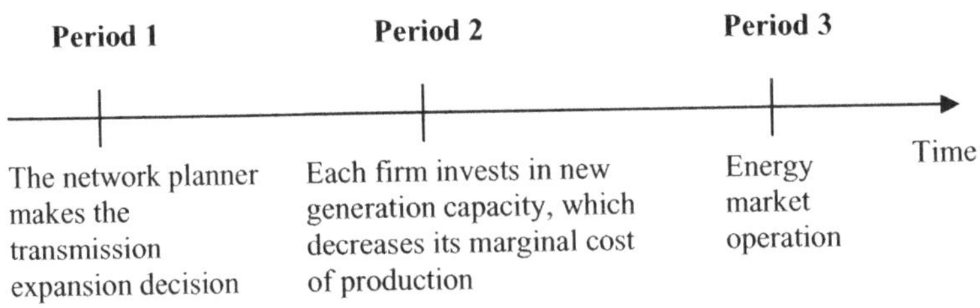

Figure 3.6. Three-period transmission investment model.

[7] A "complete- and perfect-information" game is defined as a game in which players move sequentially and, at each point in the game, all previous actions are observable to the player making a decision.

In our model, the energy market equilibrium in Period 3 is characterized as a subgame with two stages. In the first stage, nature picks the state of the world that determines the actual generation and transmission capacities as well as the shape of the demand and cost functions at each node. In the second stage, firms compete in a Nash-Cournot fashion by selecting their production quantities, while taking into consideration the simultaneous import/export decisions of the system operator whose objective is to maximize social welfare while satisfying the transmission constraints.

In the second period, each generation firm invests in new generation capacity, which lowers its marginal cost of production at any output level. For the sake of tractability we assume that generators' production decisions are not constrained by physical capacity limits. Instead we allow generators' marginal cost curves to rise smoothly so that production quantities at any node will be limited only by economic considerations and transmission constraints. In this framework, generation expansion is modeled as "stretching" the supply function so as to lower the marginal cost at any output level and thus increase the amount of economic production at any given price. Such expansion can be interpreted as an increase in generation capacity in a way that preserves the proportional heat curve or alternatively assuming that any new generation capacity installed will replace old, inefficient plants and, thereby, increase the overall efficiency of the portfolio of plants in producing a given amount of electricity. This continuous representation of the supply function and generation expansion serves as a proxy to actual supply functions that end with a vertical segment at the physical capacity limit. Since typically generators are operated so as not to hit their capacity limits (due to high heat rates and expansive wear on the generators) our proxy should be expected to produce realistic results. The return from the generation capacity investments made in Period 2 occurs in Period 3, when such investments enable the firms to produce electricity at lower cost and sell more of it at a profit. In our model, we assume that, in making their investment decisions in Period 2, the generation firms are aware to the transmission expansion from Period 1 and form rational expectations regarding the investments made by their competitors and the resulting market equilibrium in Period 3. Thus, the generation investment and production decisions by the competing generation firms are modeled as a two-stage subgame perfect Nash equilibrium.

Finally, in the first period, the network planner that we model as a Stackelberg leader in this three-period game, evaluates different projects to upgrade the existing transmission lines while anticipating the generators' and the system operator's response in Periods 2 and 3.[8] In particular, we consider here the case in which the transmission planner evaluates a single transmission expansion decision, but the proposed approach can be applied to more complex investment options.

[8]No attempt is made to co-optimize transmission expansion and redispatch decisions. We assume that the transmission planning function treats the real time redispatch function as an independent follower (even if they reside in the same organization such as an ISO or RTO) and anticipates its equilibrium response as if it was an independently controlled entity with no attempt to exploit possible strategic coordination between transmission planning and real time dispatch. One should keep in mind, however, that such coordination might be possible in a for-profit system operator enterprise such as in the United Kingdom.

Because the transmission planner under this paradigm anticipates the response by the generators, optimizing the transmission investment plan will determine the best way of inducing generation investment so as to maximize the objective function set by the transmission planner. Therefore, we will use the term *proactive network planner* to describe such a planning approach, that results in outcomes which, although still inferior to the integrated resource planning paradigm, often result in the same investment decisions. In this paper, we limit the transmission expansion decision to expanding the capacity of any one existing line according to some specific transmission-planning objective. We assume the transmission expansion does not alter the original PTDF matrices, but only the thermal capacity of the line. This would be the case if, for the expanded line, we replaced all the wires by new ones (with new materials such as "low sag wire") while using the same existing high-voltage towers. Since the energy market equilibrium will be a function of the thermal capacities of all constrained lines, the Nash equilibrium of generation capacities will also be a function of these capacity limits. The proactive network planner, then, has multiple ways of influencing this Nash equilibrium by acting as a Stackelberg leader who anticipates the equilibrium of generation capacities and induces generation firms to make better investments.

We further assume that the generation cost functions are both increasing and convex in the amount of output produced and decreasing and convex in the generation capacity. Furthermore, as mentioned before, we assume that the marginal cost of production at any output level decreases as generation capacity increases. Moreover, we assume that both the generation capacity investment cost and the transmission capacity investment cost are linear in the extra-capacity added. We also assume downward-sloping linear demand functions at each node. To further simplify things, we assume no wheeling fees.

3.4.2 Model Notation

Sets:

- N: set of all nodes
- L: set of all existing transmission lines
- C: set of all states of contingencies
- N_G: Set of generation nodes controlled by generation firm G
- $\mathcal{G}$: Set of all generation firms

Decision variables:

- q_i^c: quantity generated at Node i in State c
- r_i^c: adjustment quantity into/from Node i by the system operator in State c
- g_i: expected generation capacity of facility at Node i after Period 2
- f_l: expected thermal capacity limit of Line 1 after Period 1

Parameters:

- g_i^0: expected generation capacity of facility at Node i before Period 2
- f_ℓ^0: expected thermal capacity limit of Line 1 before Period 1
- g_i^c: generation capacity of facility at Node i in State c, given g_i
- f_ℓ^c: thermal capacity limit of Line 1 in State c, given f_ℓ
- $P_i^c(\cdot)$: inverse demand function at Node i in State c
- $CP_i^c(q_i^c, g_i^c)$: production cost function of the generation firm located at Node i in State c
- $\mathrm{CIG}_i(g_i, g_i^0)$: cost of investment in generation capacity at Node i to bring expected generation capacity to g_i.
- $\mathrm{CI}_\ell(f_\ell, f_\ell^0)$: investment cost in Line 1 to bring expected transmission capacity to f_ℓ.
- $\phi_{\ell,i}^c$: power transfer distribution factor on Line 1 with respect to a unit injection/withdrawal at Node i, in State c.

3.4.3 Model Formulation

We start by formulating the third-period problem. In the first stage of Period 3, nature determines the state of the world, c. In the second stage, for a given State c, generation firm G ($G \in \mathcal{G}$) solves the following profit-maximization problem:

$$\begin{aligned} &\text{Max}_{q^c_{i \in N_G}} \ \pi_G^c = \sum_{i \in N_G} P_i^c(q_i^c + r_i^c) \cdot q_i^c - CP_i^c(q_i^c, g_i^c) \\ &\text{s.t.} \quad q_i^c \geq 0, i \in N_G \end{aligned} \tag{3.2}$$

Simultaneously with the generators' production quantity decisions, the system operator solves the following welfare maximizing redispatch problem (for the given State c):

$$\begin{aligned} &\text{Max}_{\{r_i^c\}} \Delta W^c = \sum_{i \in N} \left(\int_0^{r_i^c} P_i^c(q_i^c + x_i) dx_i \right) \\ &\text{s.t.} \quad \sum_{i \in N} r_i^c = 0 \\ &\qquad -f_\ell^c \leq \sum_{i \in N} \phi_{\ell,i}^c \cdot r_i^c \leq f_\ell^c, \quad \forall \ell \in L \\ &\qquad q_i^c + r_i^c \geq 0, \qquad \forall i \in N \end{aligned} \tag{3.3}$$

Given that we assume no wheeling fees, the system operator can gain social surplus, at no extra cost, by exporting some units of electricity from a cheap-generation node while importing them to other nodes until the prices at the nodes are equal, or until some transmission constraints are binding.

The previously specified model assumptions guarantee that both (3.2) and (3.3) are convex programming problems, which implies that first order necessary conditions (i.e. KKT conditions) are also sufficient. Consequently, to solve the Period-3 problem (energy market equilibrium), we can just jointly solve the KKT conditions of the problems defined in (3.2), for all generation firms G, and (3.3) which together form a linear complimentarily problem (LCP), which can be easily solved with off-the-shelf software packages.

In Period 2, each firm determines how much to invest in new generation capacity by maximizing the expected value of the investment (we assume risk-neutral firms) subject to the anticipated actions in Period 3. Since the investments in new generation capacity reduce the expected marginal cost of production, the return from the investments made in Period 2 occurs in Period 3. Thus, in Period 2, the firm G solves the following optimization problem:

$$\begin{aligned} &\text{Max}_{g_i \in N_G} \sum_{i \in N_G} \left\{ E_c\left[\pi_i^c\right] - \text{CIG}_i\left(g_i, g_i^0\right) \right\} \\ &\text{s.t} \quad \text{KKT conditions of the problems defined in (3.2) for all } G \in \mathcal{G} \text{ and (3.3)} \end{aligned} \tag{3.4}$$

The problem defined in (4) is a Mathematical Program with Equilibrium Constraints (MPEC) problem and the problem of finding an equilibrium investment strategy for all the generation firms is an Equilibrium Problem with Equilibrium Constraints (EPEC), in which each firm solves an MPEC problem parametric on the other firms investment decisions and subject to the joint LCP constraints characterizing the energy market equilibrium in Period 3. Unfortunately, this EPEC is constrained in a non-convex region and, therefore, we cannot simply write down the first order necessary conditions for each firm and aggregate them into a large problem to be directly solved.

As indicated earlier, we consider here only the simple case in which the network planner makes a single transmission expansion decision that will determine which line (among the already existing lines) it should upgrade, and what transmission capacity it should consider for that line, in order to optimize its transmission-planning objective. Thus, in Period 1, the network planner solves the following optimization problem:

$$\begin{aligned} &\text{Max}_{\ell, f_\ell} \Phi\left(q_i^c, r_i^c, g_i, \ell, f_\ell\right) \\ &\text{s.t.} \quad \text{Equilibrium solution of periods 2 and 3} \end{aligned} \tag{3.5}$$

where Φ $(\cdot)$ represents the transmission-planning objective used by the network planner.

In the case where the transmission-planning objective is the expected social welfare, we have:

$$\begin{aligned} \Phi\left(q_i^c, r_i^c, g_i, \ell, f_\ell\right) = \sum_{i \in N} \Bigg\{ & E_c\left[\int_0^{q_i^c + r_i^c} P_i^c(q)\,dq - \text{CP}_i^c\left(q_i^c, g_i^c\right) \right] \\ & - \text{CIG}_i\left(g_i, g_i^0\right) \Bigg\} - \text{CI}_\ell\left(f_\ell, f_\ell^0\right) \end{aligned} \tag{3.6}$$

3.4.4 Transmission Investment Models Comparison

Now, we would like to compare the transmission investment decisions made by a *proactive network planner* (PNP) as defined above with the comparable decisions made by a *reactive network planner* (RNP), who plans transmission expansions by considering its impact on the energy market but without accounting for the generation investment response and its ability to influence such investments through the transmission expansion.

In the RNP model, the network planner selects the optimal location (among the already existing) and magnitude for the next transmission upgrade while considering the currently installed generation capacities. This case can be considered as a special case of the model described above where the generators are constrained in Period 2 to select the same generation capacity that they already have. Thus, in Period 1, the RNP solves the following optimization problem:

$$\begin{aligned} &\text{Max}_{\ell, f_\ell} \Phi\left(q_i^c, r_i^c, g_i, \ell, f_\ell\right) \\ &\text{s.t.} \quad \text{Equilibrium solution of periods 2 and 3} \\ &\qquad g_i = g_i^0, \forall i \in N \end{aligned} \tag{3.7}$$

In evaluating the outcome of the RNP investment policy, we will consider, however, the generators' response to the transmission investment (which is suboptimal) and its implication on the spot market equilibrium.

By comparing (3.5) and (3.7), we observe that, if we eliminated the 2-Period problem conditions of each problem, then both problems would be identical. Thus, there exists a correspondence from generation capacities space to transmission capacities space, $f^*(g)$, that characterizes the "unconstrained" optimal investment decisions of both the PNP and the RNP. Since the second periods of both models are identically modeled, there also exists a correspondence from transmission capacities space to generation capacities space, $g^*(f)$, that characterizes the optimal decisions of generation firms under both the PNP and the RNP approach. The optimal solution of the PNP model is at the intersection of these two correspondences. That is, the transmission capacity chosen by the PNP, f_{PNP}^*, is such that $f^*(g^*(f_{\mathrm{PNP}}^*)) = f_{\mathrm{PNP}}^*$. On the other hand, the transmission capacity chosen by the RNP, f_{RNP}^*, is on the correspondence $f^*(g)$, at the currently installed generation capacities (i.e., $f_{\mathrm{RNP}}^* = f^*(g^0)$). Thus, the optimal solution of the second period of the RNP model is on the correspondence $g^*(f)$, at transmission capacities f_{RNP}^*. Since the correspondence $g^*(f)$ characterizes the optimality conditions of the Period 2 problem in the PNP model, any pair $(g^*(f), f)$ represents a feasible solution for the PNP model. Consequently, the optimal solution of the RNP model, $(g^*(f_{\mathrm{RNP}}^*), f_{\mathrm{RNP}}^*)$, is a feasible solution of the PNP model. Therefore, the optimal solution of (3.5) cannot be worse than the optimal solution of (3.7).

Summarizing, under any transmission-planning objective, the optimal value obtained from the proactive network planner model is never smaller (worse) than the optimal value obtained from the reactive network planner model.

It is interesting to note that, although the previous result states that a RNP cannot do better than a PNP, the sign of the inefficiency is not evident. That is, without adding more structure to the problem, it is not evident whether the network planner underinvests or overinvests in transmission under the RNP model as compared to the PNP investment levels.

3.5 ILLUSTRATIVE EXAMPLE

We illustrate the results derived in the previous section with the simple three-node network displayed in Figure 3.7. We assume that each node has both local generation and local demand. Moreover, for simplicity, we consider three generation firms in the market (each firm owning the generators at a single node).

We assume that the electric characteristics of the three transmission lines of the network in Figure 3.7 are identical. For these three transmission lines, the resistance is 0.15 p.u., the reactance is 0.3 p.u., and the thermal capacity rating is 16 MVA.

The uncertainty associated with the energy market operation is classified into five contingent states, as shown in Table 3.1. Table 3.2 shows the nodal information in the normal state.

We assume the same production cost function, $CP_i(q_i, g_i)$, for all generators. Note that $CP_i(q_i, g_i)$ is increasing in q_i, but it is decreasing in g_i. Moreover, recall that we have assumed that generators have unbounded capacity. Thus, the only important effect of investing in generation capacity is lowering the production cost. We also assume that all generation firms have the same investment cost function, given by $\text{CIGi}(g_i, g_i^0) = 6\cdot(g_i - g_i^0)$, in dollars. The before-Period-2 expected generation capacity at Node i, g_i^0, is 60 MW (the same for all nodes). In our model, the choice of the parameter g_i^0 is not important because the focus of this work is not on generation adequacy. Instead, what really matters in our model is the ratio (g_i^0/g_i) since we focus on the cost of generating power and the effect that both generation and transmission investments have on that cost.

As indicated earlier, the KKT conditions for the Period 3 problem of the PNP model constitute a Linear Complementarity Problem (LCP). We solve it, for each contingent state by minimizing the complementarity conditions subject to the linear

TABLE 3.1. States of Contingencies Associated to the Energy Market Operation.

State	Probability	Type of uncertainty and description
1	0.80	Normal state: Data set as in Table 3.2
2	0.05	Demand uncertainty: All demands increase by 20%
3	0.05	Demand uncertainty: All demands decrease by 20%
4	0.05	Network uncertainty: Line 1–2 goes down
5	0.05	Generation uncertainty: Generator at Node 3 goes down

TABLE 3.2. Nodal Information Used in the Three-Node Network in the Normal State.

Data type (units)	Information	Nodes where apply
Inverse demand function ($/MWh)	$P_i(q) = 50 - q$	1
Inverse demand function ($/MWh)	$P_i(q) = 60 - q$	2
Inverse demand function ($/MWh)	$P_i(q) = 80 - q$	3
Generation cost function ($/MWh)	$CP_i(q_i, g_i) = (0.4\cdot q_i^2 + 25\cdot q_i)\cdot(g_i^0/g_i)$	1, 2, and 3.

equality constraints and the non-negativity constraints.[9] The Period 2 problem of the PNP model is an Equilibrium Problem with Equilibrium Constraints (EPEC), in which each firm faces a Mathematical Program subject to Equilibrium Constraints (MPEC).[10] We attempt to solve for an equilibrium, if at least one exists, by iterative deletion of dominated strategies. That is, we sequentially solve each firm's profit-maximization problem using as data the optimal values from previously solved problems. Thus, starting from a feasible solution, we solve for g_1 using $g_{(-1)}$ as data in the first firm's optimization problem (where $g_{(-1)}$ means all firms' generation capacities except for Firm 1's), then solve for g_2 using $g_{(-2)}$ as data, and so on. We solve each firm's profit-maximization problem using sequential quadratic programming algorithms implemented in MATLAB®.

We test our model from a set of different starting points and using different generation-firms' optimization order. All these trials gave us the same results. For the PNP model, the optimal levels of generation capacity under absence of transmission investments are $(g_1^*, g_2^*, g_3^*) = (60.9, 119.7, 80.6)$, in MW. Table 3.3 lists the corresponding generation quantities (q_i), import/export quantities (r_i) and nodal prices (P_i) in the normal state.

To solve the Period 1 problem of the PNP model, we iteratively solve Period 2 problems in which a single line has been expanded and, then, choose the expansion producing the highest expected social welfare. For simplicity, we do not consider transmission investment costs (it can be thought that the per-unit transmission investment cost is the same for each line upgrade so that we can get rid of these costs in the expansion decision). In this sense, our results establish an upper limit in the amount of the line investment cost. We tested the PNP decision by comparing the results of independently adding 16 MVA of capacity (doubling the actual line capacity) to each one of the three lines of the network in Figure 3.7. The results are summarized in Table 3.4. In Table 3.4, "Avg. L" corresponds to the average expected Lerner index[11] among all generation firms, "P.S." is the expected producer surplus of the system, "C.S." is the expected consumer surplus of the system, "C.R." rep-

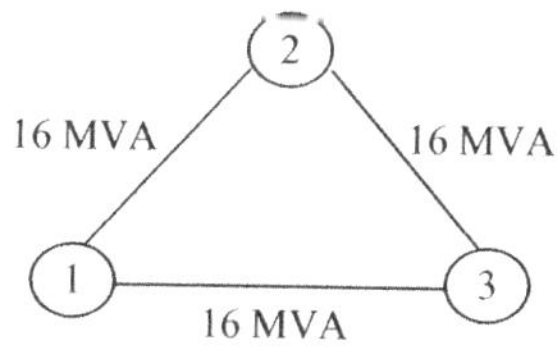

Figure 3.7. Three-node network used in our case study.

[9] Any LCP can be written as the problem of finding a pair of vectors $x, y \in R^n$ such that $x = q + M \cdot y$, $x^T \cdot y = 0$, $x \geq 0$, and $y \geq 0$, where $M \in R^{n \times n}$, $q \in R^n$. Thus, we can solve it by minimizing $x^T \cdot y$ subject to $x = q + M \cdot y$, $x \geq 0$, and $y \geq 0$. If the previous problem has an optimal solution where the objective function is zero, then that solution also solves the corresponding LCP.

[10] See (Yao et al., 2004) for a definition of both EPEC and MPEC.

[11] The Lerner index is defined as the fractional price markup i.e. (Price—Marginal cost) / Price.

TABLE 3.3. Generation Quantities, Adjustment Quantities, and Nodal Prices in the Normal State, under the PNP Model.

Node	q_i (MWh)	r_i (MWh)	P_i ($/MWh)
1	11.57	−6.91	45.34
2	22.64	−6.91	44.26
3	18.33	13.81	47.85

TABLE 3.4. Assessment of Single Transmission Expansions under the PNP Model.

Expansion Type	Avg. L	P.S. ($/h)	C.S. ($/h)	C.R. ($/h)	W ($/h)	g^* (MW)
No expansion	0.388	907.1	633.5	55.3	1595.9	[60.9; 119.7; 80.6]
16 MVA on line 1–2	0.388	907.1	633.5	55.3	1595.9	[60.9; 119.7; 80.6]
16 MVA on line 1–3	0.439	852.0	724.5	58.4	1634.9	[97.2; 116.6; 81.0]
16 MVA on line 2–3	**0.441**	**883.8**	**696.2**	**67.9**	**1647.9**	**[97.2; 99.5; 96.8]**

TABLE 3.5. Assessment of Single Transmission Expansions under the RNP Model.

Expansion Type	$\overline{\text{Avg.L}}$	$\overline{\text{P.S.}}$ ($/h)	$\overline{\text{C.S.}}$ ($/h)	$\overline{\text{C.R.}}$ ($/h)	$\overline{\text{W}}$ ($/h)
No expansion	0.280	918.8	422.4	70.2	1411.4
16 MVA on line 1–2	0.280	918.8	422.4	70.2	1411.4
16 MVA on line 1–3	**0.281**	**909.3**	**489.4**	**23.1**	**1421.8**
16 MVA on line 2–3	0.280	918.8	423.2	68.5	1410.5

resents the expected congestion rents over the entire system, "W" is the expected social welfare of the system, and "g*" corresponds to the vector of all Nash-equilibrium expected generation capacities.

From Table 3.4, it is evident that the best single transmission line expansion (in terms of expected social welfare) that a PNP can choose in this case is the expansion of line 2–3. Now, we are interested in comparing the PNP decision with the decision that a RNP would take under the same system conditions. We tested the RNP decision by comparing the results of independently adding 16 MVA of capacity (doubling the actual line capacity) to each one of the three lines of the network in Figure 3.7. The results are summarized in Table 3.5, where we use the notation $\overline{x}$ to represent the value of x as seen by the RNP.

From Table 3.5, it is clear that the social-welfare-maximizing transmission expansion for the RNP is, in this case, to expand line 1–3. Thus, the true optimal levels of the RNP model solution are: Avg. $L = 0.439$, $P.S. = \$852.0/\text{h}$, $C.S. = \$724.5/\text{h}$, $C.R. = \$58.4/\text{h}$, $W = \$1634.9/\text{h}$, and $g^* = (97.2, 116.6, 81.0)$, in MW. By comparing

Table 3.4 and Table 3.5, it is evident that the optimal decision of the PNP differs from the optimal decision of its reactive counterpart. Specifically, the PNP considers not only the welfare gained directly by adding transmission capacity (on which the RNP bases its decision), but also the way in which its investment induces a more socially efficient Nash equilibrium of expected generation capacities.

3.6 CONCLUSIONS AND FUTURE WORK

In this chapter we illustrated, through a simple radial-network example, how different planning objectives can result in divergent optimal expansions of a network. In particular, we showed that the maximization of social welfare, the minimization of local market power, the maximization of consumer surplus and the maximization of producer surplus can all result in divergent optimal expansions of a transmission network. Consequently, finding a unique politically feasible and fundable network expansion policy could be a very difficult, if not impossible, task. Accordingly, even if we agreed that a weighted sum of consumer surplus and producer surplus is the appropriated objective function to use, the weights to be used would be a controversial matter since different weights could lead to different optimal network expansions.

One of the key assumptions of the radial-network example presented in this chapter is that at least one of the generators can exercise local market power. Without considering local market power (that is, in a world where every generator faces a perfectly competitive market), the results and conclusions obtained here are not valid. However, given the prevalence of local market power in the power generation business, our results cannot be dismissed.

Motivated by the strong interrelationship between power generation and transmission investments, we have introduced a new transmission planning paradigm that attempts to capture some of the efficiency gains of integrated resource planning which is no longer feasible in an unbundled-market-based electricity industry. Our proposed approach employs a three-period model of transmission investments in which the transmission planner acts as a Stackelberg leader anticipating the effect of transmission expansion on generation investment and the subsequent energy market equilibrium. In this model, oligopolistic generation firms respond to transmission investments by interacting as Nash players in the generation investment game while anticipating the outcome of Cournot competition in the energy market.

Our future work will extend our three-period transmission investment model so that we can better characterize real-world power systems. An important extension is the analysis of our model when allowing the construction of lines at new locations (rather than upgrading existing lines). In this case, an expansion can change the electric properties of the network (and, thus, the PTDF matrices), which represents a more realistic scenario. Another valuable extension is the consideration of risk-averse generation firms. We expect to obtain more moderate generation investment levels when including risk aversion in the generation investment decisions.

BIBLIOGRAPHY

[1] Federal Energy Regulatory Commission (FERC). 2002. *Notice of Proposed Rulemaking (NOPR) on Standard Market Design and Structure RM01-12-000*, 18 CFR Part 35, Washington D.C.

[2] Leautier, T. 2001. "Transmission Constraints and Imperfect Markets for Power." *Journal of Regulatory Economics*, 19(1): 27–54.

[3] London Economics International LLC. 2002. *Final Methodology: proposed approach for evaluation of transmission investment*. Report prepared for the CAISO by London Economics International LLC. Cambridge, MA.

[4] Sheffrin, Anjali. 2005. "Gains from Trade and Benefits of Transmission Expansion for the IEEE Power Engineering Society", *Proceedings of the IEEE Power Engineering Society 2005 General Meeting (Track 5)*, San Francisco, U.S.A.

[5] Shrestha, G. and J. Fonseka. 2004. "Congestion-Driven Transmission Expansion in Competitive Power Markets", *IEEE Transactions on Power Systems*, 19(3): 1658–1665.

[6] Styczynski, Z. 1999. "Power Network Planning Using Game Theory", *Proceedings of the 13th Power Systems Computation Conference (PSCC)*, Trondheim, Norway: 607–613.

[7] Sun, H. and D. Yu. 2000. "A Multiple-Objective Optimization Model of Transmission Enhancement Planning for Independent Transmission Company (ITC)", *Proceedings of the IEEE Power Engineering Society 2000 Summer Meeting*, Seattle, U.S.A: 2033–2038.

[8] Yao, J., S. Oren, and I. Adler. 2004. "Computing Cournot Equilibria in Two Settlement Electricity Markets with Transmission Constraints", *Proceeding of the 37th Hawaii International Conference on Systems Sciences (HICSS37)*, Big Island, Hawaii, U.S.A.: 20051b.

APPENDIX

In this appendix, we present the additional computations for the example proposed in Section 3.2 of this chapter showing that the maximization of social welfare, the minimization of local market power, the maximization of consumer surplus and the maximization of producer surplus can all result in divergent optimal expansions of the transmission network. In particular, by altering the marginal cost of production at Node 1, we show here that: the transmission expansion that maximizes total producer surplus can differ from the expansion that maximizes social welfare and from the expansion that maximizes consumer surplus in the same network; and the transmission expansion that maximizes consumer surplus can differ from the expansion that maximizes social welfare in the same network.

Assume that c_1 = \$26/MWh. Then, under the SSNS, the generation firm at Node 1 optimally produces $q_1^{(SSNS)} = 120$ MWh and charges $P_1^{(SSNS)}$ = \$38/MWh. With this quantity and price, the producer surplus at Node 1 is $PS_1^{(SSNS)}$ = \$1,440/h and the consumer surplus at this node is $CS_1^{(SSNS)}$ = \$720/h. The Lerner index at Node 1 is $L_1^{(SSNS)} = 0.32$.[12] Moreover, as in the case where c_1 = \$25/MWh, under the SSNS, the firms at Node 2 optimally produce an aggregate amount $q_2^{(SSNS)} = 69.6$ MWh, and the market-clearing price is $P_2^{(SSNS)}$ = \$30.4/MWh. Also, the producer surplus at Node 2 is equal to $PS_2^{(SSNS)}$ = \$363/h and the consumer surplus at this node is equal to $CS_2^{(SSNS)}$ = \$2,420/h.[13]

[12] See footnote # 4.

[13] See footnote # 5.

Accordingly, the total producer surplus, the total consumer surplus, and the social welfare under the SSNS are $PS^{(SSNS)}$ = \$1,803/h, $CS^{(SSNS)}$ = \$3,140/h, and $W^{(SSNS)}$ = \$4,943/h, respectively.

Under the NBTCS, according to the Cournot-competition assumption, the monopolist at Node 1 optimally produces $q_1^{(NBTCS)}$ = 105 MWh while the competitive fringe at Node 2 optimally produces $q_2^{(NBTCS)}$ = 104 MWh (these output levels imply that there is a transmission flow of 39 MWh from Node 2 to Node 1). In this case, the market-clearing price is $P^{(NBTCS)}$ = \$35.6/MWh. With these new quantities and prices, the producer surplus at Node 1 is $PS_1^{(NBTCS)}$ = \$1,005/h and the producer surplus at Node 2 is $PS_2^{(NBTCS)}$ = \$807/h.[14] As well, the consumer surpluses are $CS_1^{(NBTCS)}$ = \$1,043/h for Node 1's consumers and $CS_2^{(NBTCS)}$ = \$2,076/h for Node 2's consumers. The new Lerner index at Node 1 is $L_1^{(NBTCS)} = 0.27$.

Assuming again that the transmission investment is made by an independent entity, the total producer surplus, the total consumer surplus, and the social welfare under the NBTCS are equal to: $PS^{(NBTCS)}$ = \$1,812/h, $CS^{(NBTCS)}$ = \$3,119/h, and $W^{(NBTCS)}$ = \$4,931/h—investment costs, respectively.

Comparing the SSNS and the NBTCS, we can observe that the expansion that maximizes total producer surplus is building a transmission line with "adequate" capacity (i.e., capacity greater than 39 MW). However, both the expansion that maximizes social welfare and the expansion that maximizes total consumer surplus are keeping each node as self-sufficient ($W^{(NBTCS)} < W^{(SSNS)}$, even if the investment costs were negligible, and $CS^{(NBTCS)} < CS^{(SSNS)}$). That is, in the case where we have c_1 = \$26/MWh, the construction of a non-binding-capacity line decreases both social welfare and total consumer surplus while this network expansion maximizes total producer surplus. This analysis indicates that, as in the case of the simple example presented here (with c_1 = \$26/MWh), the transmission expansion that maximizes total producer surplus in a particular network can be different from the expansion that maximizes social welfare and the expansion that maximizes total consumer surplus in the same network.

Now, assume that c_1 = \$24/MWh. Then, under the SSNS, the monopolist at Node 1 optimally produces $q_1^{(SSNS)}$ = 130 MWh and charges $P_1^{(SSNS)}$ = \$37/MWh. With this quantity and price, the producer surplus at Node 1 is $PS_1^{(SSNS)}$ = \$1,690/h and the consumer surplus at this node is $CS_1^{(SSNS)}$ = \$845/h. The Lerner index at Node 1 is $L_1^{(SSNS)} = 0.35$.[15] Moreover, as in the previous cases, under the SSNS, the generation firms at Node 2 optimally produce an aggregate amount $q_2^{(SSNS)}$ = 69.6 MWh, and the market-clearing price is $P_2^{(SSNS)}$ = \$30.4/MWh. Also, the producer surplus at Node 2 is equal to $PS_2^{(SSNS)}$ = \$363/h and the consumer surplus at this node is equal to $CS_2^{(SSNS)}$ = \$2,420/h.[16]

Accordingly, the total producer surplus, the total consumer surplus, and the social welfare under the SSNS are $PS^{(SSNS)}$ = \$2,053/h, $CS^{(SSNS)}$ = \$3,265/h, and $W^{(SSNS)}$ = \$5,318/h, respectively.

[14] See footnote # 6.

[15] See footnote # 4.

[16] See footnote # 5.

Under the NBTCS, according to the Cournot-competition assumption, the monopolist at Node 1 optimally produces $q_1^{(NBTCS)} = 119$ MWh while the competitive fringe at Node 2 optimally produces $q_2^{(NBTCS)} = 99$ MWh (these output levels imply that there is a transmission flow of 33 MWh from Node 2 to Node 1). In this case, the market-clearing price is $P^{(NBTCS)} = \$34.8/\text{MWh}$. With these new quantities and prices, the producer surplus at Node 1 is $PS_1^{(NBTCS)} = \$1{,}281/\text{h}$ and the producer surplus at Node 2 is $PS_2^{(NBTCS)} = \$729/\text{h}$.[17] As well, consumer surpluses are $CS_1^{(NBTCS)} = \$1{,}157/\text{h}$ for Node 1's consumers and $CS_2^{(NBTCS)} = \$2{,}126/\text{h}$ for Node 2's consumers. The new Lerner index at Node 1 is $L_1^{(NBTCS)} = 0.31$.

Assuming again that the transmission investment is made by an independent entity, the total producer surplus, the total consumer surplus, and the social welfare under the NBTCS are equal to: $PS^{(NBTCS)} = \$2{,}010/\text{h}$, $CS^{(NBTCS)} = \$3{,}283/\text{h}$, and $W^{(NBTCS)} = \$5{,}293/\text{h}$—investment costs, respectively.

Comparing the SSNS and the NBTCS, we can observe that the expansion that maximizes total consumer surplus is building a transmission line with "adequate" capacity (in theory, with capacity greater than 33 MWh). However, the expansion that maximizes social welfare is keeping each node as self-sufficient because $W^{(NBTCS)} < W^{(SSNS)}$, even if the investment costs were negligible. This analysis makes evident that, as in the case of the example presented here (with $c_1 = \$24/\text{MWh}$), the transmission expansion that maximizes total consumer surplus in a particular network can be different from the expansion that maximizes social welfare in the same network.

[17] See footnote # 6.

CHAPTER 4

PAYMENT COST MINIMIZATION WITH DEMAND BIDS AND PARTIAL CAPACITY COST COMPENSATIONS FOR DAY-AHEAD ELECTRICITY AUCTIONS

Peter B. Luh[1], Ying Chen[1], Joseph H. Yan[2], Gary A. Stern[2], William E. Blankson[1], and Feng Zhao[1]

[1]University of Connecticut, [2]Southern California Edison

EDITORS' SUMMARY: Currently most deregulated electricity markets use an auction mechanism that minimizes the total bid cost to select supply and demand bids and their associated power levels, but use a uniform market clearing price settlement mechanism to charge demand bids and pay supply bids, causing the total payment cost to be different from what was minimized in the auction. Studies have shown that for a given set of bids, using an auction that directly minimizes the total payment cost would lead to a reduced cost that consumers have to pay, and this is consistent with FERC's goals on standard market design. Although the discussion on the appropriate auction mechanism is still ongoing, it is clear that if payment cost minimization were adopted, there are inadequate methods to solve the problem. Building on our recent work for a market with given demand and full compensation of startup costs, this chapter solves a payment cost minimization problem for a market with demand bids and partial compensation of capacity costs. Numerical testing results demonstrate the method is effective to provide near-optimal solutions, is scalable, and provides valuable economic insights.

Economic Market Design and Planning for Electric Power Systems, Edited by James Momoh and Lamine Mili

4.1 INTRODUCTION

Deregulated electricity markets (e.g., the day-ahead market) operated by Independent System Operators (ISOs) generally use an auction mechanism to select bids and determine their associated power levels. A settlement mechanism is then used to charge or pay selected bids. There are two main auction mechanisms: Bid cost minimization where the total bid cost is minimized, and payment cost minimization where consumers' total payment cost is minimized. There are two main settlement mechanisms as well: Pay-as-Bid where selected bids are paid or charged at their bid prices, and the Pay-at-MCP where selected bids are paid or charged at a uniform Market Clearing Price. While markets are moving toward the Locational Marginal Pricing when transmission networks are considered, we shall for simplicity consider uniform market clearing pricing of the day-ahead energy market.

Currently most ISOs in the US use bid cost minimization to select bids, but use Pay-at-MCP for settlement. The auction and settlement mechanisms are inconsistent since the total payment cost is different from what was minimized in the auction (Yan and Stern, 2002; Yan et al., 2008). Studies have shown that for a given set of bids, using payment cost minimization as opposed to bid cost minimization for selection in conjunction with the pay-at-MCP settlement scheme would lead to a reduced payment cost for consumers (Jacobs, 1997; Hao et al., 1998; Alonso et al., 1999; Vazquez and Rivier, 2002; Yan and Stern, 2002; Mendes, 2002; Hao and Zhuang, 2003). This is consistent with FERC's goals on standard market design (Federal Energy Regulatory Commission, 2002) to provide a competitive environment for electricity and to lower the amount that consumers have to pay. However, no systematic methods existed before to solve such a problem. While the discussion on which auction mechanism is appropriate is ongoing, it is clear that if payment cost minimization were adopted, there are inadequate methods to solve the problem. A method has been presented in Luh et al., 2006 for payment cost minimization of an energy market with given system demand, full compensation of startup costs, no reserve requirements, and no transmission congestion.

In some day-ahead energy markets (e.g., ISO-PJM and ISO-NE), there are demand bids in addition to supply bids, and demand is determined through the auction process itself. Also, a unit's capacity costs that may contain startup costs, operation and maintenance costs, and no-load costs are not fully compensated but by the excess of its bid value over its market energy value in a day. Building on Luh et al., 2006, this chapter incorporates these two features to payment cost minimization in conjunction with pay-at-MCP, assuming that there are no reserve requirements and no transmission congestion. In the following, a review of relevant literature is presented in Section 4.2. The mathematical formulation of the problem is presented in Section 4.3, with units, demand bids, and MCPs mutually coupled. In addition, the non-additive form of the total payment cost and the complicated partial compensation formula make the objective function not additive in terms of bids, and not additive in time. To overcome the inseparability, the problem is solved by using augmented Lagrangian relaxation and surrogate optimization (Zhao, Luh, and Wang, 1999) as presented in Section 4.4. To reduce computational requirements while ensuring algorithm convergence, units and demand bids are individually solved, and when optimizing a particular unit, levels of other units are allowed to vary to satisfy the "surrogate optimality condition." The difficulty caused by non-time-additive compensation formula is overcome by re-defining compensations to be independent variables subject to linear inequality constraints. Numerical testing results in Section 4.5 demonstrate that this method is effective to provide near-optimal solutions, is scalable, and provides valuable economic insights.

4.2 LITERATURE REVIEW

Most ISOs solve the bid cost minimization problem by using traditional unit commitment and economic dispatch algorithms (e.g., Guan et al., 1992; Baldick, 1995; Carpentier et al., 1996; Jimenez and Conejo, 1999; Zhai 2002; Guan, Zhai and Papalexopoulos, 2003; and Padhy, 2004) to select supply bids and demand bids, and determine their hourly levels. With demand treated as negative generation, demand bids are regarded as negative supply bids, and the total bid cost is minimized subject to power balance and other relevant constraints. The problem is NP hard; however, due to its separability, it can be effectively solved by using the Lagrangian relaxation or other mixed-integer optimization techniques to obtain near-optimal solutions. MCPs and capacity cost compensations are then calculated at the end as by-products.

There are two methods in the literature to solve payment cost minimization problems. One was presented in Mendes 2002 based on forward dynamic programming. The author, however, admitted that the method was not suited for large problems due to its curse of dimensionality. The second was presented in Luh et al., 2006 under the assumption that system demand is given, startup costs are fully compensated, and there are no reserve requirements and no transmission congestion. The method consists of using augmented Lagrangian relaxation to overcome the difficulties of subproblem solution oscillation that would otherwise arise, and surrogate optimization (Zhao, Luh, and Wang, 1999) to overcome the difficulties caused by inseparability. The key idea is that the relaxed problem does not have to be solved optimally. Rather, approximate optimization is sufficient to update the multipliers if the "surrogate optimization condition" is satisfied. Due to inseparability, the relaxed problem as a whole is taken as a subproblem and optimized with respect to a particular bid while allowing the adjustment of other bids to satisfy the surrogate optimization condition. Numerical testing results demonstrate that the method is effective, and yields significantly reduced payment costs as compared to what is obtained by bid cost minimization for a given set of bids.

4.3 PROBLEM FORMULATION

Consider a day-ahead energy market with I units indexed by $i = 1, 2, \dots I$, and J demand bids indexed by $j = 1, 2, \dots J$. It is assumed for simplicity that there are no reserve requirements and no transmission congestion. Unit i is characterized at Time $t(1 \le t \le T)$ by its minimum and maximum generation levels denoted by $p_i^{\min}(t)$ (MW) and $p_i^{\max}(t)$ (MW); startup cost $S_i^{SU}(t)$ ($/Start) that is incurred if and only if

Unit i is turned ON from an OFF state at Hour t; operation and maintenance cost $S_i^{OM}(t)$ that is incurred if Unit i is ON, and a price curve consisting of up to 10 blocks each with an associated price. For simplicity, the price curve is considered to have a single block with a constant price of $c_i(t)$ (\$/MW). The status of Unit i at Time t is represented by a binary variable $x_i(t)$ with "1" representing "ON" and "0" representing "OFF." Its capacity-cost compensation is denoted as e_i^c (\$).

Demand Bid j at Time t is characterized by its maximum demand level $d_j^{\max}(t)$ (its minimum level is assumed to be zero), and a price curve consisting of up to 10 blocks each with an associated price. For simplicity, the price curve is considered to have a single block with a constant price of $b_j(t)$ (\$/MW). The status of Bid j at Time t is represented by a binary variable $y_j(t)$ with "1" representing "selected" and "0" otherwise, and the demand level is denoted as $d_j(t)$ (MW).

The market clearing price at Time t is denoted as MCP(t) (\$/MW). The task is to select units and bids and their associated levels to minimize the total payment cost subject to individual unit and bid constraints, energy balance constraints, and MCP definition.

Objective Function. The total payment cost to be minimized includes the MW payments and partial capacity-cost compensations, i.e.:

$$J \equiv \sum_{i=1}^{I} \left\{ \sum_{t=1}^{T} \mathrm{MCP}(t)\, p_i(t) + e_i^c \right\}. \tag{4.1}$$

According to some ISOs' (e.g., ISO-NE and ISO-PJM) practice, Unit i is compensated only when its MW payment is less than its bid value, with the compensation amount e_i^c given by:

$$e_i^c \equiv \max\left\{0, \sum_{t=1}^{T} \left[o_i^r(t)\, p_i(t) + S_i^{SU}(t) + S_i^{OM}(t) - \mathrm{MCP}(t)\, p_i(t) \right] \right\}, \forall i \tag{4.2}$$

Note that e_i^c depends on $\{p_i(t)\}$, $\{\mathrm{MCP}(t)\}$, $\{S_i^{SU}(t)\}$ and $\{S_i^{OM}(t)\}$ for all t, and the maximization operation in (4.2) makes e_i^c non-additive in time.

Individual Unit Constraints. If Unit i is OFF at Time t, its generation level should be zero. If Unit i is ON at Time t, its generation level should be within its minimum and maximum levels, i.e.:

$$p_i(t) = 0, \textit{ if } x_i(t) = 0, \tag{4.3}$$

$$\underline{p}_i(t) \le p_i(t) \le \overline{p}_i(t) \textit{ if } x_i(t) = 1, \forall i, \forall t \tag{4.4}$$

Since $\underline{p}_i(t)$ could be greater than zero, the feasible region of $p_i(t)$ may not be contiguous.

Power Balance Equations. The total generation should equal the total demand at any time (Wang et al., 2003), i.e.:

$$\sum_{j=1}^{J} d_j(t) - \sum_{i=1}^{I} p_i(t) = 0, \text{ for } t \text{ from 1 to } T \tag{4.5}$$

MCP-Offer Definition. Currently most ISOs resolve the auction by using a two-step procedure. In the first step, bid selections are determined by solving a unit commitment problem. With the given set of bids selected, the levels of these bids are then determined by solving an economical dispatch problem, and MCPs are the bid prices of marginal units. Following this and for simplicity, MCP for an hour is defined here as the maximum bid price of selected units, i.e.:

$$\text{MCP}(t) \equiv \max\{c_i(t), \forall i \text{ such that } x_i(t) = 1\}, \forall t \tag{4.6}$$

MCP-Bid Constraints. If the price of demand Bid j is higher than or equal to the MCP of a particular hour, then Bid j will be selected; and if its price is lower than MCP, then it will not be selected. The MCP and bid constraints are thus formulated as:

$$y_j(t)(\text{MCP}(t) - b_j(t)) + (1 - y_j(t))(b_j(t) - \text{MCP}(t)) \leq 0, \forall j, \forall t \tag{4.7}$$

Demand Bid Level Constraints. If Bid j is selected at Time t, its selected level should be greater than zero, and cannot exceed its maximum demand level, i.e.:

$$if\ \ y_j(t) = 1, 0 < d_j(t) \leq d_j^{\max}(t), \forall j, \forall t \tag{4.8}$$

In the above formulation, units and demand bids are coupled through power balance constraints (4.5), MCP definition (4.6), and MCP-bid constraints (4.7). This problem is complicated since the objective function contains the cross product terms of $\{p_i(t)\}$ and $\{\text{MCP}(t)\}$, with the latter being a function of units and demand bids to be selected. Consequently, the problem is pseudo-separable, and the standard Lagrangian relaxation approach requiring problem separability cannot be directly applied. Furthermore, the maximum operation in the compensation formula causes additional difficulties.

4.4 SOLUTION METHODOLOGY

Our method to solve the above problem is based on Luh et al., 2006 by using augmented Lagrangian relaxation (Wang and Shahidehpour, 1995; Bertsekas, 1999; and Al-Agtash, 2001) and surrogate optimization (Zhao, Luh, and Wang, 1999). Augmented Lagrangian is formed to improve convergence by relaxing coupling constraints and selectively adding quadratic penalty terms. In view of the inseparability of the original problem and the added penalty terms, the relaxed problem cannot be decomposed, and Surrogate optimization is used to solve the relaxed problem. The key idea of surrogate optimization is that approximate optimization of the relaxed problem is sufficient to obtain a good direction if the "surrogate optimization condition" is satisfied. In the remaining of the subsection, augmented Lagrangian is formed in Section 4.4.1, unit subproblems are formed and solved in Section 4.4.2. Demand bid subproblems are formed and solved in Section 4.4.3, and the dual problem is solved in Section 4.4.4. Heuristics to form a feasible solution is presented in Section 4.4.5, and initialization and the stopping criteria are presented in Section 4.4.6.

4.4.1 Augmented Lagrangian

The standard Lagrangian is formed by relaxing coupling constraints with multipliers. Since the cross product terms of unit levels and MCPs are in the Lagrangian, the relaxed problem cannot be decomposed into individual unit or bid subproblems. Furthermore, the linearity of the Lagrangian in term of unit and demand levels will cause subproblem solutions to oscillate. Consequently, direct application of the standard Lagrangian relaxation technique will not be effective. To overcome the difficulties, augmented Lagrangian relaxation is used, which is formed by selectively adding quadratic penalty terms associated with coupling constraints to the standard Lagrangian, leading to a quadratic relaxed problem with improved convergence. This, however, leads to additional inseparability. Surrogate optimization will be used to overcome this inseparability as well. Let multipliers $\{\lambda(t)\}$ relax power balance equations (4.5), and $\{\upsilon_j(t)\}$ relax MCP-bid constraints (4.7), and quadratic penalties are added for the equality constraints (4.6). The relaxed problem is formed as:

$$\min_{\{p_i(t)\},\{d_j(t)\}} L_c(\lambda, \upsilon, \text{MCP}, p, d), \text{ with}$$

$$\begin{aligned} L_c(\lambda, \upsilon, \text{MCP}, p, d) = {} & \sum_{i=1}^{I}\left\{\sum_{t=1}^{T}\text{MCP}(t)p_i(t) + e_i^c\right\} + \sum_{t=1}^{T}\left\{\lambda(t)\left(\sum_{j=1}^{J}d_j(t) - \sum_{i=1}^{I}p_i(t)\right)\right. \\ & \left. + \frac{c}{2}\left(\sum_{j=1}^{J}d_j(t) - \sum_{i=1}^{I}p_i(t)\right)^2\right\} + \sum_{i=1}^{J}\sum_{t=1}^{T}\upsilon_j(t)[y_j(t)(\text{MCP}(t) \\ & - b_j(t)) + (1-y_j(t))(b_j(t) - \text{MCP}(t))], \end{aligned} \tag{4.9}$$

where c is a positive penalty coefficient. The relaxed problem is subject to compensation formula (4.2), individual unit constraints (4.3)–(4.4), MCP definitions (4.6), and demand bid level constraints (4.8).

4.4.2 Formulating and Solving Unit Subproblems

Formulating Unit Subproblems. Given multipliers at the nth iteration, subproblem for Unit i is formed by collecting all terms involving Unit i from (4.9). Note that since the status of unit i, $\{x_i(t)\}$, affects MCPs as defined in (4.6), terms involving MCPs are also included in the subproblem. As a result, Unit i subproblem is formed as:

$$\min_{\{x_i(t)\},\{p_i(t)\}} L_i, \text{ with}$$

$$\begin{aligned} L_i \equiv {} & \sum_{k=1}^{I}\left\{\sum_{t=1}^{T}\text{MCP}(t)p_k(t) + e_k^c\right\} + \sum_{t=1}^{T}\left\{\lambda(t)\left(\sum_{j=1}^{J}d_j(t) - \sum_{k=1}^{I}p_k(t)\right)\right. \\ & \left. + \frac{c}{2}\left(\sum_{j=1}^{J}d_j(t) - \sum_{k=1}^{I}p_k(t)\right)^2\right\} + \sum_{j=1}^{J}\sum_{t=1}^{T}\upsilon_j(t)(2y_j(t)-1)\text{MCP}(t). \end{aligned} \tag{4.10}$$

Surrogate Optimization. In view that decision variables of other units and of demand bids are in the subproblem, optimally solving (4.10) is difficult. According to the "surrogate optimization condition" (Zhao, Luh, and Wang, 1999), the subproblem does not have to be solved optimally. Rather, its new solution only needs to be "better than" its solution in the previous iteration. In our context, Unit i is solved to satisfy the following "surrogate optimization condition:"

$$L_i\left(\lambda^n, \upsilon^n, (x_i, p_i)^n, (x_{k\neq i}, p_{k\neq i})^{n-1}, (y_j, d_j)^{n-1}\right) < L_i\left(\lambda^n, \upsilon^n, (x_i, p_i)^{n-1}, (x_{k\neq i}, p_{k\neq i})^{n-1}, (y_j, d_j)^{n-1}\right) \quad (4.11)$$

The satisfaction of (4.11) implies that the "surrogate subgradient" thus obtained forms an acute angle with the direction toward the optimal multiplier, and is thus a good direction for updating multipliers.

Re-defining Compensation. To solve the unit subproblem (4.10), Unit *i's* ON/OFF status $x_i(t)$ and its level $p_i(t)$ for each hour must be determined. Since $S_i^{SU}(t)$ in (4.10) depends on the unit's ON/OFF status at two consecutive hours t-1 and t, a natural way to solve the subproblem is to use dynamic programming (DP) where hours are stages, and ON/OFF status of Unit i for each hour are states. However, stage-wise costs and state transition costs cannot be directly identified since the maximum terms in (4.2) cause variables belonging to different hours to couple. To overcome this, compensations $\{e_k^c\}_k$ are redefined as independent variables subject to the following linear inequality constraints:

$$e_k^c \geq \sum_{t=1}^{T}\left[c_k(t)p_k(t) + S_k^{SU}(t) + S_k^{OM}(t) - \mathrm{MCP}(t)p_k(t)\right], \forall k, \quad (4.12)$$

and

$$e_k^c \geq 0, \forall k. \quad (4.13)$$

By using multipliers $\{\mu_k\}$ to relax the coupling constraints in (4.12), the subproblem is rewritten as:

$$\begin{aligned} L_i \equiv & \sum_{k=1}^{I}\left\{\sum_{t=1}^{T}\mathrm{MCP}(t)p_k(t) + e_k^c\right\} + \sum_{t=1}^{T}\left\{\lambda(t)\left(\sum_{j=1}^{J}d_j(t) - \sum_{k=1}^{I}p_k(t)\right)\right. \\ & \left. + \frac{c}{2}\left(\sum_{j=1}^{J}d_j(t) - \sum_{k=1}^{I}p_k(t)\right)^2\right\} + \sum_{j=1}^{J}\sum_{t=1}^{T}\upsilon_j(t)(2y_j(t)-1)\mathrm{MCP}(t) \\ & + \sum_{k=1}^{I}\mu_k\left[\sum_{t=1}^{T}\left(c_k(t)p_k(t) + S_k^{SU}(t) + S_k^{OM}(t) - \mathrm{MCP}(t)p_k(t)\right) - e_k^c\right], \end{aligned} \quad (4.14)$$

subject to (4.3), (4.4), (4.6), and (4.13).

Determining Power Levels. From (4.14), the state transition cost is identified as $\mu_i S_i^{SU}(t)$, and the stage-wise cost $v_i(t)$ is obtained by collecting all terms pertaining to t from (4.14) with the exception of $\mu_i S_i^{SU}(t)$, i.e.,

$$
\begin{aligned}
v_i(t) = {} & \mathrm{MCP}(t)\sum_{k=1}^{I} p_k(t) + \lambda(t)\left(\sum_{j=1}^{J} d_j(t) - \sum_{k=1}^{I} p_k(t)\right) + \frac{c}{2}\left(\sum_{j=1}^{J} d_j(t) - \sum_{k=1}^{I} p_k(t)\right)^2 \\
& + \sum_{j=1}^{J} \upsilon_j(t)(2y_j(t)-1)\mathrm{MCP}(t) \\
& + \sum_{\substack{m=1 \\ m\neq i}}^{I} \mu_m\left[c_m(t)\,p_m(t) + S_m^{SU}(t) + S_m^{OM}(t) - \mathrm{MCP}(t)\,p_m(t)\right] \\
& + \mu_i\left(c_i(t)\,p_i(t) + S_i^{SU}(t) + S_i^{OM}(t) - \mathrm{MCP}(t)\,p_i(t)\right)
\end{aligned}
\tag{4.15}
$$

With stage-wise costs and state transition costs identified above, dynamic programming can be used to solve the subproblem. One straightforward way is to discretize generation levels following Bard, 1988 and Ferreira et al., 1989. The computational requirements, however, would be prohibitive. To avoid this, our idea is to extend Guan et al., 1992 by directly deriving the optimal On-state $p_i(t)$ that minimize $v_i(t)$ by setting $\partial v_i(t)/\partial p_i(t)$ to zero, i.e.,

$$
p_i(t) = \sum_{j=1}^{J} d_j(t) - \sum_{\substack{m=1 \\ m\neq i}}^{I} p_m(t) - \frac{1}{c}((1-\mu_i)\mathrm{MCP}(t) - \lambda(t) + \mu_i c_i(t)). \tag{4.16}
$$

subject to (4.4), in which levels of other units $\{p_m(t)\}_{m\neq i}$ and of demand bids $\{d_j(t)\}$ are kept at their latest values, and MCP(t) is evaluated by using (4.6) with $x_i(t) = 1$ and the latest values of $\{x_m(t)\}_{m\neq i}$. The off-state $p_i(t)$ is directly set to zero in view of (4.3).

With $\{v_i(t)\}$ evaluated for both ON and OFF states for all t, the new solutions of $\{x_i(t),\ p_i(t)\}$ for all the hours are obtained by using dynamic programming following Guan et al., 1992. The new solutions of {MCP(t)} are obtained as by-products of the DP process following (4.6).

Determining Partial Capacity-Cost Compensation. Since the compensation e_i^c has been redefined as an independent variable, it needs to be optimized within the subproblem as well. To determine e_i^c, all terms involving e_i^c are pulled out from L_i:

$$
r_i = (1-\mu_i)\,e_i^c. \tag{4.17}
$$

Since r_i is linear in terms of e_i^c, solution may oscillate. To avoid this, (4.17) is approximated by a quadratic function following Guan et al., 1995:

$$
r_i' \equiv a_0\left(e_i^c\right)^2 + a_1 e_i^c + a_2. \tag{4.18}
$$

The solution e_i^c is then obtained by minimizing r_i' subject to (4.13) with coefficients a_0, a_1, and a_2 adaptively adjusted.

Checking Surrogate Optimality Conditions and Adjusting Variables. After the subproblem is solved, the surrogate optimality condition (4.11) is examined. If (4.11) is satisfied, then the new solution is accepted, the nth iteration finishes, and multipliers is updated as presented in Section 4.4.4. Otherwise, decision variables of other units are adjusted as presented next, and (4.11) is re-examined. If (4.11) is still not satisfied, the new solution is discarded, and another unit subproblem or demand bid subproblem is solved.

If (4.11) is not satisfied, heuristics is used to reduce L_i so that (4.11) is more likely to be satisfied. It has been observed in Luh et al., 2006 that the violation of (4.11) is mostly caused by fixing other units' variables at their previous values, since in this case, the minimization of Unit i's variables is biased toward their previous values. To overcome this bias, variables of other units are adjusted as follows, and L_i is then re-calculated by using dynamic programming.

If at Time t, Unit i was ON at Iteration n-1, and remains to be ON at Iteration n and has the highest price among all the ON units, then $v_i(t)$ for the OFF state is re-evaluated since turning off Unit i might lead to a decrease in MCP(t). In the case that turning off Unit i may cause a large decrease in supply and result in a high quadratic penalty $\frac{c}{2}\left(\sum_{j=1}^{J} d_j(t) - \sum_{k=1}^{I} p_k(t)\right)^2$, levels of other units are adjusted to satisfy the power balance constraint (4.5). This is done by sequentially increasing the levels of other ON units in the ascending order of their prices. If (4.5) still cannot be satisfied, units that are currently OFF are sequentially turned on in the ascending order of their amortized costs (bid price plus startup cost divided by the maximum generation level). The OFF-state $v_i(t)$ is then re-calculated by using the adjusted variables.

Conversely, if at Time t, Unit i was OFF at iteration n-1 and remains to be OFF at Iteration n, and its Price $c_i(t)$ is lower than MCP(t), then $v_i(t)$ for the ON state is re-evaluated in view that turning on unit i might enable turning off some other units with prices higher than $c_i(t)$, and result in a decrease in MCP(t). To avoid the excessive penalty for the violation of (4.5), levels of other ON units are adjusted in the descending order of their prices to satisfy (4.5). The ON-state $v_i(t)$ is then re-calculated by using the adjusted variables. After the whole adjustment procedure is done, L_i is re-calculated by using dynamic programming.

4.4.3 Formulating and Solving Bid Subproblems

For demand Bid j, the subproblem objective function is formed by collecting all terms involving demand Bid j from (4.9). Since the objective function thus formed is additive in time, and the levels of demand Bid j for different hours are independent of each other, the subproblem can be further decomposed into the following T subproblems, one for each hour:

$$\min_{\{y_j(t), d_j(t)\}} L_{j,t}, \text{ with } L_{j,t} \equiv \lambda(t) d_j(t) + \frac{c}{2}\left[d_j(t)^2 + 2 d_j(t)\left(\sum_{\substack{m=1 \\ m \neq j}}^{J} d_m(t) - \sum_{i=1}^{I} p_i(t) \right)\right] - \upsilon_j(t)(1 - 2y_j(t)) b_j(t). \tag{4.19}$$

To solve this subproblem, other variables $\{p_i(t)\}$ and $\{d_m(t)\}_{m \neq j}$ are kept at their previous values, and two cases are considered: if demand bid j is selected ($y_j(t) = 1$), and if demand bid j is not selected ($y_j(t) = 0$). For the case with $y_j(t) = 1$, $d_j(t)$ is determined by using the first order necessary condition on $L_{j,t}$ subject to the demand bid level constraint (4.8). For the case with $y_j(t) = 0$, $d_j(t)$ is set to zero. Then $L_{j,t}$ for the two cases are compared, and the one that yields the lower $L_{j,t}$ is selected. In view that this case yields the minimum $L_{j,t}$, its solution is always better than the previous solution, and the following "Surrogate Optimization Condition"

$$L_j\left((\lambda, \upsilon_j)^n, (p, d_{m\neq j})^{n-1}, (y_j, d_j)^n\right) < L_j\left((\lambda, \upsilon_j)^n, (p, d_{m\neq j})^{n-1}, (y_j, d_j)^{n-1}\right), \quad (4.20)$$

is naturally satisfied before convergence. Therefore, the new solution is always accepted, and there is no need to adjust other variables. After the subproblem is solved, multipliers are updated as next.

4.4.4 Solve the Dual Problem

Once a unit subproblem solution satisfying (4.11) or a demand bid subproblem solution satisfying (4.20) is obtained, a surrogate subgradient is used to update multipliers with a proper stepsize. The surrogate subgradient component associated with a multiplier is the associated level of constraint violation, e.g., the component for $\upsilon_j(t)$ is obtained based on the MCP and bid constraint (4.7) as:

$$\tilde{g}^k_{\upsilon_j}(t) = y^k_j(t)\left(\mathrm{MCP}^k(t) - b_j(t)\right) + \left(1 - y^k_j(t)\right)\left(b_j(t) - \mathrm{MCP}^k(t)\right); \quad (4.21)$$

Then $\upsilon_j(t)$ is updated as:

$$\upsilon^{k+1}_j(t) = \max\left(0, \upsilon^k_j(t) + s^k \tilde{g}^k_{\upsilon_j}(t)\right); \quad (4.22)$$

The stepsize s^n is selected based on the following "Surrogate Stepsize Condition:"

$$0 < s^n < \left(L^* - L^n_c\right) / \|\tilde{g}^n\|_2^2. \quad (4.23)$$

where $\|\tilde{g}^n\|_2$ is the L2 norm of the surrogate subgradient, L^n_c is the surrogate dual, and L^* is the optimal dual cost. Since L^* is generally unknown, it needs to be estimated. In our method, L^* is estimated as the lowest feasible cost obtained thus far based on heuristics to be presented in subsubsection 4.4.5. To reduce the computational requirements, the heuristics runs every few iterations.

Since L^* may be overestimated, the resulting stepsize may violate the surrogate stepsize condition (4.23). No theoretical results have yet been developed to guarantee the satisfaction of (4.23). Our numerical testing experience suggests using small step sizes for large n. Therefore, the following diminishing stepsize rule (Eq. (6.28) in Bertsekas 1999) is used:

$$s^n = \alpha^n\left(L^* - L^n_c\right) / \|\tilde{g}^n\|_2^2, \text{ with } \alpha^n = \frac{1+K}{n+K}, \quad (4.24)$$

where K is a fixed positive integer.

4.4.5 Generating Feasible Solutions

In the heuristics to generate a feasible solution, units with prices higher than {MCP(t)} are selected and awarded at their capacities, units with prices equal to

{MCP(t)} are also selected and their levels are first kept at the levels obtained from solving (4.10), and units with prices less than {MCP(t)} are not selected. For each demand bid, its status is adjusted to satisfy the MCP-Bid constraints (4.7), and its demand levels are adjusted to satisfy the power balance equations (4.5) and demand bid level constraints (4.8). If (4.5) cannot be satisfied after the above steps, the levels of units with prices equal to {MCP(t)} are then adjusted to satisfy (4.5). Once (4.5), (4.7), and (4.8) are all satisfied, partial capacity-cost compensations are then calculated by using the compensation formula (4.2).

4.4.6 Initialization and Stopping Criteria

According to Zhao, Luh, and Wang, 1999, the initial multipliers and decision variables should satisfy the following "Surrogate Initialization Condition:"

$$L_c^0 < L^*, \tag{4.25}$$

where L_c^0 is the augmented Lagrangian (4.9) calculated with the initial multipliers and decision variables. The initial $\{\lambda(t)\}$ are selected to be the MCPs obtained by using the priority-based commitment and dispatch. Multipliers associated with other constraints are set to zeros for simplicity. Initial unit levels, demand bid levels and compensations are also obtained from this process. If (4.25) is not satisfied, L_c^0 is reduced by adjusting variables following subsubsection 4.4.2.

The algorithm terminates when the average absolute change of the multipliers is less than a specified threshold ε_1 over a few iterations:

$$\frac{1}{N_c}\left(\left\|\lambda^{k+1}-\lambda^k\right\|_1+\left\|(\upsilon)^{k+1}-(\upsilon)^k\right\|_1+\left\|(\mu)^{k+1}-(\mu)^k\right\|\right)<\varepsilon_1 \tag{4.26}$$

and the average level of constraint violation is less than a specified threshold ε_2 over a few iterations:

$$\frac{1}{N_c}\left(\left\|g_\lambda\right\|_1+\left\|\max\left(0,\, g_\mu\right)\right\|_1+\left\|\max\left(0,\, g_\upsilon\right)\right\|_1\right)<\varepsilon_2 \tag{4.27}$$

where N_c is the number of the constraints relaxed. A feasible solution is then constructed by using the heuristics presented in subsubsection 4.4.5. The resulting total payment cost is then compared with the lowest feasible cost obtained thus far, and the solution with the lower cost is chosen to be the problem solution.

4.5 RESULTS AND INSIGHTS

The above method has been implemented in C++ on a Pentium-IV 2.67 GHz personal computer. Two examples are presented below. Example 1 uses a two-hour problem to demonstrate the effects of startup-cost compensation on bid selection, MCP, and total payment cost. Example 2 then demonstrates the scalability of our method for multiple problems with an increasing number of units and demand bids. For comparison purpose, the method of Luh et al., 2006 is used to obtain results for cases with given demand and full startup cost compensation. Operation and maintenance costs are assumed to be zeros, and capacity costs contain startup costs only.

Example 1. Consider a two-hour problem with four units as summarized in Table 1. Unit 1 is a hydro unit and has a low bid price and no startup cost. Units 2 to 4 are thermal units with various bid prices and startup costs. Assume for simplicity that all the units were initially off. System demand at Hour 1 is 200 MW, and at Hour 2 is 170 MW. The partial capacity-cost compensation and full capacity-cost compensation are compared.

Results are presented in Table 2, and demonstrate that different unit selections and MCPs were obtained for two compensation schemes. These results can be verified to be optimal by using exhaustive search. With full startup cost compensation, Unit 2 is not selected due to its high startup costs; and Units 1, 3, and 4 are selected to satisfy the system demand. The MCPs are $80/MW (price of Unit 4) for both hours. With partial capacity cost compensation, Unit 2 together with Units 1 and 3 are selected, and the MCPs are $75/MW (price of Unit 3) for both hours, $5/MW lower than those obtained with full compensation. This demonstrates that units with low prices and high startup costs are easier to be selected when partial compensation is used as compared to the case with full compensation. Results also show that the total payment cost with partial compensation is $1,460 (or 4.9%) less than that with full compensation.

TABLE 4.1. Parameters for Example 1.

	Min_MW	Max_MW	5/MW	Start Up Cost
Offer 1	5	40	10	0
Offer 2	10	60	37	5,000
Offer 3	5	100	75	50
Offer 4	0	70	80	50

TABLE 4.2. Summary of Results for Example 1.

Hour	Full Compensation			Partial Compensation		
	1 MW	2 MW	Startup cost ($)	1 MW	2 MW	Compensation ($)
Offer 1	40	40	0	40	40	0
Offer 2	0	0	0	60	60	440
Offer 3	100	100	50	100	70	50
Offer 4	60	30	50	0	0	0
	MCP(1) = $80/MW MCP(2) = $80/MW Total payment costs = $29,700			MCP(1) = $75/MW MCP(2) = $75/MW Total payment costs = $28,240		

Example 2. To test the scalability of our algorithm, problems with an increasing number of units and demand bids are tested based on semi-realistic data sets. Cases considered include 10 units and 10 bids; 10 units and 20 bids; 30 units and 10 bids; and 30 units and 20 bids. For each case, ten data sets are randomly generated using Gaussian distributions as summarized in Tables 4.3 and 4.4. Among the units, 45% of the capacity comes from nuclear units, 15% from hydro units, and 20% from thermal units with low bid prices but high startup costs, and the remaining 20% from thermal units with high bid prices but low startup costs. Each unit has identical parameters over a 24-hour period. Assume that initially nuclear units were on while thermal and hydro units were off.

The same stopping criteria are used for all the four cases, with thresholds ε_1 in (4.27) and ε_2 in (4.28) set to 0.01. The satisfaction of (4.27) implies the near convergence of the multipliers, and the satisfaction of (4.28) implies that the surrogate subgradient obtained is close to zero. These mean that the surrogate dual cost obtained is close to the top of the surrogate dual function. With the use of the effective heuristics in subsubsection 4.4.5, the solution quality is believed to be good. The average number of iterations and CPU times over ten randomly generated data sets for each of the four cases are summarized in Table 4.5. It can be seen that the

TABLE 4.3. Characteristics of Generation Offers for Example 2.

Types of Offers	Mean MW/Offer	Std. Dev. MW/Offer	Mean Price 5/MW
Nuclear	1,340	200	15
Hydro	620	50	23
Thermal	300	30	37
Thermal	300	30	80
Types of Offers	**Std. Dev. $/MW**	**Mean SU. Cost ($)**	**Std. Dev. of SU. Cost ($)**
Nuclear	2	0	0
Hydro	1.5	0	0
Thermal	2	8,000	500
Thermal	2	200	50

TABLE 4.4. Characteristics of Demand Bids for Example 2.

Mean MW/Bid	Std. Dev MW/Bid	Mean Price 5/NW	Std. Dev. 5/MW
67.5	7.5	75	10

TABLE 4.5. Summary of Results for Example 2.

(GO, DB)	(10, 10)	(10, 20)	(30, 10)	(30, 20)
ANIs	103	147	231	293
CPU time	3.6	5.2	10.3	12.1

major factor affecting the number of iterations is the number of units, with the number of demand bids playing a less significant role. CPU times are only a few minutes on a PC. This testing thus demonstrates the scalability of our method for problems with increasing numbers of units and demand bids.

GO: Total number of generation offers; DB: Total number of demand bids; ANIs: Average number of iterations; CPU time: Measured in minutes.

4.6 CONCLUSION

Currently, most ISOs in the United States conduct bid cost minimization in auctions and settle the payments with market clearing prices. An alternative auction mechanism that minimizes the consumer payment cost has been brought to recent discussions. This paper presents a systemic method to solve payment cost minimization auction with demand bids and partial capacity-cost compensation. Our method is developed by using the augmented Lagrangian relaxation and surrogate optimization framework with enhanced features to handle the difficulties caused by the coupling of units and demand bids, and the complicated compensation formula. Numerical testing results demonstrate that the method is effective and computationally efficient. This work paves the way for the further study of the payment cost minimization auction mechanism.

ACKNOWLEDGMENT

This work was supported in part by the National Science Foundation under grants ECS-0323685, ECS-0621936, and by a grant from Southern California Edison.

BIBLIOGRAPHY

[1] Al-Agtash S. Hydrothermal Scheduling by Augmented Lagrangian: Consideration of Transmission and Pumped Storage Units. *IEEE Transactions on Power Systems* 2001; 16(4):750–756.

[2] Alonso J, Trias A, Gaitan V, Alba JJ. Thermal Plant Bids and Market Clearing in an Electricity Pool: Minimization of Costs vs. Minimization of Consumer Payments. *IEEE Transactions on Power Systems* 1999; (4):1327–1334.

[3] Baldick R. The Generalized Unit Commitment Problem. *IEEE Transactions on Power Systems* 1995; 0(1):465–475.

[4] Bard JF. Short-term Scheduling of Thermal-Electric Generators Using Lagrangian Relaxation. *Operation Research* 1988; 36(5):756–766.

[5] Bertsekas DP. *Nonlinear Programming, Second Edition*, 1999, Athena Scientific, Belmont, Massachusetts.

[6] Carpentier P, Cohen G, Culioli JC, Renaud A. Stochastic Optimization of Unit Commitment: A New Decomposition Framework. *IEEE Transactions on Power Systems* 1996; 11(2):1067–1073.

[7] Federal Energy Regulatory Commission, Federal Energy Regulatory Commission Paper on Standardized Transmission Service and Wholesale Electric Market Design, 2002.

[8] Ferreira LA, Andersson T, Imparato CF, Miller TE, Pang CK, Svoboda A, and Vojdani AF. Short-term Resource Scheduling in Multi-Area Hydrothermal Power Systems. *Electric Power & Energy Systems* 1989; 11(3):200–212.

[9] Guan X, Luh PB, Yan H, Amalfi JA. An Optimization-Based Method for Unit Commitment. *International Journal of Electrical Power & Energy Systems* 1992; 14(1):9–17.

[10] Guan X, Luh PB, Zhang L. Nonlinear Approximation Method in Lagrangian Relaxation–Based Algorithms for Hydrothermal Scheduling. *IEEE Transactions on Power Systems* 1995; 10(2):772–778.

[11] Guan X, Zhai Q, Papalexopoulos A. Optimization Based Methods for Unit Commitment: Lagrangian Relaxation Versus General Mixed Integer Programming. *IEEE Power Engineering Society General Meeting July* 2003; Vol. 2, pp. 13–17.

[12] Hao, S, Angelidis GA, Singh H, Papalexopoulos AD. Consumer Payment Minimization in Power Pool Auctions. *IEEE Transactions on Power Systems* 1998; 13(3):986–991.

[13] Hao S, Zhuang F. New Models for Integrated Short-Term Forward Electricity Markets. *IEEE Transactions on Power Systems* 2003; 18(2):478–485.

[14] Jacobs JM. Artificial Power Networks and Unintended Consequences. *IEEE Transactions on Power Systems* 1997; 12(2):968–972.

[15] Jimenez N, Conejo A. Short-Term Hydrothermal Coordination by Lagrangian Relaxation: Solution of the Dual Problem. *IEEE Transactions on Power Systems* 1999; 14(1):89–95.

[16] Luh PB, Blankson WE, Chen Y, Yan JH, Stern GA, Zhao F. Optimal Auction for the Deregulated Electricity Market Using Augmented Lagrangian and Surrogate Optimization. *IEEE Transactions on Power Systems* 2006; 21(2):568–578.

[17] Mendes DP. Resource Scheduling and Pricing in a Centralised Energy Market. Proceedings of the 14th Power System Computation Conference; June 2002; Seville, Spain, pp. 1–7.

[18] Padhy NP. Unit Commitment—A Bibliographical Survey. *IEEE Transactions on Power Systems* 2004; 19(3):1196–1205.

[19] Vazquez C, Rivier M, Perez-Arriaga IJ. Production Cost Minimization Versus Consumer Payment Minimization in Electricity Pools. *IEEE Transactions on Power Systems* 2002; 17(1):119–127.

[20] Wang SJ, Shahidehpour SM. Short-Term Generation Scheduling with Transmission and Environmental Constraints Using an Augmented Lagrangian relaxation. *IEEE Transactions on Power Systems* 1995; 10(3):1294–1301.

[21] Wang J, Redondo NE, Galiana FD. Demand-side Reserve Offers in Joint Energy/Reserve Electricity Markets. *IEEE Transactions on Power Systems* 2003; 18(4):1300–1306.

[22] Yan JH, Stern GA. Simultaneous Optimal Auction and Unit Commitment for Deregulated Electricity Markets. *Electricity Journal* 2002; 15(9):72–80.

[23] Yan JH, Stern GA, Luh PB, Zhao F. Payment Cost Minimization vs. Bid Cost Minimization for ISO Markets. *IEEE Power and Energy Magazine* 2008; 6(2):24–36.

[24] Zhai Q, Guan X, Cui J. Unit Commitment with Identical Units Successive Subproblem Solving Method Based on Lagrangian Relaxation. *IEEE Transactions on Power Systems* 2002; 17(4):1250–1257.

[25] Zhao X, Luh PB, Wang J. Surrogate Gradient Algorithm for Lagrangian Relaxation. *Journal of Optimization Theory and Applications* 1999; 100(3):699–712.

CHAPTER 5

DYNAMIC OLIGOPOLISTIC COMPETITION IN AN ELECTRIC POWER NETWORK AND IMPACTS OF INFRASTRUCTURE DISRUPTIONS

Reetabrata Mookherjee[1], *Benjamin F. Hobbs*[2], *Terry L. Friesz*[3], *and Matthew A. Rigdon*[3]

[1]Zilliant, Inc., [2]Johns Hopkins University, [3]Pennsylvania State University

EDITORS' SUMMARY: This chapter puts forward a dynamic game theoretic model of oligopolistic competition in spatially distributed electric power markets having a 24-hour planning horizon. The purpose of this model is to allow quick testing of the effects of changes to the underlying electric power network. Therefore, the game is formulated as a nonlinear complementarity problem that can be solved efficiently using sequential linearization and a Lemke's type algorithm for each resulting linear complementarity problem. The underlying electric power network is represented by the widely accepted linearized DC approximation, allowing the substitution of power transmission distribution factors for Kirchhoff's energy balance and voltage laws. The model is tested on a 15-node representation of the northwest European electricity market formed by Belgium, France, Germany and the Netherlands. The effects of various infrastructure disruptions, in the form of network capacity changes, are simulated.

5.1 INTRODUCTION AND MOTIVATION

This chapter proposes a model of dynamic oligopolistic competition in electric power networks that draws on some of the recent literature on electric power market equi-

Economic Market Design and Planning for Electric Power Systems, Edited by James Momoh and Lamine Mili

librium models.[1] The model describes an electricity power market consisting of electric power generating firms competing on an underlying power system network, whose capacity is rationed by an independent system operator (ISO) using congestion pricing. The competing generating firms wish to allocate power generated at multiple locations to different markets in order to maximize their profits. The ISO's job is to efficiently clear the markets for power by setting wheeling fees for the transmission of power between locations on the network. Each individual market participant's problem is formulated as a discrete time mathematical program, in which sales by competitors and prices of transmission services are taken as exogenous. The collection of these coupled discrete time mathematical programs describe the dynamic Cournot-Nash game that we are interested in. This game is represented by a nonlinear complementarity problem (NCP) through the analysis of the necessary conditions for optimality of each participant's optimization problem. Representing the problem as an NCP allows us to make direct use of powerful commercial solvers such as PATH for efficient computation, in turn making it convenient for analyzing the impact of large disruptive events in electric power systems.

In this chapter, we think of large (or extreme) events as those events with a very large deviation from an expected or typical condition. These events usually have a negative effect on system performance and may arise due to belligerent agents or system failures. Designing and planning for extreme events is not a trivial matter. A clear tradeoff exists between the costs of designing and planning for an extreme event and the costs of managing the event if it occurs. We have witnessed such extreme events in recent years. Some examples include the 2003 electricity blackout in the northeast United States and southeast Canada, and Hurricane Katrina in 2005 that greatly affected several states in the southern United States. An earlier extreme event of interest is the 1977 blackout in New York City that was the result of lightning strikes knocking out lines in a critical transmission corridor. This brings us to another point: a localized extreme event may have a global impact in a network setting. The model proposed in this paper may be used to understand how local disruptions in an electric power system can effect the flows of power throughout the network over time, considering oligopolistic behavior of generating companies.

The model put forth in this paper is deterministic. However, a simulation type experimentation method is proposed for generating perturbations to the parameters to test the effects of extreme events. The model may also be used directly to test different scenarios including failures of transmission lines and power generating facilities. Alternatively, random outages can be generated using Monte Carlo simultion. Because the model is dynamic in nature, it is possible to see how the system reacts to a failure as well as how it returns to equilibrium when a failed component is brought back into service. This is also of interest as it allows the modeler to directly model any scenario they wish to define.

One of the major areas of application of complementarity and variational inequalities-based models of economic equilibria is electric power markets; more so

[1] A glossary of economic terms, such as "oligopolistic competition" and "equilibrium models" is provided as an appendix to this chapter.

since this economically crucial industry underwent a transition from tight regulation to intense competition subject to loose regulatory constraints. It is not our intention, nor does space permit, to list or discuss all previous models of this general type. However, Daxhelet and Smeers (2001), Day et al. (2002), and Hobbs and Helman (2004) summarize the relatively recent literature. Unlike other engineered systems, technology and cost information is widely available for power industries that facilitate modeling—in most of the cases independent system operators (ISO) publish daily load profiles, market prices and other information on a daily basis through their websites. Also, commercial databases are available that describe characteristics of generators in markets. At the same time, the unique characteristics of electricity transmission, such as Kirchhoff's power and voltage laws, present intriguing challenges to modelers and systems engineers.

This paper makes use of publicly available demand and supply information for a numerical example based upon an actual power system in northwest Europe formed by Belgium, France, Germany and the Netherlands. This system is approximated by a network that includes 15 nodes and 12 generating firms (Neuhoff et al., 2005). Of the 12 generating firms, eight are distinct firms, while four represent conglomerations of remaining (small) firms in each country. Of the 15 nodes, seven have demand and generation, and the remainder represent import points for the two larger countries (France, Germany). The data available for this network includes the power transmission distribution factors (PTDFs) for each transmission link and network node, transmission line capacities and generation costs. To the best of our knowledge, this is the first dynamic electric power equilibrium model to use this data set.

The remainder of this chapter is organized as follows. The next section (5.2) provides an overview of the modeling approach, while Section 5.3 presents the notation and the optimization problem for each market participant (generators and transmission system operator). Section 5.4 presents the resulting equilibrium model, consisting of the first-order necessary conditions for the participants' optimization problems combined with market clearing conditions. The equilibrium model is a dynamic nonlinear complementarity problem, and accounts for transmission flows and the piecewise linearity of generation cost functions. Numerical examples are presented in Section 5.5. A glossary of electricity economics terminology is provided as an Appendix defining specialized terminology from economics that is used in this paper.

5.2 SUMMARY OF MODELING APPROACH

While many existing electric power market equilibria models take a static approach, we consider a dynamic model that allows the modeling of constraints and costs that change with time. This is an important distinction as the demands realized by power generators change according to the time of the day or season of the year. Also of interest is that electric power systems are comprised of equipment that cannot instantaneously react to market changes over time. Specifically, most generating units are limited in the rate at which they can change their output. The model presented in this paper consider such bounds on the ramping rate of a generator—that is, the rate at which a generator can change its output between two successive periods.

For this model, we assume a linearized DC load flow representation of the power network. The assumptions and derivation of the general linearized model are laid out in Schweppe et al. (1988). This representation allows for the use of PTDFs to model the physical flow of power on the network, which assume that power flow on a transmission line is proportional to injections.

As mentioned previously, each firm's extremal (optimization) problem is represented as a discrete time mathematical program with linear constraints. Each firm's extremal problem depends upon the actions of their competitors and results in a set of coupled discrete time mathematical programs that define the game. The necessary conditions for this set of problems are analyzed and used to formulate a nonlinear complementarity problem (NCP.). The NCP can be solved directly using a commercial solver; alternatively, a sequential linearization scheme can be employed in which a Lemke's type algorithm is used to solve each resulting linear complementarity subproblem.

To account for the effect of disruptive events, we envision a simulation approach using the dynamic electric power model described in this paper. The model described herein may be envisioned as a day ahead model that is run by generating firms for planning and bidding purposes. The generating firms may solve this model based on their perception of the network and demands. The resulting solution will dictate their commitments to the ISO to provide energy and for what price the next day. However, as the plan is implemented the following day, the ISO is able to immediately realize variations in the system caused by various unanticipated events, and will adjust the fees charged to the generating firms to transmit their power across the network. Therefore, the generating firms will experience a different profit than was predicted by the day ahead model, and will in general adjust their generation and sales decisions.

A simulation approach could thus be used to repeatedly run the model and compare the day ahead predicted profits with the realized profits. Each simulation run will vary the parameters of the system according to some distribution.

5.3 MODEL DESCRIPTION

5.3.1 Notation

Before formally introducing the model, we give an overview of the sets, variables and parameters that are used in the exposition of the model. The notation shown in Tables 5.1 to 5.3 will be used through the remainder of this paper.

The following vector concatenations are used when applicable to simplify the notation.

$$
\begin{aligned}
q^f &: \left\{q_{i,t}^f \quad \forall i \in N, t = 0, \ldots, T\right\} \\
c^f &: \left\{c_{i,t}^f \quad \forall i \in N, t = 0, \ldots, T\right\} \\
r^f &: \left\{r_{i,t}^f \quad \forall i \in N, t = 1, \ldots, T\right\} \\
w &: \left\{w_{i,t} \quad \forall i \in N, t = 0, \ldots, T\right\}
\end{aligned}
$$

TABLE 5.1. Notations for Sets.

Sets	
F	Set of generating firms
N	Set of nodes in the power network
M	Set of nodes at which there are markets for power
A	Set of transmission lines (arc) in the network

TABLE 5.2. State Variables, Control Variables and Functions.

Variables	
$q_{i,t}^f$	Generation in MW by firm $f \in F$ at market $i \in M$ in period $t \in \{0, \ldots, T\}$
$c_{i,t}^f$	Sales (consumption) in MW by firm $f \in F$ at market $i \in M$ in period $t \in \{0, \ldots, T\}$
$r_{i,t}^f$	Ramping rate (MW/hr) of unit owned by firm $f \in F$ in market $i \in M$ in period $t \in \{0, \ldots, T\}$
$w_{i,t}$	Wheeling fee (3/MW) for market $i \in M$ in period $t \in \{0, \ldots, T\}$
$y_{i,t}$	Transfer (injection or withdrawal) of power (MW) from the hub node to the node $i \in M$ in period $t \in \{0, \ldots, T\}$
$\pi_{i,t}$	Inverse demand function (3/MW) for the market $i \in M$ in period $t \in \{0, \ldots, T\}$
$V_{i,t}^f$	Generation cost function for firm $f \in F$ at market $i \in M$ at time $t \in \{0, \ldots, T\}$
$R_{i,t}^f$	Ramping cost for firm $f \in F$ at market $i \in M$ in period $t \in \{0, \ldots, T\}$

TABLE 5.3. Model Parameters.

Parameters	
$q_{i,\max,t}^f$	Upper bound of generator for firm $f \in F$ at node $i \in M$ in period t
$r_{i,\min}^f$	Minimum ramp rate of generator owned by firm $f \in F$ at market $i \in M$
$r_{i,\max}^f$	Maximum ramp rate of generator owned by firm $f \in F$ at market $i \in M$
$a_{i,t}$	Price intercept in the linear inverse demand model for market $i \in M$ in period t
$b_{i,t}$	Quantity intercept in the linear inverse demand model for market $i \in M$ in period t
$m_{1,t}^f$	Variable cost (3/MW) of first section of two piece linear generation cost for the firm $f \in F$ at market $i \in M$ in time period t
$m_{2,t}^f$	Variable cost (3/MW) of second section of two piece linear generation cost for the firm $f \in F$ at market $i \in M$ in time period t
$b_{1,t}^f$	Fixed cost (3/MW) of first section of a two-piece linear generation cost for the firm $f \in F$ at market $i \in M$ in time period t
$b_{2,t}^f$	Fixed cost (3/MW) of second section of two-piece linear generation cost for the firm $f \in F$ at market $I \in M$ in time period t
$T_{a,t}$	Transmission capacity of arc $a \in A$ in period t
T	Total number of time periods in the planning horizon

5.3.2 Generating Firm's Extremal Problem

We are now in a position to discuss the problem faced by each generating firm. Each firm's extremal problem is formulated as a discrete time mathematical program with an objective of maximizing the profit of the firm. The strategy of each firm is dependent upon the actions of its competitors, thus we adopt the notation for variables controlled by another firm, specifically the allocation to consumption ("sales"), as c^{-f} where c^{-f}: $c^g\ \forall g \neq f$

With the preceding information, we may introduce the single generating firm f's extremal problem with c^{-f} as exogenous information

$$\max_{c^f, q^f} J_1\left(c^f, q^f; r^f; c^{-f}, w\right) = \sum_{t=0}^{N} \sum_{i \in \mathcal{M}} \left\{ \pi_{i,t}\left(\sum_{g \in \mathcal{F}} c_{i,t}^g \right) \cdot c_{i,t}^f - V_{i,t}^f\left(q_{i,t}^f\right) - w_{i,t} \cdot \left(c_{i,t}^f - q_{i,t}^f\right) \right\} \tag{5.1}$$

subject to

$$\sum_{i \in \mathcal{N}} q_{i,t}^f = \sum_{i \in \mathcal{M}} c_{i,t}^f \quad \forall t = 0, \dots, T \tag{5.2}$$

$$\pi_{i,t} = a_{i,t} - b_{i,t}\left(\sum_{g \in \mathcal{F}} c_{i,t}^g \right) \quad \forall i \in M, t = 0, \dots, T \tag{5.3}$$

$$V_{i,t}^f = \max\left(m_{1,i}^f q_{i,t}^f + b_{1,i}^f, m_{2,i}^f q_{i,t}^f + b_{2,i}^f\right) \quad \forall i \in M, t = 0, \dots, T \tag{5.4}$$

$$r_{i,t}^f = q_{i,t-1}^f - q_{i,t}^f \quad \forall i \in M, t = 1, \dots, T \tag{5.5}$$

$$0 \leq q_{i,t}^f \leq q_{i,\max,t}^f \quad \forall i \in M, t = 1, \dots, T \tag{5.6}$$

$$r_{i,\min}^f \leq r_{i,t}^f \leq r_{i,\max}^f \quad \forall i \in M, t = 1, \dots, T \tag{5.7}$$

The objective of Equation 5.1 is simply a profit maximization. Profit is the difference between the revenue at all nodes from sales minus the costs of generation and the wheeling fees. Sales are to consumers, traders, or load serving entities at individual nodes of the network; thus, we are simulating a bilateral market rather than a pool-based market (where instead all sales would be to the market operator at the location of generation).

The wheeling fee is the cost (possibly negative) of transmitting power from the network hub to the point of consumption (equal to the negative of the cost of bring power generated at that point to the hub), and is the result of the interaction of the network constraints in the ISO's model with the generators' demands for transmission services. Although the generator plays a Cournot (quantity) game against other generators, it takes the wheeling fee as exogneous, and does not believe that it will change if it changes its output or sales. Thus, the generator is playing a Bertrand (price) game against the system operator.

Equation 5.2 is a flow balance equation which states that all power produced must be allocated to sales in each period as we do not consider the storage of electricity. Equation 5.3 is the inverse demand, or market price, for electricity which depends upon the sales of all firms at that market with $a_i, b_i \in \Re_{++}^1$ for all $i \in \mathcal{M}$. Equation 5.4 is a two-piece linear generation cost function with

$m_{1,i}^f, m_{2,i}^f, b_{1,i}^f, b_{2,i}^f \in \Re_{++}^1$ for all $f \in \mathcal{F}$ and $i \in \mathcal{M}$. Equation 5.5 are discrete dynamics which define the ramping rates for the generators, while 5.7 represents the bounds on the ramping rate of a generator between two time periods with $-r_{i,\min}^f \in \Re_{++}^1$, $r_{i,\max}^f \in \Re_{++}^1$ for all $f \in \mathcal{F}$ and $i \in \mathcal{M}$. Finally, constraint 5.6 gives the bounds on generation capacity with $q_{i,\max,t}^f \in \Re_{++}^1$ for all $f \in \mathcal{F}$, $i \in \mathcal{M}$ and $t = 0, \ldots, T$. Note that the parameter for the generation capacity may change with time. This allows us to test the effects of a reduction in generation capacity due to some event for a number of time periods.

Note that each generator's extremal problem in Equations 5.1 through 5.7 is coupled to its competitiors' problems via inverse demand (Equation 5.3) and the wheeling fee, which automatically gives rise to a multiperiod Nash game. The open loop Nash equilibrium may be obtained when all the firms and the ISO simultaneously solve their respective extremal problems. A generalization of this model would represent smaller firms as "price takers", who believe that they cannot affect the price. In that case, price in 5.1 would be viewed by the firm as exogenous rather than a function in 5.3 of quantity supplied. Collectively, such price takers are referred to as a "competitive fringe."

Because we will be analyzing the necessary conditions for each firm's discrete time mathematical program in order to formulate the game as an NCP, we need to have differentiability of all constraints. Therefore, Equation 5.4 requires some special attention. Instead of using the form expressed above as:

$$V_{i,t}^f = \max\left(m_{1,i}^f q_{i,t}^f + b_{1,i}^f, m_{2,i}^f q_{i,t}^f + b_{2,i}^f\right), \tag{5.8}$$

we can instead regard $V_{i,t}^f$ as a variable and construct the following inequalities:

$$V_{i,t}^f \geq m_{1,i}^f q_{i,t}^f + b_{1,i}^f$$

$$V_{i,t}^f \geq m_{2,i}^f q_{i,t}^f + b_{2,i}^f$$

Because the problem is to maximize profit and these two equations will always have a non-negative value with a negative coefficient in the objective (assuming the amount of generation is nonnegative), the cost will always lie on the lower envelope formed by the original equation. This is because choosing a cost above that value would further reduce the the value of the objective (5.1). The resulting discrete time mathematical program after the above transformations boils down to the following:

$$\max J_1\left(c^f, q^f, V^f; r^f; c^{-f}, w\right) = \sum_{t=0}^{N} \sum_{i \in \mathcal{M}} \left\{\pi_{i,t} \cdot c_{i,t}^f - V_{i,t}^f - w_{i,t} \cdot \left(c_{i,t}^f - q_{i,t}^f\right)\right\} \tag{5.9}$$

subject to

$$\sum_{i \in \mathcal{N}} q_{i,t}^f = \sum_{i \in \mathcal{M}} c_{i,t}^f \quad \forall t = 0, \ldots, T \tag{5.10}$$

$$\pi_{i,t} = a_{i,t} - b_{i,t}\left(\sum_{g \in \mathcal{F}} c_{i,t}^g\right) \quad \forall i \in M, t = 0, \ldots, T \tag{5.11}$$

$$V_{i,t}^f \geq m_{1,i}^f q_{i,t}^f + b_{1,i}^f \quad \forall i \in M, t = 0, \ldots, T \tag{5.12}$$

$$V_{i,t}^{f} \geq m_{2,i}^{f} q_{i,t}^{f} + b_{2,i}^{f} \quad \forall i \in M, t = 0, \ldots, T \tag{5.13}$$

$$r_{i,t}^{f} = q_{i,t}^{f} - q_{i,t+1}^{f} \quad \forall i \in M, t = 1, \ldots, T \tag{5.14}$$

$$0 \leq q_{i,t}^{f} \leq q_{i,\max,t}^{f} \quad \forall i \in M, t = 0, \ldots, T \tag{5.15}$$

$$r_{i,\min}^{f} \leq r_{i,t}^{f} \leq r_{i,\max}^{f} \quad \forall i \in M, t = 1, \ldots, T \tag{5.16}$$

5.3.3 ISO's Problem

It was mentioned earlier that the wheeling fees, w, are set by the ISO in order to efficiently clear the market for transmission capacity. As we have seen, these wheeling fees are taken by generators as being exogenous to their extremal problems, and impacts their net profits. These wheeling fees are implicitly determined in the equilibrium problem by simulating the ISO as solving the following linear program for allocating scarce transmission capacity. In particular, the ISO wishes to determine the transmission flows y in order to:

$$\max_{w} J_2(t) = \sum_{i \in \mathcal{N}} y_{i,t} w_{i,t} \tag{5.17}$$

subject to:

$$\sum_{i \in \mathcal{N}} PDF_{i,a} \cdot y_{i,t} \leq T_{a,t} \; \forall a \in \mathcal{A} \tag{5.18}$$

where $\mathcal{A}$ is the arc set of the electric power network, $T_{a,t}$ is the transmission capacity on arc $a \in \mathcal{A}$ at time t, and $PDF_{i,a}$ is the PTDF that describes how much MW flow occurs through transmission line ("arc") a as a result of a unit MW injection at an arbitrary hub node and a unit withdrawal at node i. It is important to note that the parameters $T_{a,t}$ representing the transmission capacities of the arcs may change with time; this makes it possible to create scenarios where a transmission line is derated for some number of time periods. The decision variables $y_{i,t}$ denote transfers of power in MW by the ISO from a hub node to the node $i \in \mathcal{N}$ at time period t.

Note that $w_{i,t}$ is treated as exogenous in (5.17), so the ISO is modeled as if it is playing a Bertrand game against users of the network. Elsewhere, it has been shown that this representation is equivalent to the ISO setting $w_{i,t}$ so that scarce transmission capacity is allocated to those most willing to pay for it; i.e., the market clears for transmission.

To clear the market, the transmission flows $y_{i,t}$ must balance the net sales at each node. Therefore, the net transmission into a market is given by:

$$y_{i,t} = \sum_{f \in \Phi} \left(c_{i,t}^{f} - q_{i,t}^{f} \right) \quad \forall i \in M \tag{5.19}$$

Re-writing (5.17)–(5.18) we obtain the following set of linear programs, one for each time period of interest:

$$\max_{w_t} J_{2,t} = \sum_{i \in \mathcal{N}} \sum_{f \in \mathcal{F}} \left(c_{i,t}^{f} - q_{i,t}^{f} \right) \cdot w_{i,t} \tag{5.20}$$

subject to:

$$\sum_{i\in\mathcal{N}} PDF_{i,a}\cdot\left\{\sum_{f\in\mathcal{F}}\left(c_{i,t}^f-q_{i,t}^f\right)\right\}\leq T_a \quad \forall a\in\mathcal{A} \tag{5.21}$$

In this particular formulation, we ignore transmission losses; however, our model is general enough to consider non-linear losses. In the case of losses, either the ISO or the firms involved in the transaction should account for the losses and a book-keeping effort is required.

In equilibrium, the objective function (5.17) will be identical to the so-called "congestion surplus," which is the difference between the payments by consumers for electricity and the payments to generators, if payments were based on the locational marginal price at each of their locations.

5.4 FORMULATION OF NCP

We will form the nonlinear complementarity problem by analyzing the necessary conditions for both the generating firms' problems as well as the ISO's problems. Because we have only linear constraints, Abadie's constraint qualification holds and we can inspect the Karush-Kuhn-Tucker (KKT) conditions. We will first inspect the necessary conditions for the generating firms and then for the ISO. These conditions are then combined into the complete NCP representation.

5.4.1 Complementarity Conditions for Generating Firms

The mathematical program (in standard form) that we wish to examine is:

$$\min J_1\left(c^f,q^f,V^f;c^{-f},w\right)=-\sum_{t=0}^{N}\sum_{i\in\mathcal{M}}\left\{\left(a_i-b_i\sum_{g\in\mathcal{F}}c_{i,t}^g\right)\cdot c_{i,t}^f-V_{i,t}^f-w_{i,t}\cdot\left(c_{i,t}^f-q_{i,t}^f\right)\right\} \tag{5.22}$$

subject to:

$$\sum_{i\in\mathcal{N}} q_{i,t}^f-\sum_{i\in\mathcal{M}} c_{i,t}^f\leq 0 \quad \forall t=0,\ldots,T \quad \left(\zeta_t^{+f}\right) \tag{5.23}$$

$$-\sum_{i\in\mathcal{N}} q_{i,t}^f+\sum_{i\in\mathcal{M}} c_{i,t}^f\leq 0 \quad \forall t=0,\ldots,T \quad \left(\zeta_t^{-f}\right) \tag{5.24}$$

$$-V_{i,t}^f+m_{1,i}^f q_{i,t}^f+b_{1,i}^f\leq 0 \quad \forall i\in N, t=0,\ldots,T \quad \left(\gamma_{i,t}^f\right) \tag{5.25}$$

$$-V_{i,t}^f+m_{2,i}^f q_{i,t}^f+b_{2,i}^f\leq 0 \quad \forall i\in N, t=0,\ldots,T \quad \left(\eta_{i,t}^f\right) \tag{5.26}$$

$$-c_{i,t}^f\leq 0 \quad \forall i\in N, t=0,\ldots,T \quad \left(\phi_{i,t}^f\right) \tag{5.27}$$

$$-V_{i,t}^f\leq 0 \quad \forall i\in N, t=0,\ldots,T \quad \left(\delta_{i,t}^f\right) \tag{5.28}$$

$$-q_{i,t}^f\leq 0 \quad \forall i\in N, t=0,\ldots,T \quad \left(\rho_{i,t}^f\right) \tag{5.29}$$

$$q_{i,t}^f-q_{i,\max}^f\leq 0 \quad \forall i\in N, t=0,\ldots,T \quad \left(\sigma_{i,t}^f\right) \tag{5.30}$$

$$-\left(q_{i,t}^f - q_{i,t-1}^f\right) + r_{i,\min}^f \le 0 \quad \forall i \in M, t = 1, \ldots, T \quad \left(\mu_{i,t}^f\right) \tag{5.31}$$

$$q_{i,t}^f - q_{i,t-1}^f - r_{i,\max}^f \le 0 \quad \forall i \in M, t = 1, \ldots, T \quad \left(\theta_{i,t}^f\right) \tag{5.32}$$

where the variables in parenthesis are the dual variables associated with each inequality. Note that Equation 5.8 is now being represented by two inequalities (5.25) and (5.26). Also note that the ramping rates $r_{i,t}^f$ have been replaced by their representation in terms of generation $q_{i,t}^f$ as shown in Equations 5.31 and 5.32. As stated previously, Abadie's constraint qualification holds, and thus, we can inspect the KKT conditions of the above mathematical Equations 5.22 through 5.32. The KKT identities are found to be:

$$0 = 2b_{i,t}c_{i,t}^f - a_{i,t} + b_{i,t} \sum_{g \in \mathcal{F}, g \ne f} c_{i,t}^g - w_{i,t} - \zeta_t^{+f} + \zeta_t^{-f} - \phi_{i,t}^f \tag{5.33}$$

$$0 = w_{i,t} + \zeta_t^{+f} - \zeta_t^{-f} + m_{1,i}^f \gamma_{i,t}^f + m_{2,i}^f \eta_{i,t}^f - \rho_{i,t}^f + \sigma_{i,t}^f - \mu_{i,t}^f + \theta_{i,t}^f \tag{5.34}$$

$$0 = 1 - \gamma_{i,t}^f - \eta_{i,t}^f - \delta_{i,t}^f \tag{5.35}$$

with the following accompanying complementarity slackness conditions:

$$0 \le \left[-\sum_{i \in \mathcal{M}} q_{i,t}^f + \sum_{i \in \mathcal{M}} c_{i,t}^f \right] \perp \zeta_t^{+f} \ge 0 \tag{5.36}$$

$$0 \le \left[\sum_{i \in \mathcal{M}} q_{i,t}^f - \sum_{i \in \mathcal{M}} c_{i,t}^f \right] \perp \zeta_t^{-f} \ge 0 \tag{5.37}$$

$$0 \le \left[V_{i,t}^f - m_{1,i}^f \left(\sum_{i \in \mathcal{M}} c_{i,t}^f \right) - b_{1,i}^f \right] \perp \gamma_{i,t}^f \ge 0 \tag{5.38}$$

$$0 \le \left[V_{i,t}^f - m_{2,i}^f \left(\sum_{i \in \mathcal{M}} c_{i,t}^f \right) - b_{2,i}^f \right] \perp \eta_{i,t}^f \ge 0 \tag{5.39}$$

$$0 \le c_{i,t}^f \perp \phi_{i,t}^f \ge 0 \tag{5.40}$$

$$0 \le q_{i,t}^f \perp \rho_{i,t}^f \ge 0 \tag{5.41}$$

$$0 \le \left[-q_{i,t}^f + q_{i,\max}^f \right] \perp \sigma_{i,t}^f \ge 0 \tag{5.42}$$

$$0 \le \left[q_{i,t}^f - q_{i,t-1}^f - r_{i,\min}^f \right] \perp \mu_{i,t}^f \ge 0 \tag{5.43}$$

$$0 \le \left[-q_{i,t}^f + q_{i,t-1}^f + r_{i,\max}^f \right] \perp \theta_{i,t}^f \ge 0 \tag{5.44}$$

where the symbol $\perp$ denotes orthogonality of two vectors. Hence for two vectors A and B with same cardinality (i.e. $|A| = |B|$), $0 \le A \perp B \ge 0$ implies:

$$a_i \cdot b_i = 0 \quad \forall i \in |A|$$

$$a_i \ge 0 \quad \forall i \in |A|$$

$$b_i \ge 0 \quad \forall i \in |B|$$

Concatenation of the KKT identities (5.33–5.35) with the complementarity slackness conditions (5.40–5.42) respectively yields an equivalent nonlinear complementarity problem (NCP):

$$0 \le c_{i,t}^f \perp \left[2b_{i,t}c_{i,t}^f - a_{i,t} + b_{i,t} \sum_{g \in \mathcal{F}, g \ne f} c_{i,t}^g - w_{i,t} - \zeta_t^{+f} + \zeta_t^{-f} \right] = \phi_{i,t}^f \ge 0 \qquad (5.45)$$

$$0 \le q_{i,t}^f \perp \left[w_{i,t} + \zeta_t^{+f} - \zeta_t^{-f} + m_{1,i}^f \gamma_{i,t}^f + m_{2,i}^f \eta_{i,t}^f + \sigma_{i,t}^f - \mu_{i,t}^f + \theta_{i,t}^f \right] = \rho_{i,t}^f \ge 0 \qquad (5.46)$$

$$0 \le V_{i,t}^f \perp \left[1 - \gamma_{i,t}^f - \eta_{i,t}^f \right] = \delta_{i,t}^f \ge 0 \qquad (5.47)$$

We may now state the complete finite dimensional nonlinear complementarity problem for the generating firm f using Equations 5.36 through 5.47 as:

$$\begin{bmatrix} 2b_{i,t}c_{i,t}^f - a_{i,t} + b_{i,t} \sum_{g \in \mathcal{F}, g \ne f} c_{i,t}^g - w_{i,t} - \zeta_t^{+f} + \zeta_t^{-f} \\ w_{i,t} + \zeta_t^{+f} - \zeta_t^{-f} + m_{1,i}^f \gamma_{i,t}^f + m_{2,i}^f \eta_{i,t}^f + \sigma_{i,t}^f - \mu_{i,t}^f + \theta_{i,t}^f \\ 1 - \gamma_{i,t}^f - \eta_{i,t}^f \\ -\sum_{i \in \mathcal{M}} q_{i,t}^f + \sum_{i \in \mathcal{M}} c_{i,t}^f \\ \sum_{i \in \mathcal{M}} q_{i,t}^f - \sum_{i \in \mathcal{M}} c_{i,t}^f \\ V_{i,t}^f - m_{1,i}^f \left(q_{i,t}^f \right) - b_{1,i}^f \\ V_{i,t}^f - m_{2,i}^f \left(q_{i,t}^f \right) - b_{2,i}^f \\ c_{i,t}^f \\ V_{i,t}^f \\ q_{i,t}^f \\ -q_{i,t}^f + q_{i,\max}^f \\ q_{i,t}^f - q_{i,t-1}^f - r_{i,\min}^f \\ -q_{i,t}^f + q_{i,t-1}^f + r_{i,\max}^f \end{bmatrix} = F_1(z) \perp z = \begin{bmatrix} c_{i,t}^f \\ q_{i,t}^f \\ V_{i,t}^f \\ \zeta_t^{+f} \\ \zeta_t^{-f} \\ \gamma_{i,t}^f \\ \eta_{i,t}^f \\ \phi_{i,t}^f \\ \delta_{i,t}^f \\ \rho_{i,t}^f \\ \sigma_{i,t}^f \\ \mu_{i,t}^f \\ \theta_{i,t}^f \end{bmatrix} \qquad (5.48)$$

Note that the dimension of the above problem increases linearly with the number of time periods in the planning horizon as well as granularity of the time window. See Harker and Pang (1990) for a detailed discussion on the finite dimensional nonlinear complementarity problems and their relationships to the finite dimensional variational inequalities.

5.4.2 Complementarity Conditions for the ISO

Returning to the ISO's problem of interest, we have T number of linear programs each for every time period $t = 0, \ldots, T-1$:

$$\max J_{ISO,t} = \sum_{i \in \mathcal{N}} \sum_{f \in \mathcal{F}} y_{i,t}^f \cdot w_{i,t} \qquad (5.49)$$

subject to:

$$\sum_{i \in \mathcal{N}} PDF_{i,a} \cdot \left(\sum_{f \in \mathcal{F}} y_{i,t}^f \right) \le T_{a,t} \quad \forall a \in \mathcal{A}(\alpha_{a,t}) \qquad (5.50)$$

We may again inspect the KKT conditions for this mathematical program as the constraints are linear and Abadie's constraint qualification holds. The KKT identity for this problem is:

$$-w_{i,t} + \sum_{a \in \mathcal{A}} PDF_{i,a} \cdot \alpha_{a,t} = 0$$

Making use of 5.50, we obtain the following conditions:

$$w_{i,t} = \sum_{a \in \mathcal{A}} PDF_{i,a} \cdot \alpha_{a,t} \quad \forall i \in \mathcal{N} \tag{5.51}$$

$$0 \le \alpha_{a,t} \perp T_{a,t} - \sum_{i \in \mathcal{N}} PDF_{i,a} \cdot \sum_{f \in \mathcal{F}} \left(c_{i,t}^{f} - q_{j,t}^{f}\right) \ge 0 \quad \forall a \in \mathcal{A} \tag{5.52}$$

where 5.51 gives the wheeling fee and 5.52 gives the complementary conditions for the ISO. The linear complementarity problem (LCP) for the ISO is thus

$$\left[T_{a,t} - \sum_{i \in \mathcal{N}} PDF_{i,a} \cdot \sum_{f \in \mathcal{F}} \left(c_{i,t}^{f} - q_{j,t}^{f}\right) \right] = F_2(\alpha) \perp \alpha \ge 0 \tag{5.53}$$

5.4.3 The Complete NCP Formulation

The market equilibrium formulation of the dynamic game articulated above may be expressed as a single NCP by concatenating the complementarity conditions for the generating firms (5.48) for all generating firms $f \in F$ with those from the ISO (5.53):

$$\begin{bmatrix} F_1(z) \\ F_2(\alpha) \end{bmatrix} = G(y) \perp y = \begin{bmatrix} z \\ \alpha \end{bmatrix} \ge 0 \tag{5.54}$$

This NCP formulation can be represented and efficiently solved in a commercial software package such as GAMS utilizing the PATH solver. PATH uses a stabilized Newton method for the solution of the Mixed Complementarity Problem, see Dirkse and Ferris (1993). An algorithm may also be devised to solve this NCP by sequentially linearizing while solving each resulting LCP using a Lemke's type algorithm as described in Ferris and Pang (1997) and Cottle et al. (1992); note that this type of algorithm is built into PATH. Hence, PATH may be used as an off-the-shelf commercial solver.

5.5 NUMERICAL EXAMPLE

For our numerical examples, we want to test two different scenarios. The first set of examples involves the drastic decrease in the capacity of an arc from the network and the second involves the removal of a generator. In both cases, we compare the social welfare between the full capacity network and the reduced capacity network. In the former set of examples, we assume that the PTDFs do not change but because of some contingency, the system operator requires that the flow on a particular arc be drastically curtailed. The model was solved with all arcs at full capacity, and then the capacity of the arc in question was set to zero and the model was rerun to see what the effects were. The social welfare is calculated as:

$$SW = \sum_{t=0}^{\mathcal{N}} \sum_{i \in \mathcal{M}} CS_{i,t} - \sum_{f \in \mathcal{F}} V_{i,t}^{f*} \tag{5.55}$$

where *CS* is the total value accruing to consumers, represented by the integral of the demand curves:

$$CS_{i,t} = \int_{0}^{\sum_{g \in \square} c_{i,t}^{g*}} (a_{i,t} - b_{i,t} x)\, dx \tag{5.56}$$

and $V_{i,t}^{f*}$ and $c_{i,t}^{g*}$ are equilibrium values. Note that the amount paid by the consumers and the wheeling fees are not included in the calculations as they are transfers of money and cancel out in the social welfare calculation. (That is, a payment by one party is subtracted from social welfare, but the receipt of that payment by another party is added to welfare, so the net is zero.)

We are also interested in how efficiently this model can be solved. As such, we have created a numerical example based on the northwest European electricity market formed by Belgium, France, Germany and the Netherlands (Neuhoff et al., 2005). This network is comprised of 15 nodes, 28 flowgates (transmission lines) and 12 generating firms. As mentioned previously, eight of the generating firms are distinct and supply the largest fraction of power, while the other four represent conglomerations of the remaining generating firms in each country. For the purposes of the numerical example, the set of firms is $\mathcal{F} = \{1, \ldots, 12\}$, the set of nodes is $\mathcal{N} = \{1, \ldots, 15\}$ and the set of nodes at which there are markets is $\mathcal{M} = \{4, 5, 6, 8, 9, 14, 15\}$.

We consider a time horizon of one day with 24 discrete time periods. Synthetic data was created for the inverse demand parameters to represent the change in loads throughout a day. General data on daily load shapes in California was used, and was obtained from CAISO. The bounds for the ramping rates of the generators are also synthetic. The data used for the two piece linear generation cost function, generation capacitites, PTDF values and transmission line capacities was obtained from the Energy research Centre of the Netherlands (ECN) and is given in Tables 5.4 through 5.7. The network is illustrated in Figure 5.1 with the nodes and flowgates (arcs) enumerated.

The first set of results are shown in Table 5.8. The data in this table represents the social welfare for the full capacity network, as well as the difference between the social welfare of the network at full capacity and that of the network with each arc removed. The change in *SW* is computed as: change in *SW* = Full Capacity *SW* – *SW* with one or more arc capacities reduced.

We might expect that the social welfare would decrease as transmission capacity is removed from the network. However, there are cases where the removal of an arc results in an increase to social welfare; this type of observation is typically termed the Braess paradox. In this case, we see that the when arcs 6, 12, 16, 19, 20, 22, 23 and 27 are removed, the social welfare actually increases, contrary to what we expect.

To give an example of the equilibrium solution, we provide some results in Figures 5.2 though 5.7. The first of those figures shows the sales (allocation to consumption) patterns by each firm at Node 4 at equilibrium under full capacity (no derating of any arc capacity). Likewise, Figure 5.3 shows the generation rate of firms at Node 4. In Figure 5.2, we can see that Firm 6 has the highest sales throughout

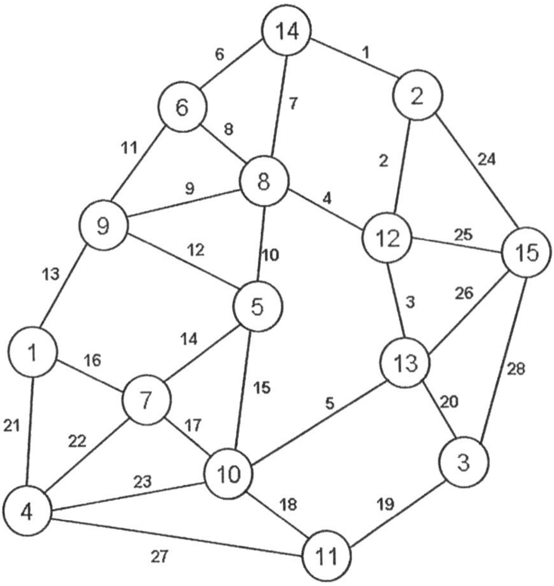

Figure 5.1. Network Illustration.

TABLE 5.4. Data for Two Piece Linear Generation Cost Function.

Firm	Node	Segment 1				Segment 2			
		LB (MW)	UB (MW)	Intercept	Slope	LB (MW)	UB (MW)	Intercept	Slope
1	5	0	0	0.0	0.000	0	1,052	3	0.046
1	9	0	0	0.0	0.000	0	894	6	0.052
2	5	0	0	0.0	0.000	0	9	42	0.000
3	5	0	0	0.0	0.000	0	28	48	0.000
5	5	0	4,066	7.3	0.001	4,066	4,320	−1125	0.280
5	9	0	2,832	6.4	0.002	2,832	5,905	−44	0.019
2	4	0	6,576	0.0	0.002	6,576	10,361	−108	0.018
6	4	0	72,617	4.0	0.000	72,617	82,909	−451	0.006
7	4	0	0	0.0	0.000	0	49	0	0.000
4	15	0	33,823	2.7	0.001	33,823	35,738	−916	0.028
8	15	0	22,541	0.6	0.002	22,541	23,247	−1703	0.077
6	15	0	0	0.0	0.000	0	54	0	0.000
7	15	0	6,550	6.2	0.001	6,550	7,862	−319	0.051
9	15	0	15,7001	7.1	0.000	15,701	20,377	−154	0.011
10	15	0	7,910	8.2	0.001	7,910	9,308	−390	0.051
3	8	0	0	0.0	0.000	0	429	6	0.070
3	6	0	0	0.0	0.000	0	2,026	5	0.021
3	14	0	0	0.0	0.000	0	1,057	13	0.018
8	6	0	448	11.2	0.005	448	1,334	4	0.021
5	6	0	0	0.0	0.000	0	94	9	0.196
5	14	0	0	0.0	0.000	0	3,237	19	0.002
11	8	0	0	0.0	0.000	0	1,138	16	0.014
11	6	0	0	0.0	0.000	0	1,301	15	0.009
12	6	0	0	0.0	0.000	0	2,421	13	0.007

TABLE 5.5. PTDF Data Computed with Node 15 as the Hub.

Node\ Line	1	2	3	4	5	6	7	8	9	10	11	12	13	14
1	−0.2068	−0.0095	−0.0449	−0.1941	−0.3034	−0.1333	−0.0726	−0.0258	−0.1025	−0.1372	−0.1502	0.0167	−0.2794	−0.1306
2	0.2769	0.1975	0.0997	−0.1300	−0.0599	0.1445	0.1323	−0.0632	0.0164	0.0601	0.0813	0.0371	0.0596	0.0695
3	−0.0904	−0.0216	−0.1155	−0.0607	−0.1036	−0.0640	−0.0366	0.0014	−0.0288	−0.0698	−0.0526	−0.0239	−0.0575	−0.0603
4	−0.1630	−0.0171	−0.0856	−0.1373	−0.3245	−0.1016	−0.0614	−0.0098	−0.0670	−0.1219	−0.1113	−0.0324	−0.1460	−0.1355
5	−0.2697	−0.0024	0.0257	−0.2939	−0.2289	−0.1639	−0.1060	0.0074	−0.0784	−0.3298	−0.1463	−0.2884	0.0637	0.2719
6	−0.3924	0.0445	0.0900	−0.3262	−0.1303	−0.3656	−0.0268	−0.3691	−0.0274	0.0434	0.2653	0.1121	0.1258	0.1047
7	−0.1897	−0.0138	−0.0624	−0.1766	−0.3372	−0.1178	−0.0719	−0.0103	−0.0767	−0.1604	−0.1282	−0.0652	−0.1397	−0.2283
8	−0.3043	−0.0035	0.1172	−0.4337	−0.1167	−0.1288	−0.1765	0.1654	0.0890	0.1464	0.0266	0.0222	0.0934	0.1161
9	−0.2977	0.0104	0.0477	−0.3089	−0.1937	−0.2103	−0.0874	−0.0896	−0.2097	−0.0980	−0.2989	0.2800	0.2115	0.1185
10	−0.1439	−0.0238	−0.1229	−0.1045	−0.4510	−0.0870	−0.0669	−0.0011	−0.0499	−0.1104	−0.0881	−0.0606	−0.0874	−0.1024
11	−0.1125	−0.0213	−0.1121	−0.0751	−0.2066	−0.0680	−0.0444	−0.0009	−0.0390	−0.0796	−0.0599	−0.0304	−0.0775	−0.0806
12	−0.0940	−0.1338	0.2474	0.1393	−0.0100	−0.0387	−0.0663	0.0610	0.0312	0.0519	0.0122	0.0086	0.0348	0.0422
13	−0.0924	−0.0421	−0.2316	−0.0086	0.1735	−0.0618	−0.0406	0.0108	−0.0178	−0.0422	−0.0410	−0.0218	−0.0370	−0.0416
14	−0.5840	0.1321	0.1001	−0.2209	−0.0846	0.2163	0.2006	−0.1002	0.0168	0.0631	0.1152	0.0620	0.0798	0.0785

TABLE 5.6. PTDF Data, Continued from Table 5.5.

Node\ Line	15	16	17	18	19	20	21	22	23	24	25	26	27	28
1	0.0101	0.4537	0.2759	0.0986	0.2966	0.0131	−0.2669	−0.0471	0.1159	−0.2153	−0.2296	−0.2716	0.1981	−0.2835
2	0.0277	0.0244	0.0605	0.0407	0.0870	−0.0250	−0.0352	−0.0234	0.0123	−0.5286	−0.2278	−0.1345	0.0462	−0.1120
3	−0.0234	−0.0128	−0.0345	−0.1434	−0.2447	0.2545	0.0447	0.0396	0.0181	−0.1120	−0.1446	−0.2527	−0.1014	−0.4907
4	−0.0188	0.0429	0.1171	0.0525	0.3751	0.0470	0.1889	0.0297	0.2788	−0.1800	−0.2058	−0.2859	0.3226	−0.3282
5	0.1108	−0.0090	0.1859	0.1007	0.2175	−0.0128	−0.0727	−0.0770	0.0328	−0.2621	−0.2657	−0.2418	0.1168	−0.2303
6	0.0508	0.0563	0.1187	0.0672	0.1510	−0.0239	−0.0696	−0.0423	0.0279	−0.3471	−0.2807	−0.1964	0.0839	−0.1749
7	0.0026	−0.2076	0.3957	0.1266	0.2975	0.0078	−0.0679	−0.1684	0.0654	−0.2034	−0.2242	−0.2826	0.1709	−0.2896
8	0.0524	0.0345	0.1089	0.0670	0.1453	−0.0306	−0.0587	−0.0419	0.0224	−0.3078	−0.3130	−0.2033	0.0782	−0.1768
9	0.0536	0.1055	0.1692	0.0843	0.1998	−0.0143	−0.1059	−0.0548	0.0452	−0.2872	−0.2716	−0.2270	0.1155	−0.2141
10	−0.0586	−0.0614	−0.1676	0.2200	0.3005	−0.0096	0.0260	−0.0037	−0.1027	−0.1677	−0.2036	−0.3186	0.0806	−0.3100
11	−0.0295	−0.0132	−0.0358	−0.2360	0.6059	0.1754	0.0643	0.0579	0.0367	−0.1337	−0.1659	−0.2708	−0.1581	−0.4295
12	0.0182	0.0111	0.0349	0.0483	0.0952	−0.0594	−0.0238	−0.0184	0.0022	−0.2277	−0.4296	−0.1980	0.0393	−0.1445
13	−0.0225	−0.0227	−0.0609	0.0574	0.0725	−0.1801	0.0143	0.0033	−0.0327	−0.1346	−0.1980	−0.4148	0.0151	−0.2526
14	0.0367	0.0333	0.0812	0.0508	0.1105	−0.0256	−0.0456	−0.0305	0.0174	−0.4518	−0.2528	−0.1591	0.0597	−0.1361

TABLE 5.7. Transmission Line Capacities.

Line	Capacity (MW)	Line	Capacity (MW)	Line	Capacity (MW)
1	2,762	8	3,329	15	3,329
2	20,222	9	20,000	16	1,282
3	20,000	10	20,000	17	896
4	20,000	11	20,000	18	1,842
5	20,000	12	20,000	19	1,326
6	1,842	13	1,207	20	1,842
7	2,971	14	267	21	1,842

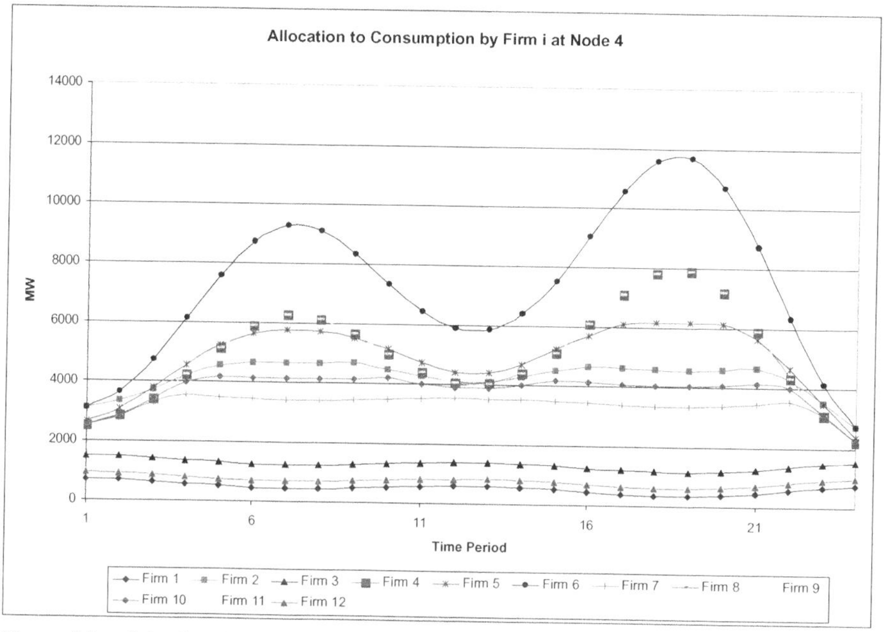

Figure 5.2. Sales by Firm *i* at Node 4.

the day. Also note that there are two predominant "humps" in the sales. These arise from the data used for the inverse demand function and represents the increase in load experienced at the beginning of the day when people arrive to work and at the end of the day when people arrive at their homes. Correspondingly, we see in Figure 5.3 that Firm 6 has the highest production at Node 4.

As mentioned previously, we also test the effects on the equilibrium of removing a generator due to some contingency for a number of consecutive time periods. Specifically, we removed the generator owned by Firm 2 at Node 4 during Time Periods 5, 6 and 7. We first solved the problem to find the equilibrium when all generators and transmission lines are working. The problem was then solved again with the said generator on forced outage and therefore removed from service. The

TABLE 5.8. Change in Social Welfare When One Arc Is Removed at a Time.

Full Capacity SW	5.020E+10						
Arc Removed	1	2	3	4	5	6	7
Change in SW	1.506E+07	0.000E+00	0.000E+00	3.362E+07	1.984E–04	–4.341E+06	5.882E+04
Arc Removed	8	9	10	11	12	13	14
Change in SW	2.585E+06	2.453E+06	5.773E+05	5.064E+05	–4.728E+05	0.000E+00	0.000E+00
Arc Removed	15	16	17	18	19	20	21
Change in SW	0.000E+00	–3.999E+05	0.000E+00	0.000E+00	–1.484E+07	–3.510E+06	6.554E+06
Arc Removed	22	23	24	25	26	27	28
Change in SW	–1.538E+06	–2.053E+07	1.743E+07	2.180E+07	1.280E+06	–2.049E+07	1.984E–04

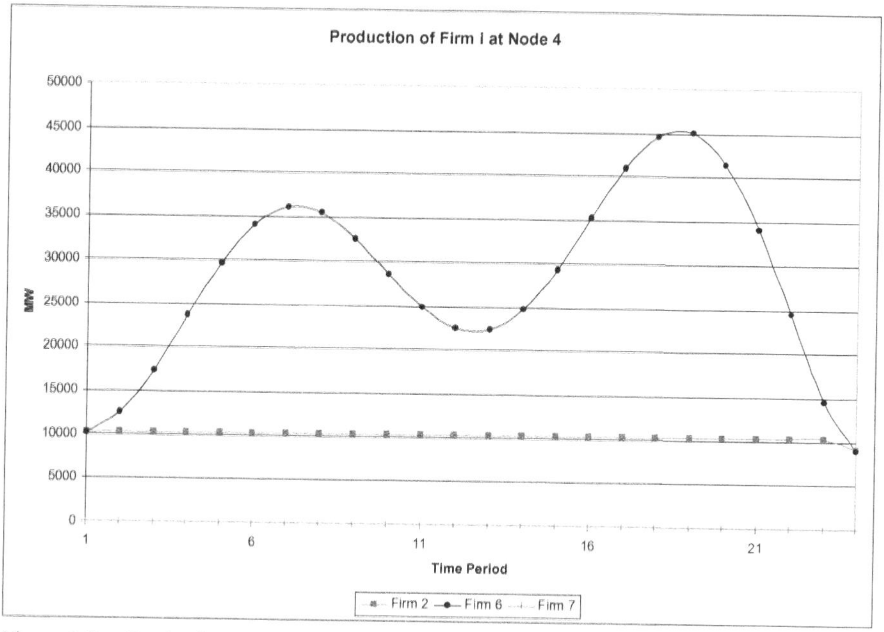

Figure 5.3. Production by Firm *i* at Node 4.

equilibriums were then compared to see what the effects of such a disruption were. The model was formulated in GAMS and solved with the PATH solver using the default settings. The solution time for this problem is under 20 seconds for one run on a 2.8 GHz Intel Pentium 4 processor machine with 512 MB of RAM.

Figure 5.4 below shows the equilibrium sales by Firm 2 for both the non-disrupted (before) and disrupted (after) cases. As we can see, the sales by Firm 2 are drastically affected by this generator failure. However, as when the generating facility becomes operational again, its sales return to the equilibrium that was established with no disruption. Meanwhile, we see in Figure 5.5 that firm 1 visibly increases its sales at Nodes 4 and 15 during the time periods during and following the disruption to the generation of Firm 2; changes in behavior can be seen at the other nodes too, though they are much more subtle. Examining the changes observed by other firms at other nodes reveals similar behavior.

We can also look at the changes in generation that are experienced during this disruption. Figure 5.6 shows the equilibrium generation pattern for Firm 2. We can see that the generator at Node 4 drops to zero for Time Periods 5, 6 and 7 and then begins to slowly ramp back up according to the ramping rate constraints. The generation level finally reaches that of the non-disrupted equilibrium around Time Period 14. We also see that Firm 2 turns on Generator 5, though it is very limited in output, during the disruption. Figure 5.7 shows the equilibrium generation of Firm 1 in both the disrupted and non-disrupted cases. It is clear that Firm 1 experiences changes to its generation at both Nodes 5 and 9 due to the disruption occuring in Firm 1's generation at Node 4. Similar results can be seen for the other generating firms as well.

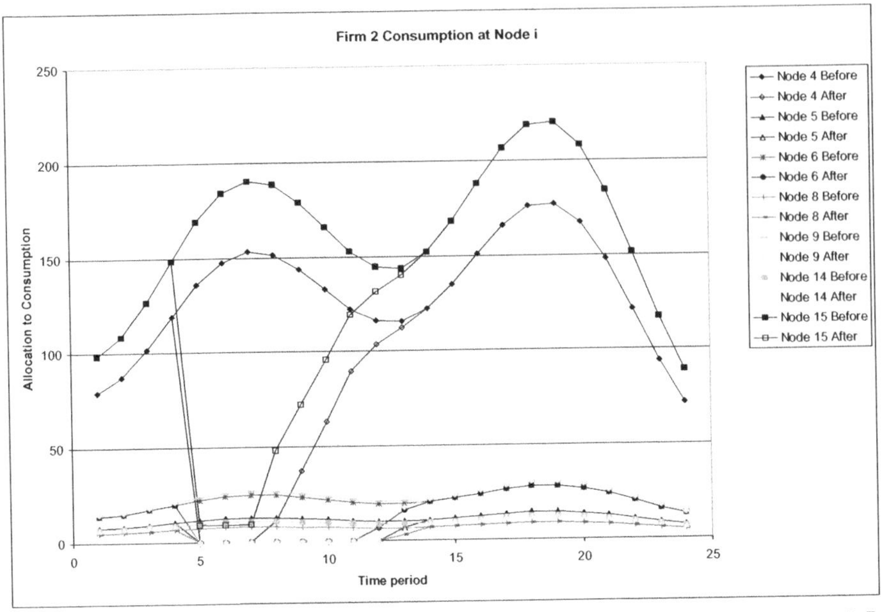

Figure 5.4. Sales by Firm 2 before and after the generator disruption in Periods 5 through 7.

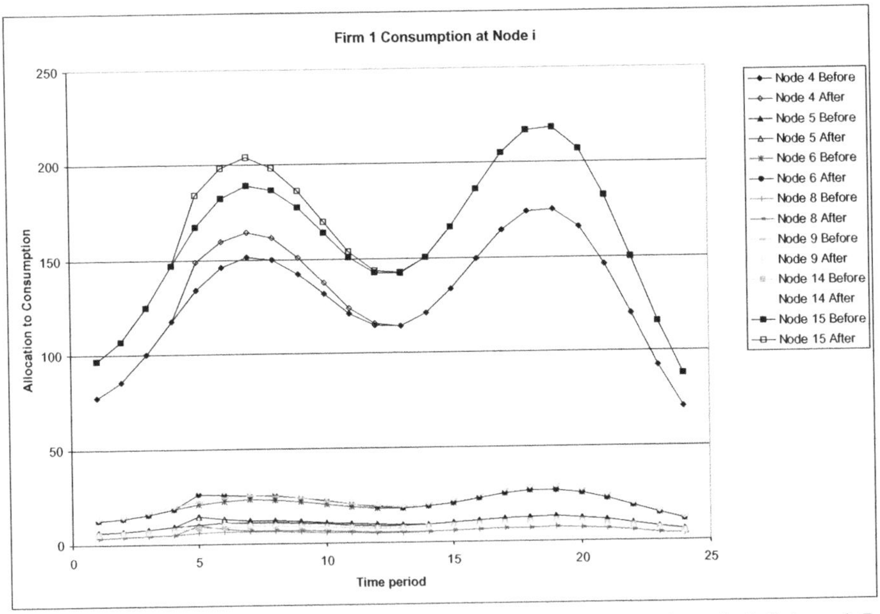

Figure 5.5. Sales by Firm 1 before and after the generator disruption in periods 5 through 7.

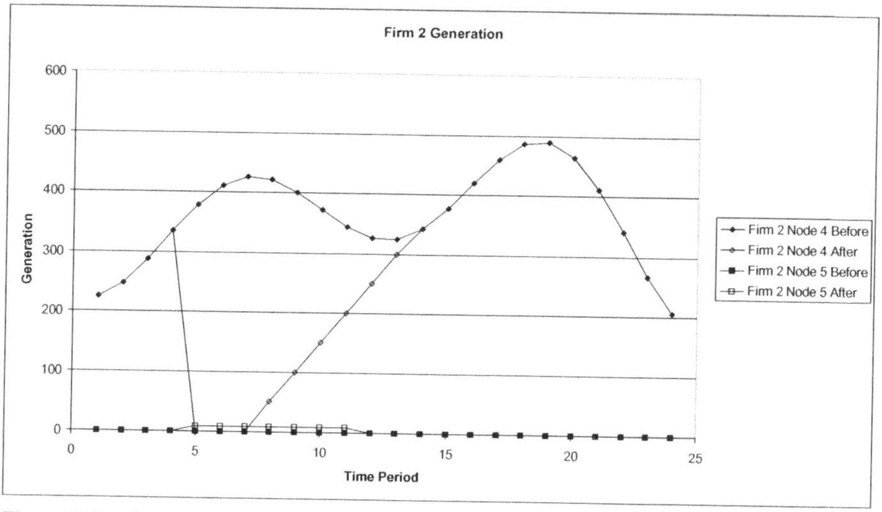

Figure 5.6. Generation by Firm 2 before and after the generator disruption.

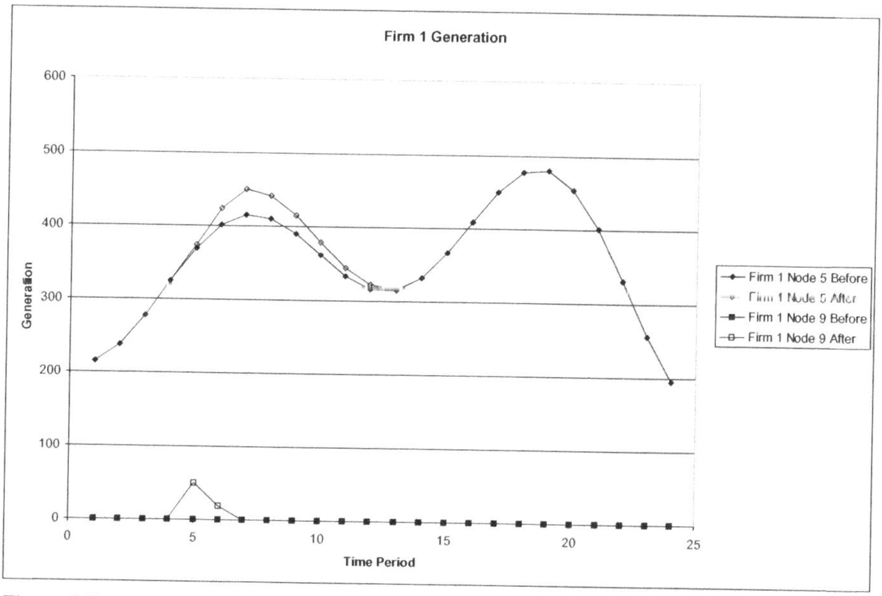

Figure 5.7. Generation by Firm 1 before and after the generator disruption.

5.6 CONCLUSIONS AND FUTURE WORK

We have proposed a dynamic game thoeretic model of oligopolistic competition in an electric power network setting. This model is formulated to lend itself to efficient computation in order to facilitate the research of extreme events in such an electric power system. We provided an outline of a possible simulation route that may be realistically carried out for the testing of extreme events. We have also shown how a local disruption in the system can have implications for operation and consumption throughout the system as is witnessed in the numerical examples.

ACKNOWLEDGMENT

This work was supported in part by the National Science Foundation under EPNES grant ECCS-0224817.

APPENDIX: GLOSSARY OF RELEVANT TERMS FROM ELECTRICITY ECONOMICS

Bilateral power sales agreement: a contract between a seller and buyer of power for delivery of a specified amount of power at a specified time for a specified duration at a specified price. "Physical" contracts may require provision from particular generators; however, in most ISO-based markets, such contracts are financial, in the sense that a provider of power may instead substitute a purchase from the spot market for physical delivery from their own facilities. This can be done by just paying the purchaser of the contract the spot price for power at that time and place.

Competitive fringe: This is a set of smaller power producers who either believe or in fact cannot significantly affect the price of power by their unilateral actions. As a result, they are "price takers," and their optimal output is that which equates their marginal cost with the local price.

Complementarity problem: Let x be a column vector of length n, and $f(x)$ be a vector-valued function of the same dimension. Then a complementarity problem is defined as the following: find x such that: $x \geq 0$, $f(x) \leq 0$, and $x^{T}f(x) = 0$.

Congestion surplus: Under an ISO-type system in which power is priced using locational marginal pricing, this is the difference between the amount of money paid to the ISO by consumers for power (quantity consumed times their local price) and the amount of money the ISO pays to producers (quantity produced times their local price). This is nonnegative, and can be strictly positive if one or more transmission constraints are binding. This is also called the "merchandizing surplus." In reality, ISOs are supposed to be nonprofit organizations, so any congestion surplus is either refunded to grid users (by lowering grid access charges) or by granting financial transmission rights.

Consumer surplus: This is the difference between total value to consumers (see definition below) and what they pay for power. Thus, this is a measure of the net benefits of power consumption.

Clear the market: If the quantity demanded in a market at a given price equals the quantity that producers are willing to provide at that same price, then the market is cleared. In the context of transmission services, this means that demand for transmission services (the flow through a particular line or other piece of equipment) does not exceed the supply (the capacity of the equipment), and the price assigned to that flow can be positive only if the flow equals the constraint.

Cournot-Nash Game: A Nash game is a game in which each player chooses its strategy subject to the assumption that no other player will change their strategy. In a Cournot-Nash game, the "strategy" is assumed to be quantity produced or sold. For example, a Cournot-Nash game among generating companies would mean that each generator believes that other generators will not alter their outputs in response to a change in its output.

Extremal problem: Also called an "optimization problem" or "mathematical program." This is a mathematical problem in which the values of decision variables are to be obtained that simultaneously satisfy a stated set of constraints and maximize (or minimize, depending on the problem statement) a given objective function. Under certain mathematical conditions, it is both necessary and sufficient for the values of the decision variables to satisfy the so-called Karush-Kuhn-Tucker conditions, which define a complementarity problem.

Game theoretic model: A model in which the outcome is determined by the interactions of players of the game, and players are aware of each other's actions when making decisions. In a market game, the outcomes are prices, quantities bought and sold, and profits for each of the players. A Cournot-Nash game is an example of such a game.

Hub node: This is a single location in the network (usually a single bus) that all transactions are assumed to pass through for the purpose of pricing transmission services. For instance, if 100 MW is sold from a generator at location A to a consumer at location B, for pricing purposes, it is broken up into two transactions: a 100 MW transfer from A to the hub, and a subsequent 100 MW transfer from the hub to B. In a locational marginal pricing system, where the price of transmission (the "wheeling fee") from A to B is defined as being equal to the price at B minus the price at A, the choice of hub node is arbitrary and does not affect the net wheeling cost from B to A.

Inverse Demand Function: This function relates the marginal willingness to pay of a consumer (the price) to the quantity demanded.

Linear complementarity problem: This is a complementarity problem in which all components of $f(x)$ are linear.

Linear program: A mathematical program in which all constraints and the objective function are linear functions of the decision variables.

Locational marginal price: This is the marginal cost, in 3/MWh, of providing another MW at a given time and location. It is the shadow price or dual variable of the power balance at a bus in an optimal power flow model. In ISO-based markets, this is also the price charged for spot purchases or sales of power.

Mathematical program: See "extremal problem," above.

Market equilibrium: This is defined as a set of market prices and quantities such that no participant in the market has an incentive to change their decisions; that is, none can increase their profit by producing, consuming, or bidding a different amount.

Mixed complementarity problem: Let x and y be column vectors of length n and m, respectively, and $f(x, y)$ and $h(x, y)$ be vector-valued functions of the same respective dimensions. Then a mixed complementarity problem is defined as the following: find $\{x,y\}$ such that: $x \geq 0$, $f(x, y) \leq 0$, $x^T f(x, y) = 0$, and $h(x, y) = 0$.

Nonlinear complementarity problem: This is a complementarity problem in which one or more of the components of $f(x)$ are nonlinear.

Oligopolistic competition: A situation in which one or more suppliers in a market can significantly affect prices for a significant length of time by their unilateral decisions and furthermore recognizes that they are able to do so. Monopoly is a special case in which there is only one such firm.

Open-loop Nash Equilibrium: This refers to the equilibrium outcome of a dynamic Nash game in which it is assumed that all players choose their strategies for all periods at once. In contrast, in a closed-loop game, decisions in one period are made correctly anticipating how they will affect decisions and outcomes in subsequent periods. It is generally much easier to solve for and prove properties of equilibria for open-loop equilibria.

Pure or perfect competition: A situation in which either no suppliers in a market can significantly affect prices for a significant length of time by their unilateral decisions, or no supplier recognizes that it is able to do so.

Social welfare: The sum of surpluses for all market participants (generators, transmission system operator, and consumer).

Surplus: Net benefits to an individual market participant. For producers, this is profit; for consumers, this is consumer surplus; and for the system operator, this is (at least in the short term) the congestion surplus.

Total value to consumers: This is the gross benefit of consumption, which is generally approximated as the integral of the inverse demand curve.

Variational Inequalities: This is a mathematical problem defined as follows. Let $f(x)$ be a vector valued function of the same dimension as the vector of decision variables x. Let X be the feasible region for x. Then a variational equality problem is: find x^* such that: $f(x) - f(x^*) \leq 0$ for all $x \in X$.

Wheeling fees: The price that users of a transmission grid pay to transmit power from one location to another. In a locational marginal pricing system, this is just the difference between the price in the two locations; however, other wheeling fee systems are also used.

BIBLIOGRAPHY

[1] Cottle RW, Pang J-S Stone, RE, (1992), *The Linear Complementarity Problem.* (Academic Press).

[2] Daxhelet O, Smeers Y, (2001), "Variational inequality models of restructured electric systems," in: Ferris MC, Mangasarian OL, and Pang J-S, *Applications and Algorithms of Complementarity* (Kluwer, Dordrecht).

[3] Day CJ, Hobbs BF, Pang J-S, (2002), "Oligopolistic competition in power networks: A conjectured supply function approach," *IEEE Transactions on Power Systems* 17(3): 597–607.

[4] Dirkse SP, Ferris MC, (1993), "The *PATH* Solver : A Non-Monotone Stabilization Scheme for Mixed Complementarity Problems," technical report CS-TR-1993-1179, University of Wisconsin Madison (also available at citeseer.ist.psu.edu/dirkse95path.html).

[5] Ferris MC, Pang J-S, (1997), Complementarity and Variational Problems: State of the Art. (SIAM Publications, Philadelphia).

[6] Harker PT, Pang J-S, (1990), "Finite-dimensional variational inequality and nonlinear complementarity problems a survey of theory, algorithms and applications," *Mathematical Programming Series B*, 48: 161–220.

[7] Hobbs BF, Helman U, (2004), "Complementarity-Based Equilibrium Modeling for Electric Power Markets," in Bunn DW, (ed.), *Modeling Prices in Competitive Electricity Markets*, Ch. 3 (Wiley, London).

[8] Neuhoff K, Barquin J, Boots MG, Ehrenmann A, Hobbs BF, Rijkers FAM, Vázquez M, (2005), "Network-constrained models of liberalized electricity markets: The devil is in the details," *Energy Economics* 27: 495–525.

[9] Schweppe FC, Caramanis MC, Tabors RE, Bohn RE, (1988), *Spot Pricing of Electricity*, Kluwer Academic Publishers (Norwell, MA).

CHAPTER 6

PLANT RELIABILITY IN MONOPOLIES AND DUOPOLIES: A COMPARISON OF MARKET OUTCOMES WITH SOCIALLY OPTIMAL LEVELS

George Deltas and Christoforos Hadjicostis
University of Illinois, Urbana-Champaign

EDITORS' SUMMARY: This chapter investigates the interaction between system availability/reliability, economic restructuring, and regulating constraints. Specifically, the effects of restructuring on electricity markets are analyzed via the interplay between market structure and the incentives provided to power generators to maintain plant availability and reliability. Here, plant reliability is defined as the probability of plant operability under the assumption that this probability is an increasing function of the plant maintenance expenditure. In this framework, three market structures are studied, which are monopoly, duopoly, and public ownership or perfect regulation where maintenance decisions are made by a regulator seeking public interest. To capture the responsiveness of the demand for a given market price, a linear downward-sloping demand curve has been assumed. Based on these assumptions, the authors derive a socially acceptable level of maintenance expenditure to achieve a given level of reliability for the three foregoing markets. For each type of market structure, the authors compute the level of plant reliability and show that the more competitive the market structure is the higher will be the level of reliability. Also, they show that a regulator seeking the public interest would prefer an even higher level of reliability than the one provided by firms with pricing (or market) power. Paradoxically, a multi-plant firm with identical plants that produce power for the same market may optimally choose different reliability level for each one of the plants.

Economic Market Design and Planning for Electric Power Systems, Edited by James Momoh and Lamine Mili

6.1 INTRODUCTION

A large and growing body of literature investigates the implications of the recent restructuring for the performance of the electricity generation market. Recent important contributions to this literature include Borenstein, Bushnell and Stoft (2000), Joskow and Tirole (2000), Borenstein, Bushnell and Wolak (2002), Wolak (2003), Luh et al. (2008), and Sauma and Oren (2008), while Garcia, Mili and Momoh (2008) in this volume nicely describes the various approaches of modeling electricity markets. This body of research focuses on the effects of restructuring on market power and prices, and the interplay between restructuring in the power markets and transmission investment. However, the effects of restructuring on environmental performance and system availability/reliability have remained largely unexplored, with a recent unpublished manuscript by Joskow and Tirole (2004) and Mookherjee et al. (2008) being among the rare contributions on this issue in the Economics literature. Note that some economic aspects of reliability as applied to specific power distribution systems have been discussed in various contexts (see, for example, Medjoudj and Aissani (2002)).

This chapter contributes to our understanding of the effects of restructuring on electricity markets by investigating the inter-relationship between market structure and incentives for power generators to maintain plant availability/reliability. The importance of availability/reliability has established reliability assessment as one of the most important aspects of any engineering design—see, for instance, Johnson (1989) and Siewiorek and Swarz (1998). The effects of availability/reliability on production costs and revenues has been particularly important in power generation and distribution systems—see, for example, Shih, Mazumdar and Bloom (1999)—and, for this reason, reliability assessment is recognized as an important problem in power generation and distribution systems [see, for example, Billinton, Allan and Salvaderi (1991), Billinton, Ringlee and Wood (1973), McGranaghan and Ignall (2002), Endrenyi (1978), and Singh and Gubbala (1994) among others)]. In this chapter we focus on the interaction between system availability/reliability, economic restructuring and regulatory constraints. Note that some of these issues have partially surfaced in earlier works, such as Gonzalez et al. (2005) and Bertling, Allan and Eriksson (2005).

In order to highlight the effects of these interactions, we consider a stylized market that consists of two power plants, each of which requires some maintenance expenditure to be operable. We define the reliability of the plant to be the probability that it is operable, and assume that reliability is an increasing function of the maintenance expenditure on that plant. We further assume that this function is concave in order to capture the existence of diminishing returns to maintenance expenditure in terms of improving reliability. Note that in this chapter reliability is equivalent to plant availability, and the two notions will be used interchangeably.

We consider three market structures:

(1) Monopoly, i.e., one firm owns both plants,

(2) duopoly, i.e., there are two firms and each owns a single plant,

(3) public ownership or perfect regulation, i.e., maintenance decisions are made by a regulator who has the public interest in mind.

The total capacity available in the market is sufficiently low that even a monopolist would never choose to withhold any output from the market *even if* all plants were operable. Even though there are time periods that a "real-life" monopolist would choose to restrict output, we ignore this possibility for three reasons. First, the marginal costs of providing power are for many firms (e.g., nuclear and coal power generators) sufficiently low that many firms would choose to keep their base-load plants going on most days. Second, a firm is not likely, in the long run, to maintain capacity that is not used sufficiently often; such capacity would not yield sufficiently high profits over variable costs to cover the initial investment cost. Third, and most important, focusing the model on markets in which the output/pricing decision is trivial allows us to highlight the effects of using reliability as a strategic variable. As we will see, such a model is in itself sufficiently rich to yield important insights. Incorporating a pricing decision in the model would increase its complexity even further. Reliability-enhancing expenditure does, of course, indirectly affect market price by affecting the supply of power in the market. However, our assumption of limited capacity implies that a supplier would choose perfect reliability if it were offered for free.

The demand side of the model is given by a linear downward-sloping demand curve which implies that the lower the capacity available in the market, the higher the price that the suppliers obtain. Such a responsiveness of the quantity demanded to the market price reflects several aspects of the current operation of many power markets. First, price in many restructured power markets is determined in power exchanges. The higher the capacity offered by suppliers in these exchanges, the lower the equilibrium price. Second, some demanders have interruptible supply contracts. When price exceeds certain levels, these demanders curtail their demand. Third, regional markets are to a large extent inter-connected. Even if total system demand in a region is not sensitive to price, the residual demand faced by suppliers in a region (after subtracting the net imports from other regions) will be declining in the equilibrium price in the regional exchange: The higher the regional price, the larger the imports from neighboring exchanges, and the lower the quantity purchased from local generators. These features of restructured power markets imply that the price in a restructured electricity market is a downward-sloping function of the capacity offered by electricity generators in that market.

In this chapter, we completely abstract from the within-day and across-days variability of demand for electricity. We also abstract completely from network capacity considerations and heterogeneity in power plants. Our demand and supply systems are, thus, highly stylized. For simplicity (and without any loss of generality) we also abstract from operating costs, other than the maintenance costs.

We find that the profit maximizing level of plant reliability of a monopoly firm is lower than the Nash Equilibrium level of reliability in a duopoly (i.e., when there are two firms in the market, each owning one plant). In turn, we find that the Nash Equilibrium level of reliability in a duopoly is lower than the level of reliability chosen by a regulator who has the public interest in mind (i.e., a regulator who equally values firm profits and consumer welfare). Finally, we show that a monopolist may choose as part of a profit maximizing strategy (under certain circumstances) to maintain a different reliability level for two identical plants, even though costs of reliability are convex in the level of reliability. In summary, we find that a decentralized market leads to under-provision of reliability relative to the socially optimal level; plant reliability is decreasing in market concentration, and; differential reliability level may be an instrument of strategic firm behavior.

These findings have policy implications, both with regards to the operation of power generators and with regards to merger guidelines. In particular, a policy that directly or indirectly encourages generators to increase plant reliability is likely to be beneficial from the social point of view. Such a policy could involve direct subsidies of maintenance expenditure, minimum maintenance expenditure requirements, or the subsidization of R&D related to fault-tolerance and plant reliability. Moreover, in reviewing proposed mergers between generators, anti-trust authorities should not only consider the implications of the merger on prices, but also its effect on system reliability.

This chapter is organized as follows. We first outline the modeling framework. This is followed by an examination of the market outcomes under a monopoly, considering the possibility that the monopolist may be able to choose different reliability levels for the two plants. We then examine market outcomes under a duopoly. Finally, we derive the optimal decisions of the regulator who makes the reliability choices with the public interest in mind. The chapter ends with a comparison of the outcomes under the three market structures, a discussion of the intuition behind the results, and some concluding remarks on policy implications and extensions to future research. As a separate and independent result from the above comparisons, this chapter also includes an illustration of the possibility that the monopolist may choose to maintain a different reliability for the two identical plans as part of a profit maximization strategy.

6.2 MODELING FRAMEWORK

Many of the results obtained in this chapter hold under quite general environments. For the sake of brevity, concreteness, and clarity of exposition, we consider here a parsimonious parametric framework. However, we note throughout on the insights obtained from the chapter's results and comment on the more general applicability of these insights.

Let the supplier price of electricity in a market be given by the linear function:

$$P = a - Q$$

where P is the price received by the suppliers (generating companies), Q is the quantity supplied by the generators in that market, and a is a parameter that indexes the strength of the market demand (the higher the value of a, the stronger the demand for electricity and the higher the price). Note that this price-quantity relationship is known in Economics as the *demand function*, and we will often refer to it as such. Underlying the assumption that price is a decreasing function of the quantity offered by the suppliers is the presumption that price is determined by some type of power exchange or other electricity market. The lower the market clearing price in such a market or exchange, the greater the demand by final consumers (as many of these consumers have interruptible supply contracts and get supplied only if price is below certain thresholds) and the lower the level of imports into that market (as producers outside that market have contracts to supply only if price exceeds certain thresholds). Therefore, the greater the supply in the power exchange or electricity market from generating companies located in that market, the lower the price will have to be so that the market clears and supply equals demand.

There are two plants in this market, each being capable of producing one unit of output when operational. Marginal costs of production are assumed to be zero (this simplifying assumption is without loss of generality). The reliability of a plant, r, is defined to be the probability that the plant is operational. This probability is a function of the maintenance expenditure, m, and a cost parameter, c. We parameterize the relationship between expenditure and plant reliability by:

$$r(m) = \max\left[\sqrt{\frac{m}{c}}, 1\right] \tag{6.1}$$

The assumption that reliability increases with the square root of maintenance expenditure is consistent with the presence of diminishing returns to increased maintenance effort. Reliability cannot exceed one, as it is defined to be the probability that the plant is operable. The cost parameter c, which is of the same units as the expenditure m, indexes the input costs of maintenance operations and/or technical change and the cost-effectiveness of the reliability technology: a 10% increase in the cost-effectiveness of the reliability technology or a 10% decrease in the price of inputs (i.e., a 10% reduction in the value of c) would allow a firm to reduce maintenance expenditures by 10% and still maintain the same level of reliability. The choice of parametric function does not affect the nature of our results (as long as it is a monotonically increasing and concave function of m); the particular choice we have made simplifies the algebra considerably and allows for derivation of closed form solutions.

With regards to parameter values, we assume that a is greater than or equal to four. This assumption ensures that if both plants were owned by a monopolist and were operational, the monopolist would choose to run them both (this, in turn, ensures that a regulator would also choose to run both plants). We do not place an upper bound on a, however, the most economically interesting range for a is the [4, 6] interval: higher values of a would likely make the construction of additional plants optimal. We also distinguish between the lower part of this range, and in particular the [4, 5] interval, from the remainder of the range of a. For values of a in [4, 5] and for some values of c between 0.5 and 1.5, it is optimal (from the profit maximization point of view) for a monopolist to choose different levels of maintenance expenditure for the two plants, even though the plants are totally identical. In most of our development, we do not consider values of a that are in the [4, 5] interval as this avoids the complexity of asymmetric solutions to the optimal choice of maintenance expenditure problem. We briefly discuss the case of a in the [4, 5] interval in Section 6.6. We make no restrictions in the range of c but the economically interesting range of c is the range for which equilibrium plant availability is not too low (relative to observed values).

If L plants are operational, then the total profits that accrue to the electricity generators are given by:

$$\Pi(L) = L \cdot (a - L) \tag{6.2}$$

This follows directly from the fact that each plant has capacity equal to one and from the price-quantity produced relationship (or demand function).

The sequence of decisions and events is as follows. The firms or the social planner choose the level of maintenance expenditure. Then, the operational state of each plant is realized. Finally, the market price is determined and profits are earned. On the behavioral side, we assume that firms and consumers (and hence, the social planner) are risk-neutral, i.e., they care about maximizing expected profit and are indifferent to risk.

6.3 PROFIT MAXIMIZING OUTCOME OF A MONOPOLISTIC GENERATOR

In principle, the monopolist can choose a different reliability level for each of the two plants. It turns out, however, that it is never optimal to choose a differential reliability level when the demand is sufficiently strong, i.e, when the demand parameter a exceeds five. This is not a general result; when demand is relatively weak and maintenance costs are moderate, it is possible that a monopolist would optimally choose a different level of reliability for the two plants even though they are in every other respect identical. As discussed in the preceding section, for expositional brevity we first consider the case for which the profit maximizing maintenance strategy of the monopolist is indeed symmetric. We then briefly consider in Section 6.7 the case for which a differential, non-symmetric, maintenance strategy is profit maximizing.

The *ex ante* expected profit of the monopolist as a function of maintenance expenditure on each plant is given by:

$$\Pi_m = r(m)^2 \cdot \Pi(2) + 2 \cdot r(m) \cdot (1 - r(m)) \cdot \Pi(1) - 2 \cdot m \tag{6.3}$$

The first term is the revenue the monopolist obtains in the event that both plants are operating, the second term is the revenue obtained when only one of the plants are operating, while the third term is the maintenance expenditure. The monopolist chooses m to maximize expected profits. Assuming an interior solution, i.e., one in which reliability is not perfect, the first order condition of profit maximization (after making the appropriate substitutions and simplifications) is given by:

$$\frac{\sqrt{c}}{2 \cdot \sqrt{m}} \cdot (a - 1) - c - 1 = 0 \tag{6.4}$$

The optimal value of m is the lower of the solution of this equation with respect to m, or the value of m, which results in perfect reliability. Solving for m, we obtain the optimal for the monopolist level of maintenance as a function of the cost of maintenance, c, and the strength of demand, a:

$$m_m^* = \min\left[c, \frac{c}{4}\left(\frac{a-1}{1+c}\right)^2\right] \tag{6.5}$$

Maintenance expenditure is capped at c because at that level of expenditure, plant reliability is 100%. It can be readily seen that the optimal level of maintenance is increasing in the strength of the demand for electricity and *decreasing* in the cost of providing reliability (provided that perfect reliability is not optimal). The fact that maintenance expenditure is decreasing in the cost parameter c is not paradoxical. As the cost of reliability increases, the monopolist finds it optimal to "purchase" less of it. If the demand for reliability is sufficiently responsive to its cost, then an increase in the cost parameter would be more than made up by a decrease in the optimal level of reliability. This in turn, would lead to a reduction in the total maintenance expenditure.

The optimal plant reliability associated with the above level of maintenance expenditure is given by:

$$r_m^* = \min\left[1, \frac{1}{2}\cdot\frac{a-1}{1+c}\right] \tag{6.6}$$

The optimal reliability is increasing in a and decreasing in c. Figures 6.1 and 6.2 graph the optimal values of m and r for $a = 5$ and for a range of c. This value of a is at the midpoint of the interesting range of a (as discussed earlier) and the range of c is chosen so that equilibrium reliability is not too low. The figures show that perfect reliability is optimal for low enough maintenance costs. When maintenance

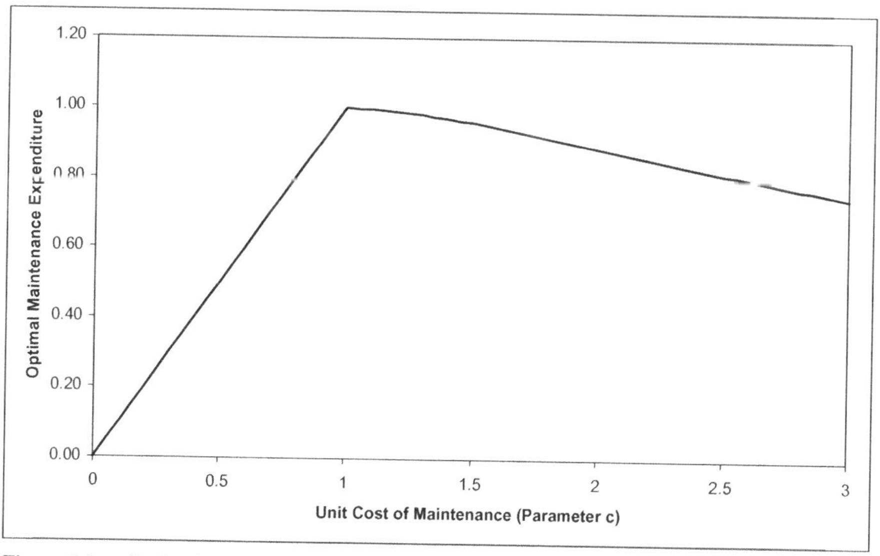

Figure 6.1. Optimal maintenance expenditure by a monopolist as a function of unit cost of maintenance. The demand parameter a has been set to five.

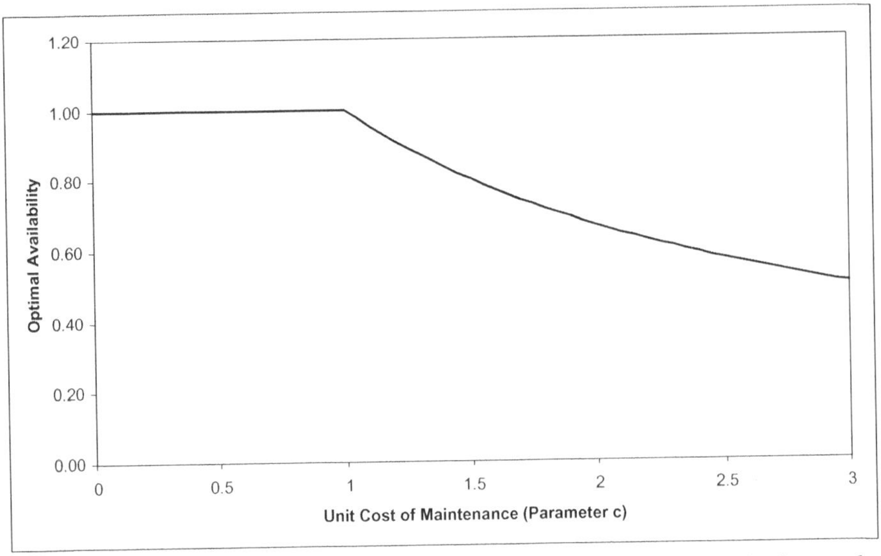

Figure 6.2. Optimal plant availability chosen by a monopolist as a function of unit cost of maintenance. The demand parameter a has been set to five.

costs are sufficiently high, the monopolist chooses less than perfect reliability. The figures also show that optimal maintenance expenditure increases linearly with the cost of maintenance when perfect reliability is optimal, but declines slowly when less than perfect reliability is optimal. The above findings are, of course, sensitive to the choice of parametric model. For example, if the incremental gain of reliability from additional maintenance expenditure went to zero as reliability goes to one, then perfect reliability would never have been optimal. We proceed next to the derivation of the market (Nash) equilibrium when each plant is owned by a different firm.

6.4 NASH EQUILIBRIUM IN A DUOPOLISTIC MARKET STRUCTURE

We now consider the possibility that there are two firms operating in this market, each owning one of two identical plants. The expected profits of Firm 1 as a function of its maintenance expenditure, m_1, and the maintenance expenditure of its competitor, m_2, are given by the expression

$$\Pi_1 = r(m_1)\cdot r(m_2)\cdot \frac{\Pi(2).}{2} + r(m_1)\cdot(1-r(m_2))\cdot\Pi(1) - m_1 \tag{6.7}$$

The first term is the revenue Firm 1 obtains when both plants are operating, the second term is the revenue it obtains when its plant is operating and the plant of its competitor is not operating, and the third term is the maintenance expenditure. Similarly, the expected profits of Firm 2 are given by the expression:

$$\Pi_2 = r(m_1)\cdot r(m_2)\cdot\frac{\Pi(2)}{2} + r(m_2)\cdot(1-r(m_1))\cdot\Pi(1) - m_2 \tag{6.8}$$

We assume that each firm chooses its maintenance expenditure to maximize its profits taking the choice of its competitor as given. Both firms make their choices simultaneously, i.e., without knowing the choices of their competitors. This is the standard set of assumptions underlying the Nash Equilibrium of a game. In our model, the Nash Equilibrium of the game played by the two generating companies is given by the solution of the system of equations obtained from First Order Conditions associated with the unilateral profit maximization of each firm's profit.

Profit maximization of Firm 1 with respect to m_1 yields the first order condition:

$$\frac{a-2}{2\cdot c}\cdot\sqrt{\frac{m_2}{m_1}} + \frac{a-1}{2\cdot\sqrt{m_1\cdot c}}\cdot\left(1-\sqrt{\frac{m_2}{c}}\right) - 1 = 0 \tag{6.9}$$

Similarly, profit maximization of Firm 2 with respect to m_2 yields the first order condition:

$$\frac{a-2}{2\cdot c}\cdot\sqrt{\frac{m_1}{m_2}} + \frac{a-1}{2\cdot\sqrt{m_2\cdot c}}\cdot\left(1-\sqrt{\frac{m_1}{c}}\right) - 1 = 0 \tag{6.10}$$

Observing that the system of the two equations is symmetric with respect to the two variables, we can utilize symmetry and impose the condition $m_1 = m_2 = m_d$. Either equation can then be written as:

$$\frac{a-2}{2\cdot c} + \frac{a-1}{2\cdot\sqrt{m_d\cdot c}}\cdot\left(1-\sqrt{\frac{m_d}{c}}\right) - 1 = 0 \tag{6.11}$$

Solving for m_d and recognizing the possibility of a corner solution (i.e., the possibility that equilibrium maintenance would result in perfect reliability) we obtain the Nash equilibrium level of maintenance expenditure. This is given by:

$$m_d^* = \min\left[c, c\cdot\left(\frac{a-1}{1+2\cdot c}\right)^2\right] \tag{6.12}$$

The equilibrium level of plant maintenance expenditure in a duopoly is increasing in the strength of the demand and decreasing in the cost of maintenance (provided that perfect reliability is not optimal). The associated level of plant reliability is given by:

$$r_d^* = \min\left[1, \frac{a-1}{1+2\cdot c}\right] \tag{6.13}$$

Observe that the equilibrium level of reliability in the duopoly is increasing in a and decreasing in the cost of maintenance, c. We postpone a comparison of the duopoly outcome with the outcome under a monopoly until Section 6, and proceed next to the derivation of the socially optimum level of reliability.

6.5 SOCIAL OPTIMUM

In this section we take up the question of what is the optimal level of maintenance expenditure from the point of view of the entire economy (or society). A regulator or a government monopoly that has the public interest in mind (also known in economic terminology as a "social planner") would choose the level of maintenance of the two plants to maximize the sum of profits and consumer surplus. Consumer surplus is the difference between the willingness of consumers to pay for electric power and the price they actually have to pay for it. It is defined as the integral of the difference between the market demand curve and the market price integrated over the quantity purchased.

For our demand specification (linear demand curve with a slope equal to minus 1), consumer surplus has a simple closed form solution. When L plants are operational, consumer surplus equals:

$$CS(L) = \frac{1}{2} \cdot (a - P(L))^2 \tag{6.14}$$

or, equivalently:

$$CS(L) = \frac{1}{2} \cdot L^2 \tag{6.15}$$

The social planner's objective function, then, is given by:

$$W(m) = r(m)^2 \cdot [\Pi(2) + CS(2)] + 2 \cdot r(m) \cdot (1 - r(m)) \cdot [\Pi(1) + CS(1)] - 2 \cdot m \tag{6.16}$$

The first term is the total surplus in the event that both plants are operational, the second term is the total surplus in the event that only one plant is operational, and the third term is the cost of maintenance. Note that we have implicitly assumed that the social planner chooses the same level of maintenance for the two plants. It can be shown that this is indeed the case for our model.

Maximizing the welfare function *W(m)* with respect to *m* yields the first order condition:

$$\frac{a - \frac{1}{2}}{\sqrt{m \cdot c}} \cdot \left(1 - \sqrt{\frac{m}{c}}\right) + \frac{a - \frac{3}{2}}{c} - 2 = 0 \tag{6.17}$$

Solving for *m* and recognizing the possibility of a corner solution we obtain the socially optimal level of maintenance expenditure:

$$m_s^* = \min\left[c, \frac{c}{4} \cdot \left(\frac{2 \cdot a - 1}{1 + 2 \cdot c}\right)^2\right] \tag{6.18}$$

The socially optimal level of maintenance expenditure behaves similarly to that of the monopolist and duopolistic market structures: it is increasing in the level of demand and decreasing in the cost of maintenance. The associated level of reliability is given by:

$$r_s^* = \min\left[1, \frac{1}{2}\cdot\frac{2\cdot a-1}{1+2\cdot c}\right] \tag{6.19}$$

We next turn to the comparison of the optimal level of reliability across the three market structures.

6.6 COMPARISON OF EQUILIBRIA AND DISCUSSION

For all three types of market structure, plant reliability is 100% for low enough costs of maintenance. For higher values of c, the equilibrium levels of reliability can be ranked. In particular, observe that:

$$r_d^* > r_m^* \Leftrightarrow \tag{6.20}$$

$$\frac{a-1}{1+2\cdot c} > \frac{1}{2}\cdot\frac{a-1}{1+c} \Leftrightarrow \tag{6.21}$$

$$2+2\cdot c > 1+2\cdot c \tag{6.22}$$

which is always true. Therefore, the monopoly level of reliability is equal to or lower than the level of reliability under a duopoly for all levels of demand and costs of maintenance (the two reliability levels are equal to each other only when c is low enough that reliability for both is 100%). Similarly, observe that:

$$r_s^* > r_d^* \Leftrightarrow \tag{6.23}$$

$$\frac{1}{2}\cdot\frac{2\cdot a-1}{1+2\cdot c} > \frac{a-1}{1-2\cdot c} \Leftrightarrow \tag{6.24}$$

$$2\cdot a-1 > 2\cdot a-2 \tag{6.25}$$

which is always true. Therefore, the socially optimal level of reliability is equal to or higher than the level of reliability under a duopoly for all levels of demand and costs of maintenance (it is equal to the reliability of the duopoly only when c is low enough that the reliability for both is 100%). Therefore, both the levels of reliability and maintenance expenditure can be ranked as follows:

$$r_s^* \geq r_d^* \geq r_m^* \tag{6.26}$$

and:

$$m_s^* \geq m_d^* \geq m_m^* \tag{6.27}$$

Figures 6.3 and 6.4 plot the optimal maintenance expenditure and reliability levels for $a = 5$ and c in [0, 3]. The ranking of reliability levels obtained in this chapter is not dependent on the particular parametric formulation of the model, and is likely to hold more generally. We explain why below.

Consider, first, the comparison of the reliability level in a monopoly with that of a duopoly. The financial incentive of a monopolist to maintain high reliability is lower than that of a single plant duopolist. If the duopolist's plant is not available for production, the market price goes up, but the duopolist cannot benefit as it has no other operational capacity. In contrast, if one of the monopolist's plants is not available for production, the resulting increase in the market price will lead to an increase in the revenue obtained from the other plant. Therefore, the monopolist values the reliability of each of the plants less than does the duopolist. Consequently, a monopolist will invest less in plant maintenance than a duopolist would. This discussion makes it clear that the result is not dependent on the functional form of market demand or the reliability-maintenance expenditure relationship. Also, our results do not depend on the assumption that the duopolist has only a single plant. All other things being equal, breaking up a four-plant monopoly into a two-plant per firm duopoly would result in an increase in plant reliability. The comparison is most meaningful if the firms choose the same level of reliability for each of their plants, as is the case in the parameter range that we examine here. Otherwise, one would have to devise some metric of average reliability.

Consider, next, the comparison of the reliability level in a duopoly with that chosen by a social planner. The duopolist chooses the level of reliability by comparing the incremental cost of improving reliability by a small amount with the incremental gain in the probability that profits are obtained by the duopolist from operating the plant. The social planner chooses the level of reliability by comparing the incremental cost of improving reliability with the incremental gain in the probability of obtaining the profits from operating the plant *plus* the consumer surplus from the plant's operation. Since the marginal benefit of a plant's operation is higher for the social planner than for the duopolist, it follows that the social planner's willingness to invest in plant maintenance will exceed that of the duopolist. Notice

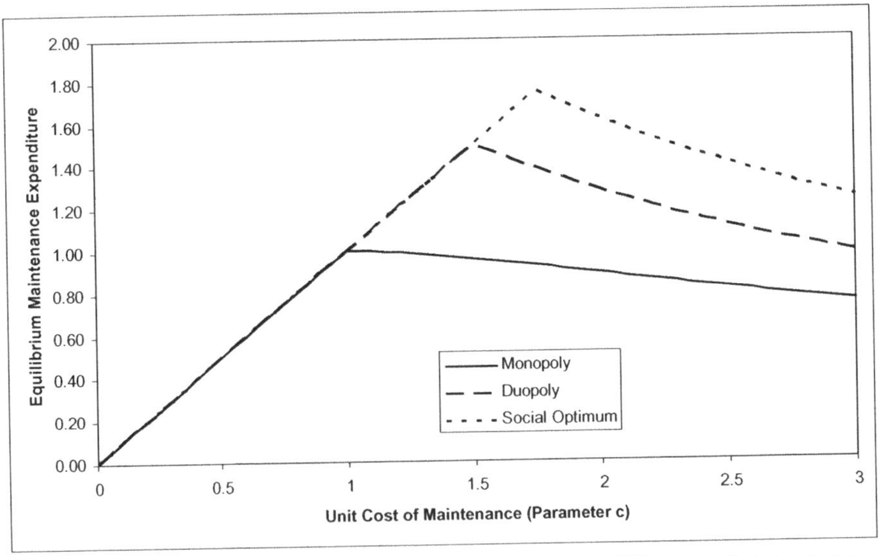

Figure 6.3. Equilibrium levels of maintenance expenditure for different market structures. The demand parameter *a* has been set to five.

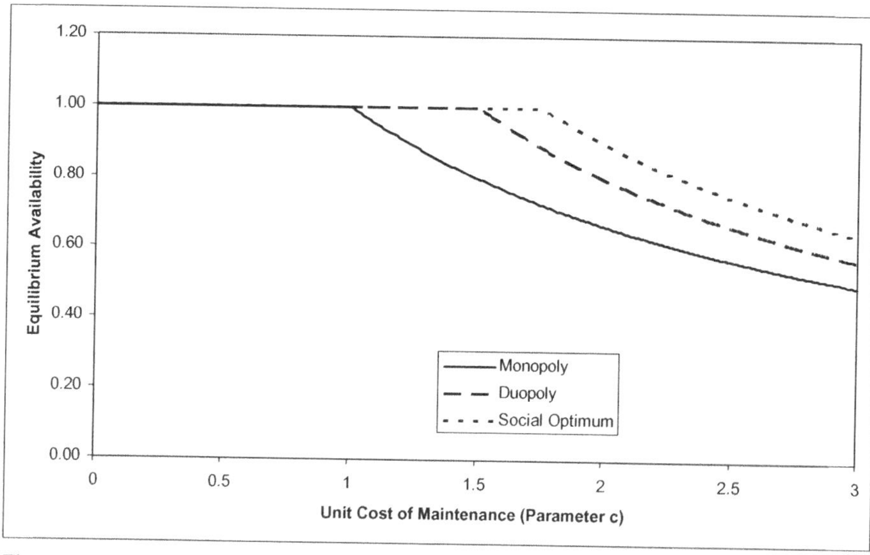

Figure 6.4. Equilibrium levels of plant availability for different market structures. The demand parameter a has been set to five.

that this argument does not rely on any functional form. Thus, any decentralized market structure would be expected to provide less reliability than is socially optimal. For the comparison to be meaningful, firms should be symmetric and choose the same level of reliability for each of their plants, and the social planner should also choose the same reliability for all the plants.

A comprehensive and general framework that investigates equilibrium maintenance expenditures and plant reliability when firms do not have symmetric capabilities, do not find it optimal to maintain the same reliability level on all of their plants, and might possibly prefer to keep operational plants idle is the subject of ongoing and future research (though the following section provides some understanding of the forces underlying the choice of asymmetric maintenance policies). However, we expect that the central insights obtained using this stylized framework will carry over to a more general framework.

6.7 ASYMMETRIC MAINTENANCE POLICIES

In the framework developed in this chapter, the monopolist and the social planner (i.e., a regulatory agency) control the maintenance expenditure of two (identical) plants. Therefore, these decision makers have the option to expend different levels of resources in order to maintain the two plants. One may think that it would not be optimal for either decision maker to maintain a different level of reliability for two plants that are identical in every respect, particularly given that the reliability of each plant is a concave function of maintenance expenditure on that plant. For a desired level of capacity, it would appear cheaper to maintain the same reliability for the two plants. Indeed, this intuition is born out for the social planner who always prefers to treat the two plants symmetrically. The monopolist, however, does not always prefer that its plants are of equal reliability. Equal reliability levels are profit maximizing for values of a that exceed five, which is the case analyzed in the preceding sections. For markets with relatively low demand (for values of a between four and five), the monopolist *may* prefer to have one plant at excellent reliability while keeping the other plant at lower reliability. The optimal (profit maximizing) availability of the two plants of a monopolist facing a linear demand parameter with intercept $a = 4$ is plotted in Figure 6.5. It can be seen that for intermediate values of maintenance cost, different levels of reliability are optimal, while for high values of maintenance cost, optimal reliability is the same for both plants.

The reason for this unexpected result is that for relatively low values of demand, the profit from having at least one plant operational is much greater than the incremental profit of having a second plant operational. Therefore, the monopolist's strategy is to ensure that one plant will definitely be able to produce, while keeping the second plant available at a much smaller fraction of the time. The reason this is not an optimal strategy is that, for relatively high values of the demand and for the same values of maintenance cost, the optimal reliability level of both of the monopolist's plants is 100% (see Figure 6.2). To complete the intuition behind asymmetric maintenance policies, one must also examine why the profit maximizing reliability is the same for the two plants when the maintenance cost is high. The reason is that when maintenance costs are high, the increase in costs from maintaining different reliability levels (due to the convexity of the cost function) outweighs any benefits arising from the ensuring that one plant is always operational.

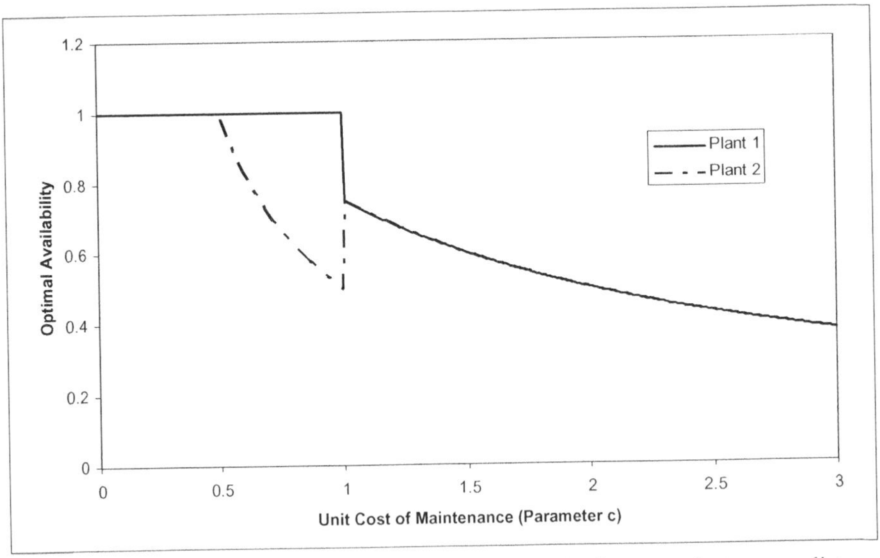

Figure 6.5. Optimal (equilibrium) levels of plant availability for a two plant monopolist. The demand parameter a has been set to four.

6.8 CONCLUSION

Our analysis has at least two policy implications. The first involves public policy towards plant reliability standards and the second involves public policy towards mergers of electricity generators. We discuss them in turn.

The findings that the level of privately provided plant reliability is lower than the socially optimal level suggests that public authorities (FERC, regional or state governmental organizations, etc.), should provide incentives for generators to increase the reliability of the units beyond the level that they would do on their own accord. The more concentrated the generation market, the more powerful the incentives need to be. These incentives can take a number of forms. One possibility is to directly stipulate a level of expenditure on maintenance operations or on system reliability. Another possibility is to subsidize maintenance expenditure. Both of these policy recommendations have their shortcomings. For example, electricity generators could try to circumvent a minimum expenditure standard by lumping much non-maintenance related expenditure together with maintenance related expenditure in an attempt to meet the standard without actually increasing bona-fide maintenance expenditure. A regulatory agency would have to undertake invasive audit procedures to ensure that only bona-fide maintenance related expenditures are counted towards the stipulated minimum. Similarly, a maintenance subsidy may lead firms to attempt to obtain this subsidy for non-maintenance related expenditure by claiming that these expenditures are relevant to increasing system reliability. Such behavior, if prevalent, would also necessitate the adoption of invasive audit procedures by the regulator.

A more effective way to increase system reliability towards the socially desirable levels might be for the regulator or other federal funding agencies to provide financial support for research activities that lead to a reduction in the cost of maintaining high reliability. In the notation of our model, subsidizing this type of research activities would eventually result in lower values of the parameter c, which in turn would lead generating companies to adopt higher levels of reliability than would otherwise be the case.

The findings that the level of privately provided plant reliability is decreasing in market concentration suggests that anti-trust authorities should consider the effects of mergers between generators on plant reliability when considering the approval of such proposed mergers. Currently, the primary area of concern for anti-trust authorities is the effect of mergers on market price. However, as we have shown in this chapter, even when firms have no incentives to curtail production in order to increase prices, a merger has an indirect effect on market prices and system reliability through a reduction in the incentives to expend resources on plant maintenance. We should note, though, that our findings are obtained by completely abstracting of network considerations. Network reliability might possibly be increasing in market concentration, providing a countervailing influence to negative relationship between plant reliability and market concentration. The examination of network effects is the subject of ongoing research.

ACKNOWLEDGMENT

This material is based upon work supported in part by the National Science Foundation under NSF/ONR EPNES Award No. 0224729. Any opinions, finding, and conclusions or recommendations expressed in this publication are those of the authors and do not necessarily reflect the views of NSF.

BIBLIOGRAPHY

[1] Bertling L, Allan R, Eriksson R (2005). A reliability-centered asset maintenance method for assessing the impact of maintenance in power distribution systems. *IEEE Trans. on Power Systems*, Vol. 20, No. 1, pp. 75–82.

[2] Billinton R, Allan RW, Salvaderi L (1991). *Applied Reliability Assessment in Electric Power Systems*. IEEE Press.

[3] Billinton R, Ringlee RJ, Wood AJ (1973). *Power-System Reliability Calculations*. MIT Press.

[4] Borenstein S, Bushnell J, Stoft S (2000). The competitive effects of transmission capacity in a deregulated electricity industry. *RAND Journal of Economics*, Vol. 31, pp. 294–325.

[5] Borenstein S, Bushnell J, Wolak FA (2002). Measuring market inefficiencies in California's restructured wholesale electricity market. *American Economic Review*, Vol. 92, pp. 1376–1404.

[6] Endrenyi J (1978). *Reliability modeling in electric power systems*. Wiley and Sons Press, New York, NY.

[7] Garcia A, Mili L, Momoh J (2008). Modeling Electricity Markets: A Brief Introduction. *Economic Market Design and Planning For Electric Power Systems*. Mili L. and Momoh J., editors. Wiley and Sons Press, New York, NY.

[8] Gonzalez C, Juan J, Mira J, Prieto FJ, Sanchez MJ (2005). Reliability analysis for systems with large hydro resources in a deregulated electric power market. *IEEE Trans. on Power Systems*, Vol. 20, No. 1, pp. 90–95.

[9] Johnson BW (1989). *Design and Analysis of Fault-Tolerant Digital Systems*. Addison-Wesley Press, Reading, MA.

[10] Joskow P, Tirole J (2000). Transmission rights and market power on electric power networks. *RAND Journal of Economics*, Vol. 31, pp. 450–487.

[11] Joskow P, Tirole J (2004). Reliability and Competitive Electricity Markets. Unpublished manuscript, MIT and U. of Toulouse.

[12] Luh PB, Chen Y, Yan JH, Stern GA, Blankson WE, Zhao F (2008) Payment Cost Minimization with Demand Bids and Partial Capacity Cost Compensations for Day-Ahead Electricity Auctions. *Economic Market Design and Planning For Electric Power Systems*. Mili L. and Momoh J., editors. Wiley and Sons Press, New York, NY.

[13] McGranaghan M, Ignall R (2002). Measuring power system performance for high-reliability applications. *EC&M Magazine*, March 2002.

[14] Medjoudj R, Aissani D (2002). Economic aspects of distribution power system reliability: Application to a 30KV network of Bejaia, Algeria. *Proceedings of the 3rd International Conference on Mathematical Methods in Reliability*.

[15] Mookherjee R, Hobbs BF, Friesz TL, Rigdon MA (2008). Dynamic Oligopolistic Competition in an Electric Power Network and Impacts of Infrastructure Disruptions. *Economic Market Design and Planning For Electric Power Systems*. Mili L. and Momoh J., editors. Wiley and Sons Press, New York, NY.

[16] Sauma EE, Oren SS (2008). Alternative Economic Criteria and Proactive Planning for Transmission Investment in Deregulated Power Systems. *Economic Market Design and Planning For Electric Power Systems*. Mili L. and Momoh J., editors. Wiley and Sons Press, New York, NY.

[17] Shih F, Mazumdar M, Bloom JA (1999). Asymptotic mean and variance of electric power generation system production costs via recursive computation of the fundamental matrix of a Markov chain. *Operations Research*, Vol. 47, pp. 703–712.

[18] Siewiorek DP, Swarz RS (1998). *Reliable Computer Systems: Design and Evaluation*. A. K. Peters Press, Natick, MA.
[19] Singh C, Gubbala N (1994). Reliability evaluation of interconnected power systems including jointly owned generators. *IEEE Trans. on Power Systems*, Vol. 9, No. 1, pp. 404–412.
[20] Wolak F (2003). Measuring unilateral market power in wholesale electricity markets: the California market, 1998–2000. *American Economic Review*, Vol. 93, pp. 425–430.

CHAPTER 7

BUILDING AN EFFICIENT RELIABLE AND SUSTAINABLE POWER SYSTEM: AN INTERDISCIPLINARY APPROACH

James Momoh, Philip Fanara, Jr., Haydar Kurban, and L. Jide Iwarere

Howard University

EDITORS' SUMMARY: This chapter introduces economic, technical, modeling and performance indices for reliability measures across boundary disciplines. The concept is being used to analyze outages of a typical power system. This chapter proposes new tools for handling probabilistic contingencies in electric power systems. It introduces a combination of new indices such as expected socially unserved energy with load loss that allow the planner to measure social impacts of contingencies. Furthermore, these indices are designed to accommodate both engineering models and public perception based on economic and social factors. The results indicate that an efficient, reliable, and sustainable power system can be built using an interdisciplinary approach that has the potential to address large-scale systems.

7.1 INTRODUCTION

In this chapter, we present opportunities for improvement in technical modeling of Flexible AC Transmission System (FACTS) devices and Distributed Generation (DG) Technologies for enhancing the efficiency and sustainability of high performance electric power systems. Mathematical models of the system components are included herein. Reliability measures and other performance analyses are described and representation of Public Perception and other attributes that cut across

Economic Market Design and Planning for Electric Power Systems, Edited by James Momoh and Lamine Mili

boundaries of disciplines is included in the analysis. Technical and economic limitations of current power systems are highlighted [1–3]. In addition, the technical as well as economic improvements brought as a result of this study are presented.

7.1.1 Shortcomings in Current Power Systems

7.1.1.1 Technical Limitations Modern electric power systems are becoming overly stressed due to increased hourly loading and slow generation expansion that often affects system stability and reliability of power delivery. The efficiency of the electric power network is also affected by natural and/or forced contingencies that have undesirable impacts on the normal operation of the network.

Research work done to date embarks on deterministic control schemes to mitigate against contingencies such as loss of critical load sets, generation, or transmission lines. Such work exhibits shortfalls in the ability to handle uncertainties of load and topology change as well as the social impacts of deploying appropriate control technology. In this chapter, we focus on the development of data models for improving the test-bed used in EPNES research.

7.1.1.2 Economic Shortcomings In addition to these technical issues, we need to address the economic shortcomings of the current system. These shortcomings include the lack of an efficient emission market and a system of market incentives, which encourage optimum investment and reliability, as well as derivatives markets, which provide protection. While these efficient and optimum market related issues are important they have been addressed by other researchers. However, this paper addresses public perception, an issue which some consider the most neglected aspect economic shortcoming. This latter point must include the public perception of reliability and failures such as outages. Traditionally private costs have been the main determinant of enhancements to the power system, while social costs/public perceptions have been largely neglected. From the economics perspective, the major focus of our research is incorporation of social costs/public perceptions into the decision-making process for designing an efficient, reliable, and sustainable power system.

7.1.2 Our Proposed Solutions to the Above Shortcomings

7.1.2.1 Technical Improvements Technically the scheme will be able to increase the loading capability of lines to their thermal capabilities, including short term and seasonal. Through inclusion of FACTS devices, we hope to be able to increase the system security through raising the transient stability limit, limiting short-circuit currents and overloads, managing cascading blackouts and damping electromechanical oscillations of power systems and machines. The scheme will also lead to secure tie line connections to neighboring utilities.

The new tools and indices developed will be able to handle uncertainty in the system and real time operations and controls. The Weighted Probability Index (WPI) is able to handle uncertainties as well as probabilitistic contingencies inherent in the electric power system.

The performance index incorporating Available Transfer Capability (ATC) and Expected Socially Unserved Energy (ESUE) with Load Loss provides an excellent combination of indices for measuring power system security, as well as social impacts of contingencies.

The results of these exercises are a well-defined criterion for managing the contingencies and losses associated with them and utilizing social and economic considerations in the planning and operation of an electric power system.

7.1.2.2 ***Incorporation of Public Perceptions, Private and Social Costs*** A central feature of our work is the creation of a Public Perception Index, which is used to derive a measure of Expected Socially Unserved Energy (ESUE). In order to achieve this, we incorporate social costs/public perceptions in the determination of the desired level of reliability; we construct and estimate an index of social and economic factors. A consumer's sense of security is a function of many factors, including the economic and social factors of their current environment [4, 11].

The index, integrates public perceptions based on spatial variations of economic and social factors. In turn, it is used in designing a reliable and sustainable power system. The factors used in this index included unemployment rates, a measure of social strife, crime rates, and measures of financial strength of state and local government.

The public perception index is used to assist in the cost and benefit analysis of adding new technologies, and control systems for the improvement of technical performance of an existing power system. It can also be employed in the design of new electrical power systems.

Analysis of major outages indicated that public perception or reaction to incidences varies based on local social and economic conditions, as well communications regarding the source of outages. These reactions indicate the level of reliability desired by the public. They also enable us to accurately estimate indirect and direct costs associated with the reliable and sustainable system. Our index is designed to accurately summarize public perception, and can readily be operationalized and incorporated into the technical engineering models for such things as contingency screening, selection of control systems, and any analysis, which utilized a cost and benefit approach.

7.2 OVERVIEW OF CONCEPTS

7.2.1 Reliability

The *reliability* of a bulk power system is the degree of assurance in providing customers with continuous quality service within accepted standards. Overall, electric power system reliability is used to evaluate the ability of a system to supply the load demand, taking into account the random effects of equipment outages, loss of lines or other network components, islanding of subsystems, load variations, and other factors that affects the energy generation-consumption equilibrium.

In general, power system reliability is divided into two groups: bulk power system reliability for the generation and transmission networks with point loads and the high voltage buses, and distribution reliability for the subsystems of the point loads. Overall, probabilistic reliability measures are most commonly expressed in terms of indices reflecting the system degree of service capability.

7.2.2 Bulk Power System Reliability Requirements

The National Energy and Reliability Council (NERC) define the reliability of the interconnected bulk electric systems in terms of two basic functional aspects:

- *Adequacy*—The ability of the electric system to supply the aggregate electrical demand and energy requirements of customers at all times, taking into account scheduled and reasonably expected unscheduled outages of system elements.
- *Security*—The ability of the electric system to withstand sudden disturbances such as electric short circuits or unanticipated loss of system elements. This issue also relates to the ability of the power systems to respond to dynamics or transient disturbances arising within the system.

The challenge of meeting these two requirements in any electric power system is important in the planning and operation of changing electric utilities. This has led to the integration of economic rationale and technically meaningful basis for recommended operational policies.

In addition, adequacy indicators reflect various factors such as system component availability and capacity, load characteristics and uncertainty, system configuration and operational conditions, etc. In reliability assessment, historical events on these factors can be used to identify weak areas (areas that need reinforcement or modifications) that degraded the system reliability. Absolute or relative reliability indices for a particular set of system data and conditions can therefore be used in a cost/benefit framework with the goal of economically supplying power—on demand—while minimizing productions costs as depicted in Figure 7.1.

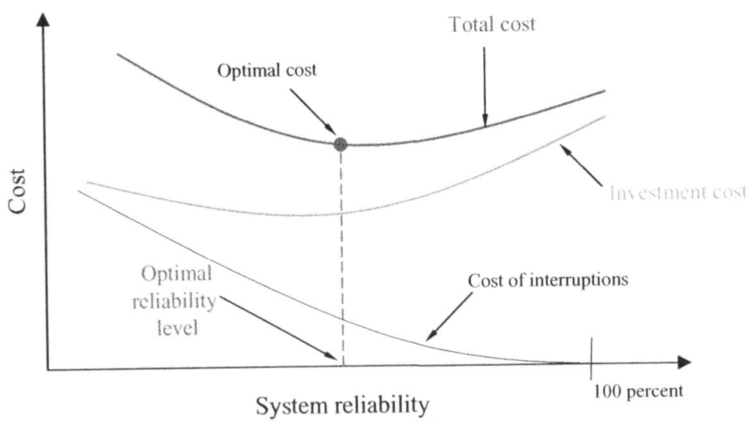

Figure 7.1. Meeting system reliability at optimum cost.

In general, several factors affect reliability assessment of the bulk power and distribution systems and some of these factors are summarized in the next section.

7.2.3 Public Perception

One recent report by EPRI [6] has stated, "... there is a major disconnect between the public's perception of the electricity sector's circumstances and its reality. The connection between the electricity system and the goods and services the public depends on, let alone those they aspire to, is vague to nonexistent in the minds of most. Even those who are engaged on the periphery of the business are generally indifferent to its future as long as the lights stay on [5]." This statement stresses the necessity for an operational definition of public perception. In our work, we define public perception as a measure of consumer satisfaction or dissatisfaction with the flow of services in the realm of electric power. Public perception is based on the premise that public desires "precautionary approaches" for risk management. In modern behavioral economic literature, it is well documented [12, 13], that consumers have aversion towards loss; therefore, they prefer "minimum risk." A power system must be designed to be reliable and sustainable. The maximum reliability level is determined by the most advanced available technology and control systems. Reliability must also be a function of the condition of "minimum risk." As a result, a reliable system should be interpreted as the one with the minimum risk level defined by consumer perception. A major question then becomes one of defining those factors that determine consumer perception.

Consumer perception and attitudes toward risk have to be studied in detail because the very premise of deregulation is based on trading the risks through a competitive market system where the participants are assumed to have different risk taking behaviors. After all, an efficient economic system is defined as the one that satisfies the Pareto Optimality Condition.

7.2.4 Power System / New Technology

The utility system called Western Systems Coordinating Council (WSCC) is used for the research study. It is scalable to other utility and military (Navy) systems.

7.2.4.1 Description of WSCC According to PSERC, The Western Systems Coordinating Council (WSCC) region is the largest and most diverse of the ten regional reliability councils of the North American Electric Reliability Council (NERC). It encompasses all or part of fourteen western states, two Canadian provinces, and portions of northern Mexico. It has characteristics that are distinct from the other three North American Interconnections. The WSCC divides into four geographic sub-regions: California, Northwest, Arizona/New Mexico/Southern Nevada, and Rocky Mountain. About sixty percent of the WSCC load is located in the coastal regions. A significant portion of the generation that serves these load centers is located inland, so transmission over long distances is needed. As a result, significant portions of the WSCC network are stability limited.

The WCSS model is given in Figure 7.2. The WSCC power system consists of several components (lines, generators, transformers, variety of loads and controls). As in most other utility systems, most of the loads are aggregated and have common load points called load centers or buses [14, 15]. In addition, the generators are grouped into various operational units and are usually committed at less than full

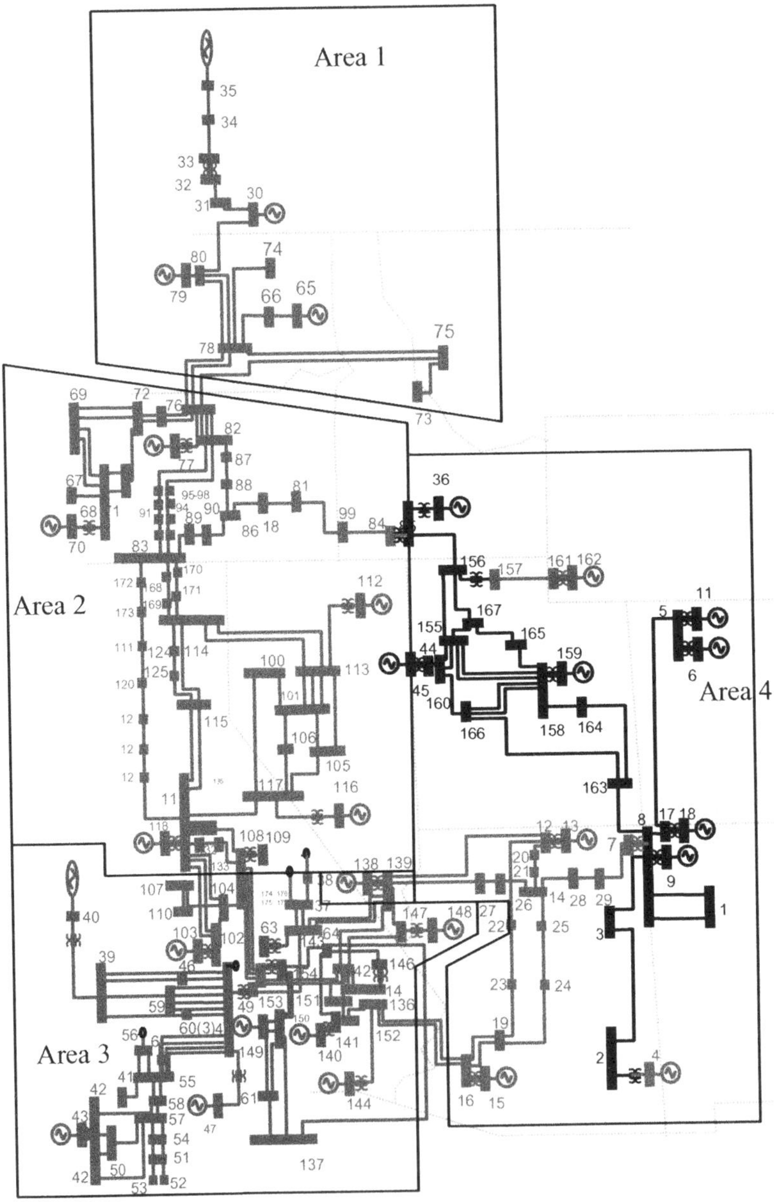

Figure 7.2. WSCC Power System Network.

capacity in each unit to allow for reserve or safety margin for security. Overall, the WSCC network consists of 205 transmission lines, 58 transformers, 29 generator buses and 104 load buses.

7.2.4.2 Components of Utility Power System Model

7.2.4.2.1 Induction Motor Load Modeling The induction machine load is considered using the steady state model equivalent circuit [10] as shown in Figure 7.3.

The input impedance of the equivalent circuit shown in Figure 7.3 is:

$$Z = \frac{\frac{r_s r_r'}{s} + \left(\frac{\omega_e}{\omega_b}\right)^2 (X_M^2 - X_{ss} X_r') + j\frac{\omega_e}{\omega_b}\left(\frac{r_r'}{s} X_{ss} + r_s X_{rr}'\right)}{\frac{r_r'}{s} + j\frac{\omega_e}{\omega_b} X_{rr}'} \tag{7.1}$$

where:

$$s = \frac{\omega_e - \omega_r}{\omega_e} \tag{7.2}$$

$$X_{ss} = X_{ls} + X_M \tag{7.3}$$

$$X_{rr}' = X_{lr}' + X_M \tag{7.4}$$

$$\omega_b = \omega_e = 2\pi f \tag{7.5}$$

s = machine slip

r_s = stator resistance

r_r = rotor resistance referred to the stator side

X_{ls} = stator leakage inductance

X_{lr} = rotor leakage inductance referred to the stator side

Since,

$$\vec{I} = \frac{\vec{V}}{Z} \tag{7.6}$$

Then the power consumed by the induction motor is

$$S = \vec{V}\,\vec{I} \tag{7.7}$$

$$S = P + jQ \tag{7.8}$$

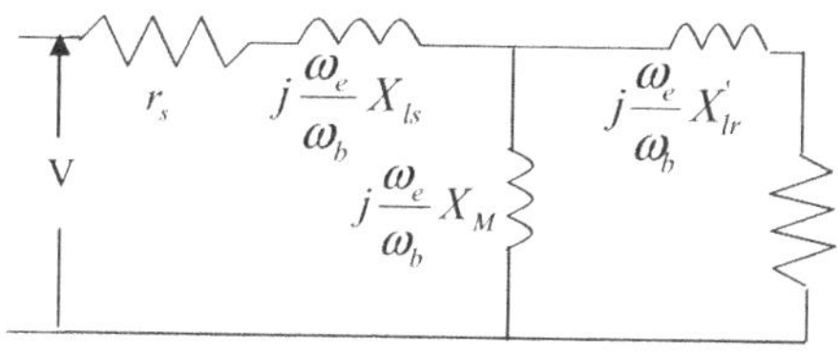

Figure 7.3. Equivalent Circuit for steady state operation of a symmetrical induction machine.

The simplified dynamic model of the induction motor can be given as

$$\dot{S} = \frac{1}{2H}(T_m(s) - T_e(S, V)) \tag{7.9}$$

Where: S = slip

T_e, T_m: Electromagnetic and mechanical load torques respectively

H: Moment of inertia

As with other utility systems, WSCC network comprises many control blocks and dynamics of load. Also included are FACTS devices and distributed generation units. The following is a brief description of the various network component models used in the utility systems.

7.2.4.2.2 Frequency Dependent Load Model This can be represented as an exponential voltage frequency dependent load as follows:

$$P = \frac{k_p}{100}\left(\frac{V}{V_0}\right)^{\alpha p}(1+\Delta\omega)^{\beta p} \tag{7.10}$$

$$Q = \frac{k_q}{100}\left(\frac{V}{V_0}\right)^{\alpha q}(1+\Delta\omega)^{\beta q} \tag{7.11}$$

where:

k_p : active power percentage

α_p : active power voltage coefficient

β_p : active power voltage coefficient

k_q : reactive power percentage

α_q : reactive power voltage coefficient

β_q : reactive power voltage coefficient

$\Delta\omega$: frequency deviation

7.2.4.2.3 Generic Dynamic Load Model Structure and Parameters Generally, in response to a step change in voltage, loads undergo a step change in real and reactive power demand. The load will then recover, over some time, to a steady state value, which may be different from its pre-disturbance value. Important characteristics of this dynamic behavior are the initial step change, the final value, and the rate of load recovery. A generic model, which captures these characteristics, is given in equations (7.12)–(7.15) below.

$$T_p \dot{x}_p = P_s(V) - P_d \tag{7.12}$$

where T_p is the load motor torque and $\dot{x}_p$ is given by:

$$x_p = P_d - P_t(V) \tag{7.13}$$

A similar model can be used for reactive power load. The functions $P_t(V)$, $P_s(V)$ define the initial step response, and the final value of power demand respectively. A convenient form for these functions is,

$$P_s(V) = P_o(V/V_o)^{n_{ps}} \tag{7.14}$$

$$P_t(V) = P_o(V/V_o)^{n_{pt}} \tag{7.15}$$

where V_o, P_o are the nominal voltage and the corresponding real power demand respectively, and n_{ps}, n_{pt} are the steady state and transient voltage indices. Reactive power functions $Q_s(V)$, $Q(V)$ can be defined similarly, but with voltage indices n_{qs}, n_{qt} respectively. The time constants T_p, T_q describe the rate of recovery of the real and reactive power loads.

7.2.4.3 Modeling of FACTS Devices

7.2.4.3.1 Brief Overview of Facts Devices The primary purpose of FACTS devices such as Thyristor Controlled Series Capacitor (TCSC) and Static VAR Compensator (SVC) is to ensure power system stability and improvement. The primary uses of TCSC are to enhance the angle stability of the power system, and to mitigate the sub-synchronous resonance by regulating real power and maximizing transient synchronizing torque between the interconnected power systems. The prototype control systems are interface with Power system simulation software to test its effectiveness through nonlinear simulations using the Central European-CIS interconnected power system as the study case.

7.2.4.3.2 Benefits of FACTS inclusion in Power Systems Operation

1. Control of power flow as ordered. The use of control of the power flow may be to follow a contract, meet the utilities' own needs, ensure optimum power flow, ride through emergency conditions, or a combination thereof.
2. Increase the loading capability of lines to their thermal capabilities, including short term and seasonal. Overcoming their limitations and sharing of power among lines according to their capability can accomplish this. It is also important to note that the thermal capability of a line varies by a very large margin based on the environmental conditions and loading history.
3. Increase the system security through raising the transient stability limit, limiting short-circuit currents and overloads, managing cascading blackouts and damping electromechanical oscillations of power systems and machines.
4. Provide secure tie line connections to neighboring utilities and regions, thereby decreasing overall generation reserve requirements on both sides.
5. Reduce reactive power flows, thus allowing the lines to carry power that is more active.
6. Reduce loop flows.
7. Increase utilization of lowest cost generation.

One of the principle reasons for transmission interconnections is to utilize lowest cost generation. When it is not possible to be performed, it follows that there is not enough cost-effective transmission capacity. Cost-effective enhancement of capability will therefore allow increased use of lowest cost generation. The TCSC model due to its suitability and the added advantage over others as highlighted is employed

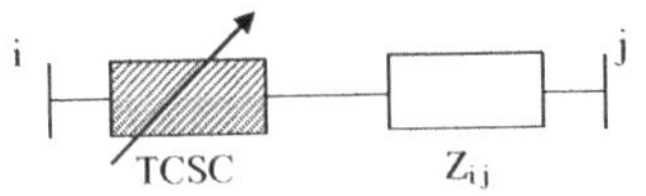

Figure 7.4. The transfer admittance matrix for the TCSC.

in this study. Below is a detailed analytical representation of the TCSC model (Figure 7.4).

7.2.4.3.3 Representation of TCSC

$$\begin{bmatrix} \Delta Ii \\ \Delta Ij \end{bmatrix} = \begin{bmatrix} -Bii & jBij \\ jBji & -Bjj \end{bmatrix} \begin{bmatrix} Vi \\ Vj \end{bmatrix} \tag{7.16}$$

For the reactance:

$$Bij = Bji = -1/x \tag{7.17}$$

The real and reactive power injections required for the simulations of TCSC can be written as:

$$\Delta Pisc = Bij|Vi||Vj|\sin\delta ij \tag{7.18}$$

$$\Delta Pjsc = -Bij|Vi||Vj|\sin\delta ij \tag{7.19}$$

$$\Delta Qisc = Bij|Vi|\{|Vi| - |Vj|\cos\delta ij\} \tag{7.20}$$

$$\Delta Qjsc = Bij|Vj|\{|Vi|\cos\delta ij + |Vj|\} \tag{7.21}$$

7.3 THEORETICAL FOUNDATIONS: THEORETICAL SUPPORT FOR HANDLING CONTINGENCIES

7.3.1 Contingency Issues

The electric power system is subject to several contingencies. These may be in terms of a disturbance resulting in line outage, loss of load or generation with damage to equipment. The technical, social, and economic impacts of these contingencies can be enormous.

Up until recently, power system disturbances have resulted in prolonged duration of clearing time and hence power outages [7–10]. This situation is not desirable and requires immediate attention since the longer it takes to clear the fault, the more damage to suppliers, customers, and all other stakeholders. This situation calls for the need for new indices for power system stability and reliability analysis, and system restoration strategies. In the past, several preventive measures and corrective actions have been taken to minimize the frequency and the extent of power outages.

Notable disturbances have been found to occur at random with peculiar associated problems requiring different solution methodologies. Some of the disturbances have led to sudden increase in load or to loss of an on-line generator, others lead to over-voltages, switching transients, harmonic and power quality problems, and over- and under-excitation of generators. Most recently recorded major disturbances have lead to complete blackout for several minutes, hours, and even days.

Unfortunately, most of the current reliability and stability analysis techniques used do not adequately account for uncertainty in the system.

Our approach is to introduce new indices that can account for these uncertainties and hence more effectively mitigate the effects of the contingencies. These contingencies include instability, duration, loss of load, equipment damage, and complete power outage. In this section, several indices have been developed for determining the stability and reliability of the power system network while accounting for uncertainty in the system and the probability of occurrence of disturbance in the network. These indices also incorporate the social and economic perspectives of the power systems operation. These indices have been tested on the IEEE 30-bus and WSCC power system networks with very convincing results. Some of these indices are as described below.

7.3.2 Foundation of Public Perception

We study public perception within the welfare economics framework, which states that competitive market outcomes are Pareto Optimal, i.e., efficient and optimum if private and social costs are identical. When there are externalities, private costs fail to reflect the true social costs of any action. Therefore, in order to attain outcomes that are both, socially and privately optimum or desired, we must rationally incorporate those neglected social costs. Public perception takes on greater significance in a deregulated power market.

A major aspect of the regulated power sector was a high premium on reliability. The problem of customer education is made more complex by the fact that power has always been available and reliable for the U.S. customer. Indeed, reliability, and high quality of service has never been listed as one of the shortcomings of regulation. Certainly, the old regime of rate base rate-of-return regulation has been a feature in secure and reliable power. Whether one argues that the Averch-Johnson effect results in a bias toward capital intensive technologies, or one takes the less scientific, but perhaps similarly valid viewpoint that it results in rate base padding—it still meant that there was an incentive to expand the capital base, and adopt the latest technologies. Thus, status quo reliability was due to excess capacity. Consumers depended upon and became accustomed to a highly reliable system supported by excess capacity without having complete information on the cost of the electrical power system. Deregulation requires consumers to make choices among alternative plans offered by suppliers based on imperfect information about the costs and benefits. Deregulated power markets partly rely on the consumers' attitudes toward risk. Because of the unforeseen risks, the transactions in the market will be based on incomplete information [19, 21]. Compared to the regulated system, the deregulated market is subjected to higher degrees of externalities and therefore higher social costs.

In order to introduce public perception to our model, we have created an index of economic and social factors. We use this index to capture the socio-economic effect of an outage. The components of the index include the unemployment (*Unem*) rate in the local area, the inverse of the local area's bond rating (1/*Bondrating*), a measure of social strife in the local area (*SocialStrife*), and the crime rate in the local area (*Crime*). All of these social and economic factors taken together provide a basis for quantifying public perceptions. We postulate that the higher the unemployment rate in the local area, the greater the level of social dissatisfaction. Bond rating is used as a measure of the financial strength of the local area. The lower the bond rating, the less is the ability of the local area to provide local public goods and services. This, therefore, will have an impact on the level of social dissatisfaction and further worsen the impact of the outage. Social strife measures the degree of racial segregation of the local area. If the minority population with less energy usage or income are evenly distributed across the local area, the social strife measure would be zero. Therefore, the higher the social strife index, the higher the level of dissatisfaction. Our final component in our index is the crime rate in the local area. The higher the crime rate, the higher the dissatisfaction and the worse the social impact of the outage. Each component of this index is standardized by dividing the point estimates by their respective system-wide standard deviations. This new index can be utilized in the ranking of contingencies, and various technology and control system choices.

We define our index mathematically as follows. The factor I_i, the index of economic and social factors in local Area i, is used to create U_i, our measure of the level of public perception as follows:

$$U_i = Pop_i * e^{I_i} \tag{7.22}$$

$$I_i = \sqrt[4]{\frac{Unem_i}{\sigma_{unem}} \times \frac{1}{\dfrac{Bondrating_i}{\sigma_{Bondrating}}} \times \frac{SocialStrife_i}{\sigma_{socailstrife}} \times \frac{Crime_i}{\sigma_{crime}}} \tag{7.23}$$

$$f_{Ui} = \frac{1-e^{-kU_i}}{1+e^{-kU_i}} \tag{7.24}$$

where U_i or our public perception concept measures the level of dissatisfaction at the time of the outage occurrence in the local Area i;

Pop_i is the population in the local Area i;

I_i is the index of economic and social factors in the local Area i;

$Unem_i$ is the level of unemployment in the local Area i affected;

$Bondrating_i$ is a measure of the economic condition of the local Area i government;

$SocialStrife_i$ is a measure racial isolation in the local Area i;

$Crime_i$ is a measure of crime rate in the local Area i;

σ_{unem}, $\sigma_{bondrating}$, $\sigma_{socailstrife}$, and σ_{crime} are the system-wide standard deviations of unemployment, bond rating, social strife, and crime for the analysis.

7.3.3 Available Transmission Capability (ATC)

According to the North American Electric Reliability Council (NERC) definition, available transmission capability (ATC) is a measure of the transfer capability remaining in the physical transmission network for future commercial activity over and above already committed uses. The ATC is the viable increase in real power transfer from one point to another in a power system. It is a useful index of power transfer margin. The ATC is limited by thermal limits of transmission lines and transformers, voltage stability analysis for voltage limits, and transient stability analysis for stability limits.

ATC can be expressed as ATC = TTC − TRM − CBM − ETC

where:

TTC = the total transfer capability

TRM = the transmission reliability margin

CBM = the capacity benefit margin

ETC = the existing transmission commitments

The steps for determining the ATC is as follows:

1. Establish and solve the base case power flow for the period.
2. Select a transfer case.
3. Use continuous power flow (CPF) to make a step increase in transfer power.
4. Establish a power flow problem consisting of the base case modified by the cumulative increases in transfer power from step 3. Solve the power flow problem and check the solution for violations of operational physical limits.
5. If there are violations, decrease the transfer power to the minimum amount necessary to eliminate them.
6. Compute the ATC from the interface flows in the adjusted solution.
7. Return to Step 2 for the next transfer case.

7.3.4 Reliability Measures/Indices

Reliability assessment lends itself to the study of the generation/transmission systems and the distribution system. In the former case, also called the bulk power system, typical reliability indices include loss of load probability, and expected unserved energy. Table 7.1a below summarizes the indices related to the generation system. Table 7.1b summarizes some basic and derived reliability measures for the composite generation and transmission system. In the latter case, distribution system reliability indices are determined based on component failure rates, customer interruption statistical records, load-point failure rate, load-point outage duration, load-point annual unavailability, and several other factors. A summary of these indices is presented in Table 7.2. The next sections summarize common reliability measures used to assess energy delivery efficiency and load serving of the power system [17].

7.3.4.1 Reliability Indices in Generating Systems Adequacy Indices in Generating Systems are calculated using Monte Carlo methods and other approaches. The basic indices in generating system adequacy assessment include expected unserved energy, loss of load expectation, and several others as summarized in Table 7.1a.

TABLE 7.1a. Reliability Indices in Generation Systems.

Index	Definition
$EUE = \sum L_{a(i)} U_i$	**The Expected Unserved Energy (EUE):** the total energy not supplied by the system. Here, $L_{a(i)}$ is the average load connected to load point i.
$LOLE = \sum_{i \in S} p_i T$	**LOLE (days/year or hr/year)** where p_i is the probability of system state i and S is the set of all system states associated with loss of load. When the LOLE is expressed in days/year, pi depends on daily peak load and the available generating capacity. When it is in hr/year, p_i depends on a comparison between the hourly load and the available generating capacity. The LOLE index does not indicate the severity of the deficiency nor the frequency nor the duration of loss of load.
$LOEE = \sum_{i \in S} 8760 C_i p_i$	**LOEE (MWh/year)** where C_i is the loss of load for system state i. The LOEE index is the expected energy not supplied by the generating system due to the load demand exceeding the available generating capacity. The LOEE incorporates the severity of deficiencies in addition to the number of occasions and their duration, and therefore the impact of energy shortfalls as well as their likelihood is evaluated.
$LOLF = \sum_{i \in S} (F_i - f_i)$	**LOLF (occ./year)** where F_i is the frequency of departing system state i and f_i is the portion of F_i which corresponds to not going through the boundary wall between the loss-of-load state set.
$LOLD = \dfrac{LOLE}{LOLF}$	**LOLD (hr/disturbance)** Frequency and duration are a basic extension of the LOLE index in that they identify the expected frequency of encountering a deficiency and the expected duration of the deficiencies.

7.3.4.2 Reliability Indices in Generation/Transmission Systems Table 7.1b summarizes reliability indices for the generation/transmission (bulk power) system. The derived indices in Table 7.1b are useful when comparing adequacies of systems of different sizes. Overall, these indices can be calculated at the peak load and expressed as an annualized index, or by considering the annual load duration curve.

7.3.4.3 Reliability Indices in Distribution System Evaluation Continuous electric service has customarily meant meeting the customers' electric energy requirements as demanded. In order to calculate the cost of reliability, the cost of an outage must be determined, and computation of the unreliability index based on service interruptions and component failure rates at the distribution level is needed. As mentioned before, the three basic load-point factors in distribution system adequacy assessment relates to load-point failure rate, load-point outage duration, and load-point annual unavailability. These are used to formulate basic distribution reliability indices as shown in Table 7.2.

TABLE 7.1b. Reliability Indices in Generation/Transmission Systems.

Index	Definition
$PLC = \sum_{i \in S} p_i$	**PLC: Probability of Load Curtailment** where p_i is the probability of system state i and S is the set of all system associated with load curtailment.
$ENLC = \sum_{i \in S} F_i$ where $F_i = p_i \sum_{k \in N} \lambda_k$	**EFLC: Expected Frequency of Load Curtailment (occ./Year)** The ENLC is the sum of occurrences of load curtailment states and therefore an upper bound of the actual frequency index. (λ_k is the departure rate of the component corresponding to system state i and N is the set of all possible departure rates corresponding to state i.)
$EDLC = PLC * 8760$	**EDLC: Expected Duration of Load Curtailments (hr/Year)**
$ADLC = EDLC / EFLC$	**ADLC: Average Duration Load Curtailments (hr/year)**
$ELC = \sum_{i \in S} C_i F_i$	**ELC: Expected Load Curtailments (MW/Year)** where C_i is the load curtailment in system state i
$EDNS = \sum_{i \in S} C_i p_i$	**EDNS: Expected Demand Not Supplied (MW)**
$EENS = \sum_{i \in S} C_i F_i D_i = \sum_{i \in S} 8760 C_i p_i$	**EENS: Expected Energy Not Supplied (MWh/year)** where D_i is the duration of system state i.
$BPII = \sum_{i \in S} C_i F_i / L$	**BPII: Bulk Power Interruption Index (MW/MW-year)** Where L is the annual system peak load in MW. This index can be interpreted as the equivalent per unit interruption of the annual peak load. One complete system outage during peak load conditions contributes 1.0 to this index.
$BPECI = EENS/L$	BPECI: Bulk Power/Energy Curtailment Index (MWh/MW-year)
$BPACI = ELC/EFLC$	**BPACI: Bulk Power Supply Average MW Curtailment Index (MW/disturbance)**
$MBPCI = EDNS/L$	**MBPCI: Modified Bulk Power Curtailment Index (MW/MW)**
$SI = BPECI * 60$	**SI: Severity Index** This index can be interpreted as the equivalent duration in minute of the loss of all loads during the peak load conditions.

7.3.5 Expected Socially Unserved Energy (ESUE) and Load Loss

We created an index Y_i (a normalized dissatisfaction function equation) to measure the economic and social effects of an outage. Y_i is used to create our measure of the expected socially unserved energy (ESUE) [16]. Moreover, the real power loss on the transmission line can be derived from the power flow calculation.

TABLE 7.2. Reliability Indices in Distribution Systems.

Indices	Definition
$SAIFI = \frac{\sum_{i \in R} \lambda_i N_i}{\sum_{i \in R} N_i}$	**SAIFI: System Average Interruption Frequency Index (interruptions/system customer/year)** where λ_i and N_i are the failure rate and the number of customers at load point i respectively; R is the set of load points in the system.
$SAIDI = \frac{\sum_{i \in R} U_i N_i}{\sum_{i \in R} N_i}$	**SAIDI: System Average Interruption Duration Index (hr/system customer/year)** where U_i is the annual unavailability or outage time (in hr/year) at load point i
$CAIFI = \frac{\sum_{i \in R} \lambda_i N_i}{\sum_{i \in R} M_i}$	**CAIFI: Customer Average Interruption Frequency Index (interruptions/customer affected/year)** where M_i is the number of customers affected at load point i. The customers affected should be counted only once, regardless of the number of interruptions they may have experienced in the year.
$CAIDI = \frac{\sum_{i \in R} U_i N_i}{\sum_{i \in R} \lambda_i N_i} = \frac{SAIDI}{SAIFI}$	**CAIDI: Customer Average Interruption Duration Index (hr/customer interruption)**
$ASAI \frac{\sum_{i \in R} 8760 N_i - \sum_{i \in R} U_i N_i}{\sum_{i \in R} 8760 N_i}$	**ASAI: Average Service Availability Index**
$ASUI = \frac{\sum_{i \in R} U_i N_i}{\sum_{i \in R} 8760 N_i}$	**SAUI: Average Service Unavailability Index**
$ENS = \sum_{i \in R} p_{ai} U_i$	**ENS: Energy Not Supplied (kWh/year)** where p_{ai} is the average load in (kW) connected to load point i and U_i is the annual outage time (hr/year) at the load point.
$AENS = \frac{ENS}{\sum_{i \in R} N_i}$	**AENS: Average Energy Not Supplied (kWh/customer/year)**
$ACCI = \frac{ENS}{\sum_{i \in R} M_i}$	**ACCI: Average Customer Curtailment Index kWh/customer affected/year)**

The dynamic nature of Y_i is depicted in Figure 7.5. As observed in Figure 7.5, dissatisfaction curves differ from one local area to another. The level of dissatisfaction increases at a decreasing rate with respect to outage time, and the area with the highest dissatisfaction curve has a greater level of dissatisfaction. For example, Area 3 consumers in Figure 7.5 have a lower negative sensitivity to power outage than Area 1 and Area 2.

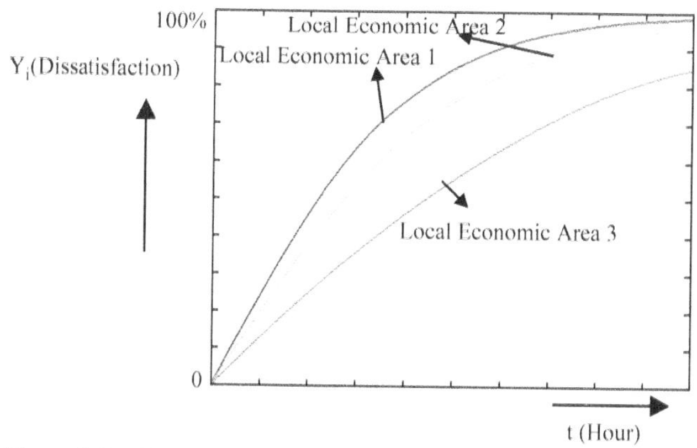

Figure 7.5. The dissatisfaction function.

Equation (7.25) shows the dissatisfaction of the selected city, and (7.26) is the ESUE of the selected city.

$$Y_i = \frac{1 - e^{-kt \cdot e^{l_i}}}{1 + e^{-kt \cdot e^{l_i}}} \tag{7.25}$$

$$ESUE_i = S_i \times \left(1 + \int_0^t Y_i dt\right) \tag{7.26}$$

where Y_i is the measure of normalized dissatisfaction,
t is the outage duration time,
S_i is the load level of area i,
$ESUE_i$ is the expected socially unserved energy of area i.

Real power loss is a basic computation in power systems, and it can be presented as following:

$$Ploss = \Delta P_{slack} \tag{7.27}$$

where $Ploss$ is the total real power loss of the system
ΔP_{slack} is the derivation of the slack bus

The impacts of contingencies can be compared from different aspects, technical and non-technical, and the new contingency screening and ranking index can be estimated. This new index incorporates social and economic factors.

7.3.6 System Performance Index

In order to capture various aspects of the impact of contingencies on power system and society, we developed an overall performance index. This index is created by weighting the line loss (Ploss), the available transmission capability (ATC) and expected socially unserved energy (ESUE) associated with the expected contingency. The higher the level of our performance index F_i, the greater the consideration needed. Our performance index is presented as follows:

$$F_i = w_1 Ploss_i - w_2 ATC_i + w_3 ESUE_i \tag{7.28}$$

where F_i is the overall performance index of contingency i,
$Ploss_i$ is the line loss of expected contingency i,
ATC_i is the ATC value of expected contingency i,
$ESUE_i$ is the expected socially unserved energy value of expected contingency i,
w_1, w_2, and w_3 are the weights for the indices respectively.

To choose the best weights, we calculate the weights of each area using linear programming, and choose the one that make the sum of F for the total system minimum.

$$\text{Min} \sum_{i \in C_j} F_i,\ C_j \text{ is the contingency set of area j} \tag{7.29}$$

$$\begin{aligned} \text{s.t.} \quad & Ploss_i w_1 - ATC_i w_2 + ESUE_i w_3 = F_i \\ & w_1 + w_2 + w_3 = 1 \\ & w_1, w_2, w_3 > 0 \end{aligned}$$

Here w_1, w_2, and w_3 are variables. Ploss, ATC, and ESUE are constants for each area.

7.3.7 Computation of Weighted Probability Index (WPI)

The weighted probability index (WPI) is an index used for ranking different scenarios and selecting the important and unimportant contingencies for further decision and control actions. The weighted probability index has been used in [17–18] for ranking voltage stability margin.

$$WPI_{ij} = w_i p_{ij} \tag{7.30}$$

where: w_i are weights reflecting the relative market, system, and social values of any particular system configuration.
p_{ij}: Power flow from bus i to bus j.

The *WPI* indices plays a major role in determining the need for voltage stability assessment in a network under stress or disturbance. The process involves comparing WPI against a set value of security threshold ρ. Depending on the result of comparison, the following decisions are made:

1. $WPI_{ij} > \rho$: Critically important (Stability computation is critically needed).
2. $WPI_{ij} = \rho$: Important (Stability computation is not necessary).
3. $WPI_{ij} < \rho$: Unimportant ones (Stability computation is not required).

A relationship is developed between WPI and a voltage stability index called expected voltage stability margin (EVSM), of a power system network. EVSM is a voltage stability index developed for evaluating the expected region of voltage stability in a power system. More concisely, the expected voltage stability margin (*EVSM*) can be defined as a mean value of the voltage stability margin determined for each probable important contingency and load level in the system. The *EVSM* concept expresses a general voltage stability "fitness" of the system for selected equipment outages and load levels.

Mathematically, VSM can be defined as follows:

$$EVSM = P_B + \sum_{i=1}^{n}\sum_{j=1}^{m} p_{ij} VSM_{ij} = P_B + \sum_{i=1}^{n}\sum_{j=1}^{m} p_i p_j VSM_{ij} \tag{7.31}$$

where:

B = base case values of the network parameters.

P_B = common probability of occurrence.

p_{ij} = Power flow from bus i to bus j.

VSM = Voltage stability margin, the definition of which can be found in [17].

A new definition of EVSM incorporating WPI is given as:

$$EVSM = EVSM + p_i \times p_j \times VSM_{ij} \quad \text{if} \quad WPI_{ij} > \rho \tag{7.32}$$

Obtaining the EVSM gives the region of stability of the system under stress and therefore a decision can be made as to whether or not the voltage stability margin should be increased for increased security.

The index WPI was developed at CESaC and has been used in several applications including ranking for expected voltage stability margin as highlighted.

7.4 DESIGN METHODOLOGIES

The flowchart of the design procedure for this study is given in Figure 7.6 in a modular form. It is broadly divided into three parts: (1) Modeling of components including FACTS Devices, (2) Contingency Evaluation, and (3) Impact Study/Analysis using the various new indices introduced in the course of this work. These indices include ESUE, Load Loss, WPI, EVSM, Performance index, and a Public Perception index. The test systems used for this study are WSCC and IEEE 30-bus system networks.

As seen in Figure 7.6 above, the first task here is to model the power system components for the utility power system. For the WSCC, the components are similar to those of the IEEE Test System; also in some cases, the modeling approach is scalable to those of the military (Navy) system models. The model includes Flexible AC Transmission System (FACTS) devices. These devices are potential control components in the network. The next step is to perform base-case load flow analysis to establish the operating limits of all network components. We then set contingencies depending on the type of impact study desired; the contingency can be in form of line loss, loss of load, or loss of generation. For each type of contingency, there is an associated effect on the overall network. This can be in form of stability, security, and reliability. For each contingency, we perform an impact analysis using the newly created indices. These indices are capable of handling uncertainties in the system and enable us to incorporate social and economic factors.

The results of these impact studies are used to evaluate the system performance under different contingencies and are used to recommend what type(s) of control action or operational planning needed:

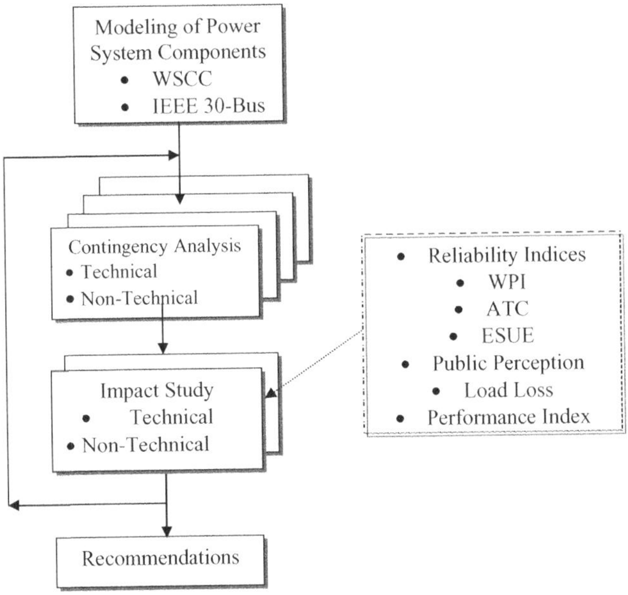

Figure 7.6. Implementation Flowchart.

7.5 IMPLEMENTATION APPROACH

7.5.1 Load Flow Analysis with FACTS Devices (TCSC) for WSCC System

This study includes solving the base-case load flow for the WSCC power system network. The aim is to observe the systems normal operating condition and the behavior under perturbed condition (under disturbance or contingency scenario). The efforts here also highlight the advantages of including FACTS devices in the power system network.

The following simple steps are followed in carrying out this analysis:

1. Convert the WSCC data from the IEEE format to PSAT format.
2. Run the base-case load flow and note the operational limits of all system components.
3. Perturb the system by increasing the load levels by 5%.
4. Note the types and locations of violations resulted from this load increase.
5. Based on the type and location of violation, carefully select the type of FACTS device needed for improvement.
6. Determine the location of the FACTS device needed to achieve the level of improvement desired.
7. Determine the amount of control variable of the FACTS device required to achieve the desired level of improvement in the network under the load change.
8. Evaluate and compare the performance with and without FACTS device.

7.5.2 Performance Evaluation Studies on IEEE 30-Bus and WSCC Systems

The series of steps involved in computing the overall performance index for the system is given below:

1. Compute the base-case load flow to establish the operating limits of the network components.
2. Perform contingency analysis (such as losing a line or generator or load, or load increase or generation decrease).
3. Perform contingency filtering to determine which one leads to violation of limits.
4. For those contingencies leading to violation, compute the total power loss.
5. Compute the available transmission capacity (ATC).
6. Compute the expected socially unserved energy (ESUE).
7. Compute the overall performance by solving the resulting optimization problem of Equations 7.28 through 7.29 repeated here for convenience.

7.6 IMPLEMENTATION RESULTS

7.6.1 Load Flow Analysis with FACTS Devices (TCSC) for WSCC System

The aim of this effort is to illustrate the advantages of including FACTS devices in a power system. The suitability of FACTS devices depend largely on the impact of the contingency on the system under study. Thus, a careful selection of FACTS device will help in achieving the objective in one area while not creating a different type of problem in another area. In this work, the load flow analysis was performed for the WSCC (slightly modified) network. The result of the base-case load flow was obtained with all components operating within their established limits. The system was then perturbed by increasing the system load by 5%. The load flow for this load level resulted in voltage limit violations at 30 buses and reactive power limit violations at 8 buses. Figure 7.7 below shows the bus voltages for the normal case and that of 5% load increase. The "acceptable region" indicates the region within which the bus voltages are within set limits. The upper bound is 1.1 p.u and the lower bound is 0.9 p.u. Outside of this range the bus voltage is said to have violated its limit. It is advisable to keep the bus voltages within the limit to avoid instability and other problems that are voltage related. For base case results, we see that all bus voltages are within the limits and thus the network is in good operating condition. However, with 5% load increase on all load buses, all the voltages have overstepped their boundaries. The increase or decrease in the voltages is either due to the excessive increase of reactive power compared to the real power or vice-versa.

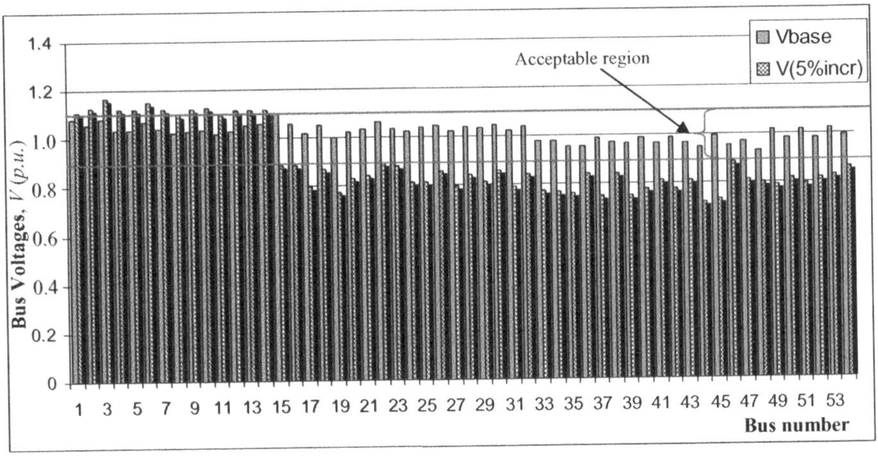

Figure 7.7. Bus voltages for the normal case and 5% load increase.

TABLE 7.3. Summary of the Load Flow with the Base Case, 5% Load Increase and 5% Load Increase with TCSC.

	Base Case Results	5% Load Increase on all Load Buses	5% Load Increase with TCSC on Line 128–129
Total Real Power Generated	615.6225	643.6423	647.3604
Total Reactive Power Generated	128.0566	220.8201	195.8978
Total Real Load Served	607.3733	629.5117	631.7046
Total Reactive Load Supplied	153.5126	132.6325	134.7087
Total Real Power Loss	8.2492	14.1306	15.6558
Total Reactive Power Loss	16.7016	129.9595	103.0261
Total No of Buses with Voltage Violations	0	30	17

In order to improve the network performance, a TCSC device was inserted in series with line 128–129 of the WSCC power system. The value of the control parameter for the TCSC was varied until no further improvement could be made. At this point, the number of bus voltage violations has reduced from 30 to 17. In addition, the active and reactive power losses have reduced from (14.1306 + j129.9595) to (15.6558 + j103.0261). By placing one or two more carefully selected FACTS devices in carefully selected locations in the system, we can remove all the violations and restore the system completely to normalcy. Table 7.3 shows the summary of the load flow with the base case, 5% load increase and 5% load increase with TCSC.

In conclusion, this segment of the work demonstrates the importance of FACTS devices in alleviating problems arising from contingencies in an electric power system. Some of these problems could be voltage stability/instability, reactive power generation, real and reactive power losses and transient (angle) stability analysis of the power system. FACTS devices when carefully selected and put in the right location can improve the performance of the system under stress largely.

7.6.2 Performance Evaluation Studies on IEEE 30-Bus System

In this section, the most urgent contingencies are selected using the newly created performance index. Figure 7.8 is the modified IEEE 30-bus test system. It has 41 branches, five generators, four phase shifters and 37 switches.

From different areas, expected contingencies are selected. In Area 1, Line 15–18, 10–21, and 12–14 are outages. In Area 2, Line 1–3, 6–7, and 2–6 are outages. In Area 3, Line 27–29, 8–28, and 6–8 are outages.

For this analysis, the data used are not the standard IEEE 30-bus system data. Therefore, we make the new assumptions displayed in Table 7.4. These assumptions can be changed for different areas in the system. In accordance with the above assumptions, the factor vectors for different contingencies in different areas are shown in Table 7.5. The last column in Table 7.5 is the weighted sum of the factors. Here we assume that all the weights are the same. The values of temperature, average wage level, and the measure of dissatisfaction are presented by f_t, f_R, and f_U

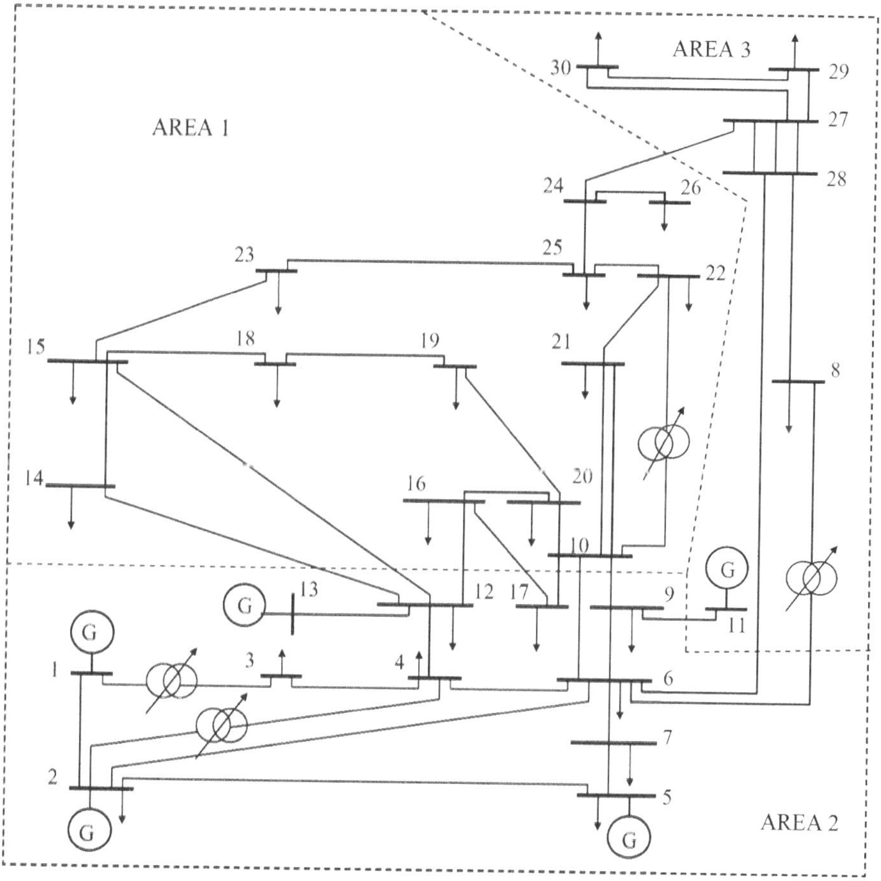

Figure 7.8 Modified IEEE 30-Bus test system.

respectively. Using the approach in [16], the values of f_t, f_R, and f_U are obtained as shown in Table 7.5.

From Table 7.5, we can rank the contingencies in the same areas. In this analysis, we found that, the higher the sum of the factors, the more valuable the contingency. After calculating the factors and ranking the contingencies, load flow was performed under the prioritized contingencies. Using equation (7.29), the weights we use in equation (7.28) are determined. The weights and the overall performance index, F, are presented in Table 7.6. From Table 7.6, the weights in Area 2 were chosen as the optimal weights since they minimize the sum of the $F_i's$. Therefore, these weights are chosen as the optimum weights to be used in Eq. (7.19). The real power losses, ATCs, and ESUEs of expected contingencies are shown in Table 7.7.

TABLE 7.4. Assumptions for IEEE 30-bus Calculations.

	Abnormal Days	Ri	Population (Million)	Sensitivity	Load Level(p.u.)
Area 1	>200	>1.2 & <0.8	<0.5	<0.3	<1
Area 2	>100 & <200	>1.2	>3	>0.3 & <0.7	>1.8
Area 3	<100	<0.8	>0.5 & <3	>0.7	>1 & <1.8

TABLE 7.5. The Factors of Contingencies.

Contingencies		f_t	f_R	f_U	Sum
Area 1	15–18	0.0649	0.0050	0.0719	0.1418
	10–21	0.0500	0.0100	0.1076	0.1676
	12–14	0.0549	0.0100	0.0790	0.1439
Area 2	1–3	0.0500	0.0300	0.1955	0.2755
	6–7	0.0350	0.0200	0.2041	0.2591
	2–6	0.0400	0.0200	0.2127	0.2727
Area 3	27–29	0.0400	0.0050	0.1253	0.1703
	8–28	0.0450	0.0060	0.1150	0.1660
	6–8	0.0500	0.0075	0.1357	0.1932

TABLE 7.6. Comparison of Weights and Overall Performance index, F.

Areas	w_1	w_2	w_3	F
Area 1	0.6533	0.2467	0.1000	75.6187
Area 2	0.1000	0.6490	0.2510	−104.5715
Area 3	0.6529	0.2471	0.1000	75.2168

TABLE 7.7. Results Comparison.

Contingency		P_{loss}(MW)	ATC(MW)	ESUE(MW)	F(MW)
Area 1	15–18	17.7668	134.8845	216.7620	–31.3561
	10–21	18.0279	92.6068	219.7230	–3.14855
	12–14	17.8831	93.3864	243.3270	2.255613
Area 2	1–3	27.4012	70.0000	206.8950	9.240765
	6–7	19.3532	85.1320	212.3550	–0.01424
	2–6	20.3738	81.9170	204.9840	0.324231
Area 3	27–29	18.0411	91.4413	124.7040	–26.2406
	8–28	17.6549	93.6677	116.1840	–29.8627
	6–8	18.6311	86.5466	113.6880	–25.7699

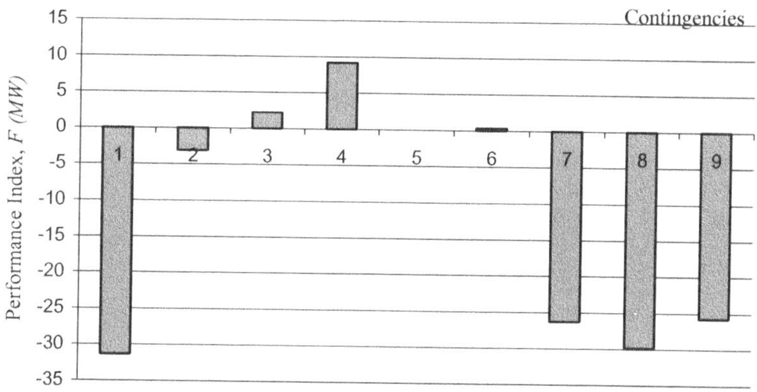

Figure 7.9. Overall performance index comparison for the 9 contingencies on the IEEE 30-Bus System.

The F of expected contingency 1–3 is largest, which means this contingency has the most technical and non-technical impact on the system when real power loss, ATC, and ESUE are utilized as indices to evaluate the expected contingencies. The plot of F for the different contingencies is given in Figure 7.9.

7.6.3 Performance Evaluation Studies on the WSCC System

As stated earlier, performance evaluation is obtained as a function of Power Transmission Loss (Ploss), Available Transmission Capability (ATC) and Expected Socially Unserved Energy (ESUE) associated with the expected contingency. That is $F_i = w_1 Ploss_i - w_2 ATC_i + w_3 ESUE_i$.

The test system used for this evaluation is the WSCC system described earlier. The Western Systems Coordinating Council (WSCC) system has 179 buses and 263 branches. For this analysis, it is split in four areas as shown in Figure 7.10. We break three lines in each area as contingencies.

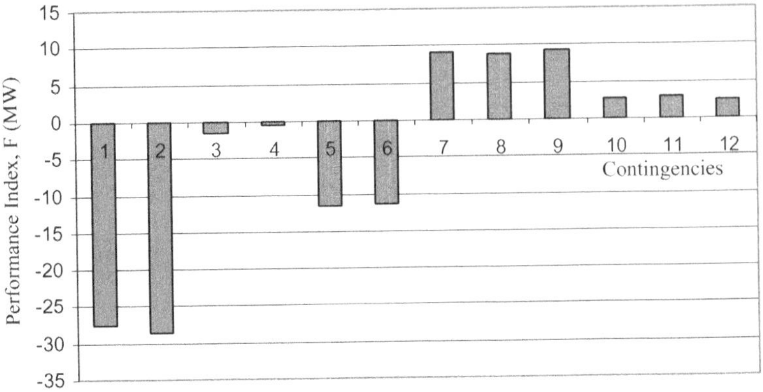

Figure 7.10. *F* comparison for WSCC network.

TABLE 7.8. The Factors of Contingencies.

Contingency		f_I	f_R	f_U	Sum
Area 1	78–80	0.216804	0.037433	0.02899	0.2832263
	74–78	0.310366	0.05933	0.059906	0.4296024
	75–78	0.113706	0.099519	0.060104	0.2733298
Area 2	89–90	0.296018	0.043223	0.04055	0.3797917
	117–119	0.376822	0.041426	0.049609	0.4678568
	100–101	0.245577	0.051903	0.442136	0.7396164
Area 3	50–57	0.147467	0.033937	0.056798	0.2382021
	48–62	0.511604	0.045718	0.029161	0.5864833
	150–154	0.17144	0.04911	0.110869	0.3314196
Area 4	163–164	0.17144	0.038681	0.024332	0.2344533
	14–24	0.245577	0.051903	0.022267	0.3197474
	158–166	0.341598	0.054147	0.023281	0.4190257

TABLE 7.9. The Comparison of Weights and F.

	W_1	w_2	w_3	F
Area 1	0.1100	0.7900	0.1000	–57.7154
Area 2	0.1322	0.7513	0.1165	55.6154
Area 3	0.102	0.7980	0.1000	9.962658
Area 4	0.1101	0.7890	0.1009	12.59068

The performance factors are obtained as in Table 7.8 following the approach of [16]. This table is used in ranking the weights. The weights are shown in Table 7.9. From Table 7.9, we find that the weights in Area 1 have the minimum F value. Using these weights, a load flow study was performed for different contingencies. The results are in Table 7.10. Figure 7.10 shows the plot of F for different contin-

TABLE 7.10. Results Comparison.

Contingency		Ploss(MW)	ATC(MW)	ESUE(MW)	F(WM)
Area 1	78–80	584.26	152.2794	283.9994	−27.6327
	74–78	574.92	152.1777	283.744	−28.6048
	75–78	734.90	140.0000	282.8257	−1.47843
Area 2	89–90	619.98	144.4758	453.1252	−0.62556
	117–119	577.39	152.5410	453.8898	−11.6055
	100–101	575.77	152.1341	454.8164	−11.3696
Area 3	50–57	574.80	152.4247	663.5158	9.164067
	48–62	575.81	152.4102	660.2736	8.962402
	150–154	574.49	151.9307	661.2247	9.291117
Area 4	163–164	581.98	148.9807	589.6479	2.785196
	14–24	584.83	152.1486	588.2118	2.955086
	158–166	578.39	152.4102	589.1058	2.336086

gencies. The overall performance for loss of 150–154 is much more serious than other contingencies. Expected contingencies 50–57 and 48–62 are also more serious but less serious than 150–154. In this case, we assume the decision maker is more concerned about public perception and assigns more weight to that factor. In other words, the social effect of the contingency is given.

A new index for ranking contingencies without a load flow study was developed in this section. It incorporates social and economic factors to rank any set of expected contingencies. The uniqueness of our approach stems from the inclusion of specific social and economic measures such as unemployment rate, social strife and crime as well as the economic conditions of the affected area. This method allows decision makers to grasp the most important contingencies. Our test results on sample power systems demonstrate the implementation of our method. The factors used to construct the index are by no means exhaustive. Other factors in constructing a similar index for ranking and evaluating the contingencies may be employed.

7.7 CONCLUSION

In this chapter, we model various power system components and FACTS devices. A load flow analysis was performed for WSCC for normal case and with 5% increase in the load level. With TCSC included, the load flow was repeated for the system with 5% increase in load level. The results show a marked improvement on the voltage levels and reactive power losses in the system. This effort was to demonstrate the importance of FACTS devices in solving problems due to contingencies in an electric power system using the WSCC as an example. Furthermore, we developed a new index for ranking contingencies, which incorporates economic and social factors, recognizing that different areas have different economic and social conditions. These conditions must be accounted for in any impact analysis.

As a proxy for economic losses, we use a number of factors such as relative wages, bond rating, social strife index, crime rate, and unemployment rate. The purpose of this index is to reflect public acceptance and cooperation during power outages. While the data we used was based on real data from cities and states in the WSCC, our analysis is yet experimental.

Due to the lack of micro data at the contingency level, the economic data could not be systematically matched to each contingency. In our analysis, each contingency was therefore matched with data from randomly selected cities. In future research, micro data will be developed at the contingency level.

Contingency evaluation was performed on the IEEE 30-bus and WSCC using the new performance index which incorporates all the social and economic measures (unemployment rate, social strife, crime rate and the economic conditions of the affected area). This performance index provides decision makers a unique way of handling contingencies in a prioritized manner for improved efficiency and effectiveness of decision making as needed.

ACKNOWLEDGMENTS

The work was supported in part by the National Science Foundation (NSF) under grant #0224873 and by a grant from the Office of Naval Research (ONR) grant #N-00014-04-1-0255.

BIBLIOGRAPHY

[1] Barquin, J., Gomez, T., and Luis F. P., "Estimating the Loading Limit Margin Taking Into Account Voltage Collapse Areas," *IEEE Transactions on Power Systems*, vol. 10, Nov. 1995.

[2] Chiang, H. D., Wang, C., and Flueck, A. J., "Look-ahead voltage and load margin contingency selection functions for large-scale power system," *IEEE Transactions on Power Systems*, vol. 12, pp. 173–180, Feb. 1997.

[3] Chiang, H. D., Li, H., and Sauer, P. W., "Toward the development of an on-line ATC evaluation tool," in *Proc. Of 1998 North American Power Symposium*, pp. 311–316.

[4] Chichilnisky, G., and Heal, G., "Global Environmental Risks," *Journal of Economic Perspectives*, vol. 7, pp. 65–86, 1993.

[5] EPRI, The Electric local area Sector Framework for the Future: Achieving the 21st Century Transformation, 2003.

[6] EPRI, and BPA, Framework for Stochastic Reliability of Bulk Power System with Stability Constraints, EPRI, Palo Alto, CA, and Bonneville Power Administration, Vancouver, WA: 2004. 1009506.

[7] Ejebe, G. C., Irisarri, G. D., Mokhtari, S., Obadina, O., Ristanovie, P., and Tong, J. "Methods for Contingency Screening and Ranking for Voltage Stability Analysis of Power Systems," *IEEE Transactions on Power Systems*, vol. 11, pp. 350–356, Feb. 1996.

[8] Fouad, A. A., and Vittal, V., Power System Transient Stability Analysis Using The Transient Energy Function Method, Prentice Hall, 1992.

[9] Fu, C., and Bose, A., "Contingency Ranking Based on Severity Indices in Dynamic Security Analysis," *IEEE Transactions on Power Systems*, vol. 14, pp.980–986, Aug. 1999.

[10] Feng, Z., Ajjarapu, V., and Long, B., "Identification of Voltage Collapse through Direct Equilibrium Tracing," *IEEE Transactions on Power Systems*, Vol. 15, No 1, February. 2000.

[11] Heal, G. M., "Interaction between Economy and Climate: A Framework for Policy Design under Uncertainty," in *Advances in Applied Microeconomics*, Greenwich: J.A.I. Press, 1984, pp. 151–158.

[12] Hanemann, M. W., "Willingness to Pay and Willingness to Accept: How Much Can They Differ?" *American Economic Review*, Vol. 81, No.3 635–647, 1991.

[13] Kahneman, D., Knetsch, J. L., and Richard H. T., "Anomalies: The Endowment Effect, Loss Aversion, and Status Quo Bias" *Journal of Economic Perspectives*, Vol. 5, No. 1, 193–206, 1991.

[14] Kang, S. W., and Meliopoulos, A. P., "Contingency Selection Via Quadratized Power Flow Sensitivity Analysis," in *proc. 2002 IEEE Power Engineering Society Summer Meeting*, vol. 3, pp. 1494–1499, July. 2002.

[15] Li, W., and Bose, A., "a Coherency based Rescheduling Method for Dynamic Security," *Proceeding of the 20th International Conference on Power Industry Computer Applications*, Columns, Ohio, pp. 254–259, 1997.

[16] Momoh, J. A., Fanara Jr., P., Kurban, Haydar, Iwarere, L. J., and Zhang, Y., "Social Impact Based Contingency Screening and Ranking," Working paper EPNES project, Submitted to Journal of Critical Infrastructure, 2005.

[17] Momoh, J. A., Makarov, Y. V., Elfayoumy, M., and Mittelstadt, W., "Expected Voltage Stability Margin," Proc. of the Large Engineering Systems Conference on Power Engineering, Halifax, Nova Scotia, Canada, June 7–9, 1998.

[18] Momoh, J. A., Makarov, Y. V., and Mittelstadt, W., "A Framework of Stability Assessment in Power System Reliability Analysis," *IEEE Transactions on Power Systems*, vol. 14, pp. 454–491, May. 1999.

[19] Pindyck, R. S., "Irreversibility, Uncertainty, and Investment," *Journal of Economic Literarure*, XXIX: 3, pp.1110–1148, September 1991.

[20] Solow, R. M., *Sustainability: an Economist's Perspective*, the Eighteenth J. Seward Johnson Lecture, Woods Hole Oceanographic Institution, Woods Hole, Massachusetts, 1992.

[21] Schmalensee, R., "Option Demand and Consumer's Surplus: Valuing Price Changes under Uncertainty," *American Economic Review*, 62:5, pp. 813–824, December. 1972.

[22] Sutherland, R. J., *Estimating the Benefits of Restructuring Electric local area Markets: An Application to the PJM Region, Version 1.1* Center for the Advancement of Energy Markets, October. 2003.

CHAPTER 8

RISK-BASED POWER SYSTEM PLANNING INTEGRATING SOCIAL AND ECONOMIC DIRECT AND INDIRECT COSTS

Lamine Mili[1] *and Kevin Dooley*[2]

[1]Virginia Tech, [2]Arizona State University

EDITORS' SUMMARY: This chapter investigates the decision making processes associated with the risk assessment and management of bulk power transmission systems under a unified methodological framework of security and survivability objectives. First, it is presented as a partitioned multiobjective risk method aimed at finding tradoffs between N-1 security and survivability to catastrophic events, namely between various levels of resiliency ranging from low to high damage severity while minimizing the cost of the design. In addition, a method is proposed that assesses the risk of cascading events using a probabilistic algorithm that pinpoints the weak links of the network by simulating the propagation of the failures throughout the transmission network due to hidden failures in relays. These weak branches are the system components in which special controllers are to be installed. Because the risk of a system failure, viewed as an event, is defined as the probability times the consequence of this event, an assessment of the consequence of this event must be carried out, for instance by means of its costs. In this chapter, these costs are defined as consisting of technical, business, and social costs, and a method for estimating them proposed. In order to examine social costs, a novel form of text analysis is proposed. This method examines media coverage of two high-profile power failures—the California crises and of the 2003 U.S. blackout. It is found that in both cases power system failure stories do not generate sustained public interest regardless of their magnitude, implying that technical costs thus do not have to be explicitly taken into account when doing risk assessments.

Economic Market Design and Planning for Electric Power Systems, Edited by James Momoh and Lamine Mili

8.1 INTRODUCTION

Critical infrastructures can be regarded as the backbone of the economy of a country since they provide the material support for the delivery of basic services to all the segments of a society. These services include fresh water supply, fuel and electric energy supply, communication routes, telecommunication services, the Internet, to cite a few. These are complex, large-scale, networked systems characterized by a strong reliance on each other via interdependencies that are mainly induced by the growing usage of computer networks for data processing and for command and control functions. Consequently, the vulnerability and security of these stratified networks are raising major concerns worldwide. For instance, the normal operation of water, telecommunications and banking systems is maintained only if there is a steady supply of electric energy. On the other hand, the generation and delivery of electric power cannot be ensured without the provision to the power plants and the power networks of fuel, water, and various telecommunication and computer services for data transfer and control purposes.

Evidently, the interdependencies that exist among critical infrastructures may turn a local disturbance in one of them into a large-scale failure via cascading events, which may have a catastrophic impact on the whole society. One example is the loss in May 20 1998 of the Galaxy IV telecommunications satellite, which not only resulted in the outage of about 90% of all pagers in the U.S., but also disturbed credit card purchases and ATM transactions [39]. Another example is the series of blackouts that struck the North American power systems during the last ten years, which include the blackout of July 1996 that hit the western part of the United States, leaving 2.2 million customers without electricity; the blackout of August 1996 that affected eleven U.S. Western states and two Canadian provinces [19]; and the Northeaster blackout of August 2003 that deprived more than 50 millions of customers from electric power. It is interesting to note that in all these examples, a system failure consists of a sequence of cascading local failures that originates from the faulted branch and spreads sequentially from one location to another over an increasingly larger region of the system. Incidentally, the failures will propagate further when the system or one of its parts is operated near the limit of its capacity. In this case, the outage or the congestion of few branches will induce the loss and/or the congestion of a growing number of system components in a cascading manner.

Catastrophic failures are low probability/high consequence events, termed extreme events. Designing the critical infrastructures to be resilient to these extreme events call for the execution of a two-level hierarchical design approach as depicted in Figure 8.1. The first level consists of the design of a robust structure of the system and the design of a monitoring and control scheme during emergency conditions. The second level will integrate these two designs into a coherent scheme. One important question that the designer needs to address is the following: Where to reinforce a networked system and which real-time control actions should be taken to confine the failure to a small region? It is clear that one good decision would be to reinforce, during the design phase, each system component (i.e. link or node) whose failure most likely leads to a catastrophic system disturbance. Such a component is termed a weak point of the network, known also as a critical point or hot spot. Another good decision would be to design and implement real-time control schemes that act on the system during an emergency state, the occurrence of which has been detected by a dedicated alarm mechanism. For example, these controllers could alleviate the system stress by disconnecting a certain number of customers either on a voluntary basis or from a priority list established beforehand. They could also bar the failure from propagating to healthy regions. In power systems, this is achieved by breaking up the system into disconnected islands, whose frontier may change from case to case as the location of the failure and the operating point of the system change. In computer networks, a data packet infected by a virus or a worm may be confined to a sub-network through fire walls.

Monitoring and reacting to risk events, while necessary, is a poor strategy by itself compared to a more proactive strategy of planning for risk and developing safer systems. The basic design investment decision is: will the additional costs associated with enhancing system safety be less than the expected costs associated with failure if such enhancements are not made? Because of the integrated nature of the system, design decisions must consider the broad impact of catastrophic failure when analyzing different design alternatives. At the equipment level, components can be chosen that are more or less reliable, more or less easy to detect imminent failure within, and more or less easy to bring back into operation once failed. Likewise, at the architectural level, different system designs will respond to failure differently, depending on redundancies that have built into the system design and how transparent cause and effect are [35]. Additionally, safety systems may actually mask the root cause of failure, making the system actually less reliable than if no safety system were there at all [40]. All of these complexities compel us to examine the decision processes associated with risk assessment of power systems.

The first part of this chapter shall present a partitioned multiobjective risk method that will allow the power system planner to achieve risk-based tradeoffs between conflicting reliability and survivability objectives, namely between resiliency to frequent failures of small and moderate damage severity (reliability issue) and resistance against rare cascading failures of high damage severity (survivability issue) subject to a upper bound on the cost of the expansion. Specifically, the methodology that we propose assesses the risk of cascading events leading to blackouts by explicitly modeling the unduly actions of the protection systems due to hidden failures in a probabilistic manner [32, 46]. This alleviate the major drawback of the current practice in power system planning that typically disregards from the analysis the occurrence of multiple contingencies because their investigation is perceived as being impossible to achieve due to the huge number of cases that need to be investigated. In addition, the Expected Loss-of-Load (ELOL) index on which the

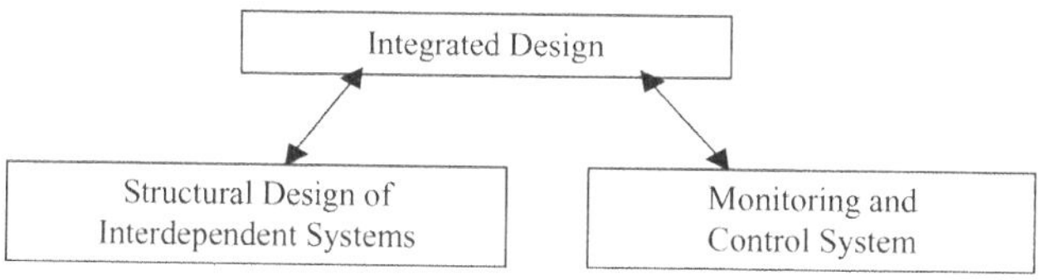

Figure 8.1. A two-level hierarchical design of a critical infrastructure aimed at providing a given level of resiliency against extreme events.

expansion is based is an average index that does not account for the small probability but high consequence associated with extreme events. Consequently, the risk of blackouts is neither assessed nor mitigated.

The second part of the chapter examines how we can estimate the costs associated with power system failure, which is a necessary data component within the partitioned risk assessment method. Failure costs are formulated to consist of three components: technical, business, and social. As has been discussed, the business costs associated with failure may far outweigh the direct technical costs; for example, it is commonly estimated that large high-tech manufacturing firms lose up to one million dollars per minute of downtime induced by energy shortages. We shall discuss how these business costs can be associated to the more readily-available metric of lost load. Specifically we shall show that business costs are associated with the scope of the failure and the time to recovery, and that these are directly related to total lost load, i.e. how much electrical load was not delivered due to the failure. Finally, we shall examine whether social costs scale linearly or nonlinearly with technical and business costs. Social costs are unknown and largely unknowable; if social costs scale linearly with the other costs, then they do not have to be explicitly estimated, but if social costs scale nonlinearly with technical and business costs, then additional parameters may be needed within the decision-theoretic framework in order to make reasonable judgments about failure costs. We will use a novel form of text analysis, Centering Resonance Analysis, to examine media coverage of two failures—the U.S. blackout in 2003 and the California energy crisis during 1999–2002—and from the text analysis infer whether social costs scale linearly or nonlinearly with the scope and time of the failure.

8.2 THE PARTITIONED MULTIOBJECTIVE RISK METHOD [31]

A partitioned multiobjective risk method allows us to design cost-effective resilient power-communication infrastructures to man-made or natural hazards. To this end, we assume that the failure rates of system components (including the probabilities of failures that are exposed only during a fault, termed hidden failures) are known since they can be estimated from historical data. Component failure rates are routinely estimated by the companies and agencies operating critical infrastructure for reliability assessment purposes. However, frequency estimation of some man-made hazards (e.g.: intentional sabotage, non-intentional human error) cannot be extrapolated into the future because the nature and magnitude of these hazards depend heavily on the unique political, social, economic and organizational environment in which they occur. It is then reasonable to assume that the probabilities of man-made hazards are unknown. Furthermore, due to the climate changes taking place throughout the globe, the frequency estimation of the extreme natural hazards from historical events is plagued with large uncertainties. Consequently, decision analysis is carried out in a situation where some of the nature-generated failures have known probabilities, while extreme natural hazards and man-made failures have unknown probabilities.

Because the services provided by critical infrastructures are vital to the economy and health of a nation, and their interruptions might have adverse large-scale social impacts, it is reasonable to assume that the decision-maker is risk averse and will adopt a mini-max strategy aimed at preventing the worst-case scenario. The goal of this strategy is to minimize the maximum conditional risk of a catastrophic failure over all possible actions that the decision-maker might take subject to limits on the costs of the design. Conditional risk is here defined as the product of the conditional probability of a cascading failure by its severity. Formally, we have:

$$\begin{aligned} &\min_i \, (\max_j u_{ij}), \\ \text{subject to} \quad & c_i \leq b \qquad \text{for } i = 1, \ldots, I, \end{aligned} \tag{8.1}$$

where u_{ij} is the conditional risk of the jth catastrophic failure under the ith action, which has a cost c_i, and where I is the total number of actions.

While we may define a catastrophic failure as a failure whose severity is larger than a given threshold, fixed a priori by the decision-maker, it is not clear how to measure the severity of a failure. This calculation presents a research challenge because, unlike the direct impacts of a failure, such as damage to equipment and fatalities, the indirect impacts are not easily quantifiable, especially if they include impacts such as business interruptions and human suffering due to psychological stress or adverse health impacts. Some costs are not borne nor considered by the decision maker. In addition, the minimax criterion is aimed only at providing robustness against extreme events. As a result, it does not guarantee good performance of the design under events with moderate severity or under normal operating conditions of the system. Therefore, an appropriate trade-off between conflicting objectives must be determined. We use a partitioned multiobjective risk method to address this issue.

The frequencies of most natural hazards can be reasonably estimated from historical data and extrapolated, at least in the short-term. Hazards include normal equipment failures and short-circuits on power substations or lines, to cite a few. While they may, under the right conditions, result in cascading failures, the minimax criterion should not be used for these routine events since it produces an overly conservative system design. In fact, as proposed by Haimes [18], a better criterion would be to minimize the conditional expected-value risk functions, f_i, $i = 1, \ldots, n$, where n is the number of the partitioned regions of the damage severity. As depicted in Figure 8.2, we typically consider for the design d_j, three ranges of severity, namely, low, moderate, and extreme ranges, which are denoted by S_{1j}, S_{2j}, S_{3j}, respectively. Assuming that the severity is a random variable X with a cumulative probability distribution function $P(x; d_j)$ and a probability density function $p(x; d_j)$, the partition of the severity maps a similar partition of the exceedance probability defined as $1 - P(x; d_j)$. Consequently, the conditional risk functions are written as:

$$f_i(d_j) = E[X | p(x; d_j), x \in S_{ij}], i = 1, 2, 3; j = 1, \ldots, m. \tag{8.2}$$

We can define the unconditional expected-value risk function $f_4(d_j)$ for the design d_j as the weighted sum of the conditional risk functions, $f_1(d_j), f_2(d_j), f_3(d_j)$, with positive weights w_1, w_2, and w_3, respectively. Hence, we have:

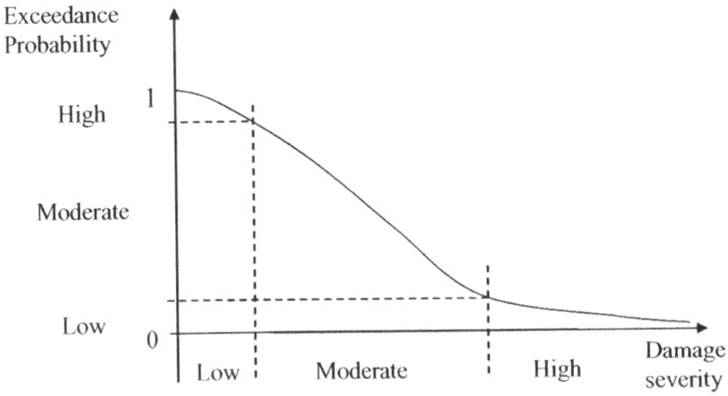

Figure 8.2. Design based on the partitioning of damage severity method.

$$f_4(d_j) = w_1\ f_1(d_j) + w_2\ f_2(d_j) + w_3\ f_3(d_j),\ \text{subject to } w_1 + w_2 + w_3 = 1 \quad (8.3)$$

By letting $f_5(d_j)$ denote the cost of the design d_j, the optimal design is then defined as the solution to a multiobjective optimization problem expressed as:

$$\min[f_i, f_5],\ i = 1, \ldots, 4 \quad (8.4)$$

As indicated by Haimes [18], the advantage of the optimization given by (8.4) over other alternatives is that it allows the decision maker to use the weights w_1, w_2, and w_3, to evaluate tradeoffs between the marginal costs associated with unit increments of the risk functions. Due to the high sensitivity of the expected catastrophic risk to the chosen partitioning of the damage severity, the final decision cannot be reached without carrying out a sensitivity analysis. The latter requires the modeling of the tails of the probability distribution, making use of the statistics of extremes [2, 17, 25]. Note that the tails of these distribution are much longer than those of the Gaussian distribution as shown for example by a study carried out by Doyle of California Institute of Technology (see Amin [3]) based on the analysis of the major disturbances in electric power systems reported in NERC [33].

8.3 PARTITIONED MULTIOBJECTIVE RISK METHOD APPLIED TO POWER SYSTEM PLANNING [31]

The partitioned multiobjective risk method is being applied to electric power systems planning. Presently, power generation and transmission expansions are typically carried out in two sequential phases, namely the adequacy phase followed by the security analysis phase. In the adequacy phase, system planners investigate several scenarios of generation expansion under the assumption that the transmission network has an infinite capacity. Upon completion of the investigation, they pick the scenario for which the generation expansion meets the load growth up to the planning horizon with a sufficiently low conditional expected probability of not meeting the load for each of the years being considered, termed the Expected Loss-of-Load (ELOL) index. This index may be regarded as the average number of days per year where the total system load is not met should a single outage of any of the generating units occurs. Once the adequacy phase is completed, the security phase is then executed. The latter consists in verifying whether the transmission network, to which is connected the planned generating units, is able to withstand the loss of a single piece of equipment (a line, a transformer, a generating unit, a power electronic device, to cite a few). Termed N-1 security analysis, this study obviously does not account for the occurrence of cascading failures leading to blackouts or brownouts. Both phases may be carried out simultaneously through the so-called composite reliability study using Monte Carlo methods [4].

Until recently, large-scale blackouts are considered as being sufficiently rare events to be disregarded from the analysis. However, at least in the U.S., ideas are evolving in this respect, prompted by the increasing number of major incidents that plagued the U.S. power systems since mid-nineties. The frequency of major blackouts, which was about one per decade until 1996, start to grow at an alarming rate since then [3, 19], and culminating in the 2003 North East blackout. The latter affected 50 million people located in eight U.S. states and the Canadian province of Ontario (US-Canada Task Force report [47]). Several causes have been identified. First, only very slow expansion of the high voltage transmission grid occurred during the last decades due to stringent regulations put forward in response to environmental concerns [43, 44]. Secondly, profound structural reforms in the power industry are geared toward creating and consolidating competitive energy markets [20, 22]. In these markets, load serving entities seek to exploit the variability of electricity prices, resulting in a growing amount of bulk power being transferred over long distances through the transmission grid (termed wheeling). This phenomenon worsens the shortage of reserve margins in transmission that have prevailed since the mid eighties.

Interestingly, a third cause of the growing number of large-scale blackouts in North America, was identified in a 1984–1988 study of major disturbances conducted by the North American Electric Reliability Council of 1988 [33, 46]. This cause is induced by the over-reaction of the grid protection systems under faults, termed relay-hidden failures. More specifically, the NERC study showed that a small fraction of relays (the name given to the protection devices) tend to unduly disconnect unfaulty pieces of equipment that they are in charge of protecting. By weakening the transmission network, over-tripping tends allows a perturbation to propagate further, especially when the power system is stressed or operated close to its maximum loadability. One peculiarity of relay-hidden failures is that they cannot be detected *a priori*, that is, they cannot be exposed before the system is perturbed. Routine maintenance testing may not detect them or even worse, may induce them by damaging relay components as was the case in the 1977 New York blackout [32].

Presently, methodologies together with algorithms are being developed that assess the risk of cascading events leading to catastrophic failure of an electric power network. Here, a catastrophic failure is defined as one that results in the outage of a sizable amount of load (commonly 10% of peak). Following the pioneering work of Thorp et al. [46], Mili et al. [32] proposed an algorithm that calculates the risk of cascading failures in an electric power system for a given production and load configuration. The approach is based on graph-theoretic breadth-first-search method that implements a probabilistic fault tree of events [26, 48] to calculate the conditional probability of a system failure, P_{Si}, written as:

$$P_{Si} = P_{Fi}(1-P_{Bi})\{P_{Pi}(P_{HPi}+P_{LPi})+(1-P_{Pi})[P_{Ri}(P_{HTNi}+P_{LTNi}) +(1-P_{Ri})(P_{HTRi}+P_{LTRi})]\}, \quad (8.5)$$

where P_{HTNi} and P_{HTRi} denote the conditional probabilities of a system failure due to relay hidden failures given the probability of a fault, P_{Fi}, on the ith line, the probability that this fault is permanent, P_{Pi}, the probability that the associated relay is reclosing, P_{Ri}, and the probability that the associated circuit breaker remains stuck in a closed position, P_{Bi}. Similarly, P_{LTNi} and P_{LTRi} denote the conditional probabilities of a system failure due to line overloads.

We extend the capability of the foregoing algorithm to minimizing the multiobjective risk function defined by (8.4) in Section 8.2 while accounting for cascading failures throughout the network under a wide variability of load and generation profiles. The risk functions $f_i(d_j)$, $i = 1, \ldots, 3$, involved in (8.3) for a design d_j, are defined as the conditional expected loss of load (ELOL) times the severity of the system failure. The integration of social and economic concerns into this risk function will be described in Section 8.4.

The algorithm being developed proceeds as follows. For a given design scenario of generation and transmission expansion denoted by d_j, the algorithm first identifies the outage sequence of lines, generating units, and loads for a given initiating fault, which is typically a short-circuit applied to a line or a node of the network. Then, for this sequence of events, it calculates the total loss of load and the total socio-economic cost associated with it. Finally, it calculates the total risk of the system failure by multiplying its estimated conditional probability by the socio-economic cost. This risk is then classified as low, medium or severe. The algorithm repeats these calculations for all the initiating faults and calculates the unconditional expected-value risk function, $f_4(d_j)$ given by (8.3) and the dollar value of the design d_j denoted by $f_5(d_j)$. The whole procedure is then repeated for each of the n selected design scenarios of generation and transmission expansion, d_j, $j = 1, \ldots, n$. The best design will be the one that minimizes the vector-valued risk-cost function given by (8.4).

Obviously, this is a formidable combinatorial optimization problem. We propose to solve it via variance-reduction Monte Carlo methods. A general account of these methods is provided by Rubinstein [38], and Cochran [6]. In power systems, Thorp et al. [46] advocate the use of importance sampling. Oliveira et al. [34] showed that control variates significantly reduce the number of samples in Monte Carlo-based composite power system reliability evaluation. Billinton and colleagues applied importance sampling (Savaderi and Billington [42]), stratified sampling (Kahn and Billington [24]), and antithetic variates (Kahn and Billington [24]; Sankarakrishan and Billinton [41]) to efficiently calculate various reliability indices for composite power systems. They found that antithetic variates outperform the other methods. Marnay and Strauss [27] compare anthetic and stratified sampling when estimating chronological production hourly marginal cost. They recommend the use of a procedure that combines antithetic sampling and proportional stratification. The latter method is being investigated and adapted to our problem.

8.4 INTEGRATING THE SOCIAL AND ECONOMIC IMPACTS IN POWER SYSTEM PLANNING

A comprehensive analysis of the costs of catastrophic failure in power systems should go beyond merely identifying the immediate costs associated with the replacement of failed technical components. If we ignore the broader social and economic impacts associated with the cascading effects of multiple contingencies, we are likely to grossly underestimate failure costs and thus under-invest in risk mitigation. The difficulty in estimating these costs should not be an excuse for them to be ignored. There are several dimensions that must be considered when examining the social and economic impacts of power system failures:

- Explicitness of impact. Some costs can be quantified in a precise way (e.g. cost of replacement equipment), while others cannot be quantified in any precise way (e.g. impact of loss of consumer confidence).
- Target of impact. Some costs are the burden of the energy providers (e.g. recovery costs), while other costs are seen at the societal level (e.g. loss of business revenue due to downtime) and do not directly, immediately impact the provider. Thus, when calculating the cost of impact, one needs to clarify whether the analysis is to consider provider costs, societal costs, or both.
- Uncertainty of impact. Some consequences can be predicted with some confidence, making them more explicit, while other consequences cannot be even fathomed a priori.

The failure cost is formulated as follows:

$$\text{Failure Cost} = \text{Recovery Cost} + \text{Business Costs} + \text{Social Costs}$$

Recovery Costs are probably largely taken on by providers, and tend to be explicit in nature. They include equipment replacement and the cost of additional labor and/or contracting workforce needed for recovery. Repair data can be used to make estimates of such costs, and historical cases can be used as a basis for prediction. Recovery costs are expected to scale linearly, e.g. a recovery of size 2 X is likely to cost twice as much as a recovery of size X. On the other hand, *Business Costs* can be substantial. To the energy provider, there are the losses associated with lost revenue, which is directly related to the amount of power not delivered and its associated retail cost. There will also be loss of customer goodwill, which can result in lost business and thus lost revevenue; additionally, the provider may incur regulatory costs in terms of either fines or required expenses to respond to regulatory action. The provider is not the only business that loses during a power outage however. To businesses within the provider network, the losses associated with business costs can be extremely large. It was estimated that the blackouts during the California energy crisis cost some high-tech manufacturing firms up to one million dollars per minute of downtime; likewise, overall lost business revenue from the 2003 Northeast U.S. blackout is estimated at 710 billion (McClure [28]). Finally, *Social Costs* may or may not be substantial, depending on the nature of the failure event. Some social costs include social inconvenience, contingent losses (e.g. loss of medical response), and decreases in public opinion, which in turn may spurn other side effects.

A more precise estimate of the overall Business Cost can be generated by determining, for every business *J* affected by the event, the expected revenue per (e.g.) minute for that firm *R*(*J*) and the amount of (e.g.) minutes the business was affected *T*(*J*), and then find *S R*(*J*) * *T*(*J*). If we do not require a precise answer but simply wish to rank design alternatives (e.g. compare on relative v. actual cost), then a surrogate measure can be used. If we assume that the mix of load drawn by businesses versus residential consumers is constant, then we could simply use lost load, which should scale linearly with actual lost business revenues.

ICF Consulting [21], an energy consulting firm, claims to have performed research on "customer willingness to pay" as a means to estimate what we are calling Business Costs and Social Costs. They argue that customer preferences for what they are willing to pay to recover lost power are an indicator of these societal costs. Their data yields an estimate that customers are willing to pay one hundred times the retail cost of energy in order to recovery it. For the 2003 Northeast Blackout, this corresponds to 78 billion, which is close to the 710 billion estimate made by EPRI (McClure [28]). If the loss to each individual is 100 times the lost load (retail value) to them, then the total Business and Social Cost could be estimated as 100 times the retail value of the total lost load. Note that if we do not require an exact estimate, then the scaling factor of 100 is moot—we can simply use lost load, which should scale linearly to the real Business and Social Costs.

8.5 ENERGY CRISES AND PUBLIC CRISES

In Section 8.4, we discussed how social and economic issues might be considered when evaluating the severity of a particular failure scenario. The Recovery and Business Costs can be considered "linear" costs, as impact tends to increase in a linear fashion as a function of the number of people impacted and the length of time they are impacted for (the lost load). If we are to consider extreme, worst-case scenarios, then we must also consider the possibility that there are "hidden" Social Costs associated with a power failure that go beyond what can be articulated and explicitly observed, and that such hidden costs might be "nonlinear" in their impact (Goldstein [15]).

8.5.1 Describing the Methodoly for Economic and Social Cost Assessment

The unarticulated component of the Social Cost equation is public perception. Like the complex cascading failures discussed in Section 3, public perceptions percolate through social networks in rapid and unpredictable ways (Rogers [37]). Why should public perception be considered important to consider when evaluating the severity of a failure? First, public perception of entities associated with the delivery and management of the energy (e.g. utility companies) is likely to erode, which may lead to negative financial consequences to such firms. Second, critical and sustained downturns in public attitudes in one area (e.g. energy) can in turn trigger cascading of public attitudes in other areas (e.g. consumer), which can expand the severity of the event. Third, such a dynamic can induce frustration with the political systems surrounding such institutions, inducing political instability. In developing countries where such instability already exists, and where energy is typically delivered (at least in part) by government entities, this simply contributes to the continued chaos; chronic energy crises imply chronic social crises. For example, continued power outages and shortages during the U.S. occupation of Iraq eroded Iraqi civilian support for the occupying force.

Obviously, public attitude will be affected by a system failure. A system pushed to crisis (disequilibrium), however, becomes unpredictable, so it is difficult to see how such costs could be made explicit (into a monetary value). However, as long as the goal is to rank designs by their risk, and not determine exact costs, then such social costs do not need to be explicitly considered if they are also linear. In other words, if social costs are linear in terms of the size and duration of the outage, then they will simply change the final numbers, but not the relative rankings of the designs, since other costs being considered also scale as a function of size and duration.

Therefore a key question becomes: Does the public respond to energy crises in a linear or nonlinear way? In order to answer this question, we chose two high-profile cases, the 2000–2002 California Energy crisis, and the 2003 power Blackout in the Northeast U.S. These two are very different, "extreme" cases, as detailed in Table 8.1.

Because we do not have access to polls of public perceptions during these crises, we must be creative in determining whether public perceptions began to change in a linear or nonlinear fashion. First, we make the premise that the media produces news that people wish to consume (Gans [13]). Second, individuals will tend to wish to read and hear about topics that are "in mind" (Greenberg [16]; Gantz [14]; Dooley and Corman [9]). Third, if there is a public "buzz" about a topic, it will be tend to be "in mind" of many individuals, and these individuals will in turn wish to consume news about that topic. This leads to an "increasing returns" dynamic: (a) media reports event, (b) public attitudes decline rapidly, (c) issue is "on mind" of many people, (d) media increases reporting on issue, (e) public attitudes further decline.

If this is true, then significant changes in media coverage around a topic should be coincident with significant changes in public attitudes about that topic. In other words, if an energy crisis "has legs" as a media story, it indicates there is such a

TABLE 8.1. Energy Crisis Cases.

	2003 U.S. Blackout	California Energy Crisis
Number affected	Many directly	Some directly; many indirectly
Duration	2 weeks	2 years
Dominant failure mode	Technical	Technical, Political, Economic
Dominant effect to consumer	Loss of power	Prices

hidden social dynamic—energy crises can become public crises and induce nonlinear effects. If an energy crisis is simply a story to fill the news pages and does not garnish much reader interest, then we can posit that energy crises do not lead to public crises (by themselves), and therefore such hidden social costs can be ignored in computing design-risk costs.

In order to analyze media coverage of these two events, we implemented the following four-step procedure. First, we searched the Lexis-Nexis news database to find articles related to the two search terms "California energy" and "Northeast blackout." This yielded 387 articles for the 2003 U.S. blackout case and 4635 cases for the California energy crisis case. Note that data for the latter case was collected back to 1997, in order to form a baseline prior to the crisis. Secondly, we performed computerized text analysis on each text using Centering Resonance Analysis, termed CRA for short (Corman et al. [7]). This method is described in Section 8.5.1. Thirdly, we created multiple time series for each case. In the 2003 U.S. blackout case, we selected a time unit of one day, as the event was largely over in about ten days; in the California energy crisis case we selected a time unit of one month, as we had over a seven year period and the event lasted nearly two years. Finally, we performed time series analysis to examine the dynamical patterns for each of the time series, and drew conclusions relative to the research question.

For each time unit in each case of the third step, we calculated the number of articles published, then we averaged the tone of those articles. Tone is defined as the ratio of words with positive connotation to words with negative connotations, and is scaled from −1.0 to 1.0 (Dooley and Corman [10]). A specific, tested taxonomy is used to make such word classifications. Word influence values are used as weights in these calculations. In order to interpret the relative size of our statistical values, we used a baseline of some 250,000 articles published by Associated Press. According to our logic, public crises should be reflected by increasingly negative tone in the media. Then, we averaged the intensity of those articles. Intensity is defined as the proportion of words with positive or negative connotation to total words (0–1), and measures the degree of emotionality of a text. According to our logic, public crises should be reflected by increasing intensity in media coverage.

8.5.2 The CRA Method

CRA is a representational method, a form of network text analysis. CRA produces stand-alone representations of a text that do not depend on the sorts of dictionaries, semantic networks, or ontologies mentioned above; its representations may then be employed in analyses aimed at positioning or inference. Unlike other network text analysis methods, CRA is based on a theory of communicative coherence. Specifically, CRA draws on centering theory (McKoon and Ratcliff [29]) in assuming that competent authors/speakers generate utterances that are locally coherent by focusing their statements on conversational centers. Centers are noun phrases constituting the subjects and objects of utterances, and are generally entities such as objects, events or persons. In a written text, for example, each sentence (except the first) has a backward-looking center that refers to a preferred forward-looking center expressed in the previous utterance. The author/speaker also establishes an ordered set of forward-looking centers to which the next utterance can coherently refer. A given utterance is made locally coherent by connecting the backward-looking center in a predictable way to previous forward-looking centers. Under the assumptions of centering theory, then, communicators speak or write coherently by creating utterances that deploy a stream of centers—more specifically, noun phrases—in a strategic way, ultimately creating a semantic structure of centers.

CRA consists of four steps: selection, linking, indexing, and mapping. CRA categorizes texts in terms of a pattern of connections between words that are crucial to the centering process. Compiling these connections in all utterances in a text yields a CRA network representing the text. During selection, CRA parses an utterance into its component noun phrases. A noun phrase is a noun (plus zero or more additional nouns and adjectives) that serves as the subject or object of a sentence. Since the centering process operates through noun phrases, this step acts as a filter, retaining only those words relevant to the centering process. The second step, linking, converts the sequences into networks of relationships between centering tokens. All centering tokens in the utterance are linked sequentially, and then all possible pairs of tokens within the noun phrase are linked. In indexing, the network of centering token associations is analyzed to determine the relative influence of each node. To the extent that a CRA network is structured, some nodes are more influential than others in channeling flows of meaning (McPhee et al. [30]). We operationalize this idea of influence as centrality of nodes in the CRA network. The final step in CRA processing is concept mapping wherein the network or networks are appropriately visualized.

8.5.3 Data Analysis of the California Crises and of the 2003 U.S. Blackout

Figures 8.3 and 8.4 show time series for the number of articles for the 2003 U.S. blackout and the California energy crisis cases, respectively. With the 2003 U.S. blackout, the media response is immediate and very short-lived; the number of articles decreases significantly after only two days—this despite the fact it being the worst power failure in U.S. history. While there are not enough samples in this time

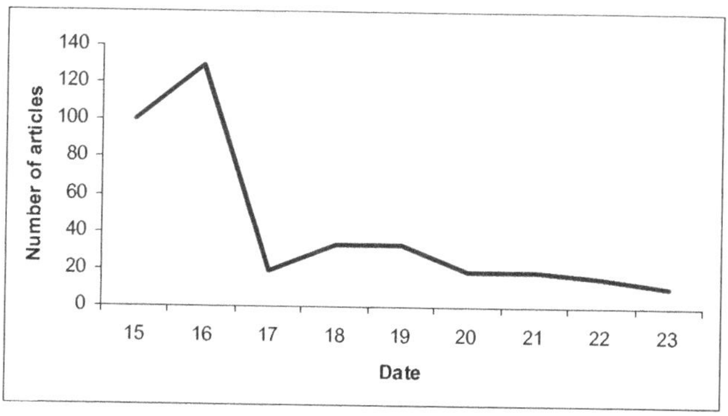

Figure 8.3. Number of articles each day, 2003 Northeast blackout case.

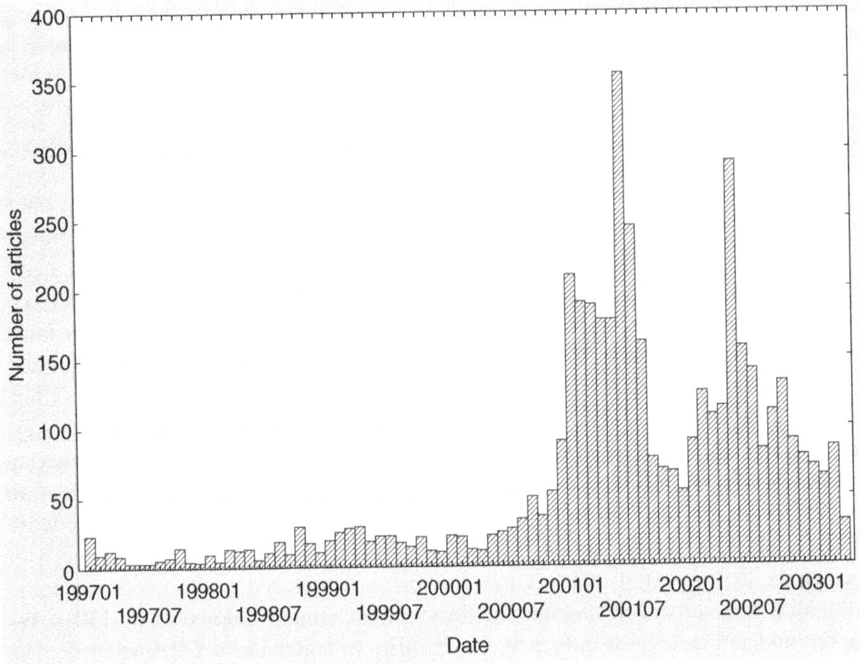

Figure 8.4. Number of articles per month, California energy crisis case.

series to do modeling, a visual inspection of the time series suggests the following. If we view the 2003 U.S. blackout event as an "impulse function" and the number of articles as an impulse response function for the media, then we conclude that the media response is extremely damped. Given no other "excitation," this story quickly died down. We posit that part of the reason the story died after two days was because it did not engender a large social impact and response; its consequences were direct, and the "story" was not of interest to the public.

With the California energy crisis, we see that 2001 brought about an increase in the number of articles about California Energy, as the crisis began to percolate. ARMA time series models were fit to the data (Poole et al. [36]), and the best fit model was an autoregressive order one (AR(1)) model with parameter sets equal to 0.91, a standard error of 0.09, and a root mean squared value of 0.82). This indicates a system with strong memory, but only across two time periods (i.e. months) (Dooley and Van de Ven [11]). Unlike the 2003 U.S. blackout case where a single impulse impacted the media system, we conceptualize a stream of impulses (the random input component of the AR model) hitting the media system and the media reacting to it. Because we only see simple first order behavior, we conclude: (a) the strong correlation month-month indicates that "stories" within the larger history come and go for short periods of time, (b) the sustained level of media activity is due to a sustained meta-event, i.e. series of failure events on multiple levels (Taylor and Van Doren [45]), and (c) there is no evidence of non-linearity, hence no evidence that the sustained nature of the story was necessarily due to public interest in it.

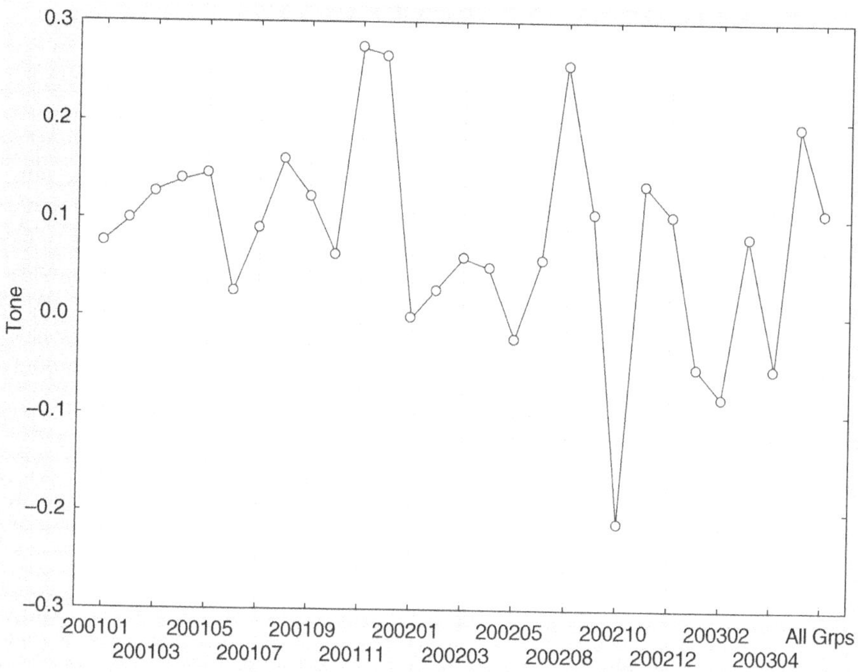

Figure 8.5. Average article tone per month for the California energy crisis case.

As further evidence, Figure 8.5 shows the tone of articles from the California energy crisis case over the critical two year period. We see that it is stable over this time, and has a mean of 0.08; compared to our baseline data, this actually makes it relatively neutral in tone. Additionally, we found the tone of articles in the 2003 U.S. blackout case also to be stable over that shorter period of time, with a mean of 0.04. Likewise, in both cases, intensity values were low, indicating the use of largely non-emotional language. Thus, media coverage of the energy events was not negative or emotional. This is further evidence that energy stories are not engendering a public response—the media is writing about them in a neutral way, and the story remains a story only as long as "events" (whether they be technical or social or economic failures) continue to occur over time.

Thus we conclude that there are no hidden, nonlinear costs associated with how public perceptions may be affected by energy-related risk events. The social and economic impact of energy failure events can be treated as functional and rational—the losses associated with an energy failure (Recovery, Business, Social) can be estimated by their observable impact; we do not need to worry about how human emotion and consequent dynamics might lead to unpredictable consequences. To the degree there is an attitudinal and/or emotional response, it is directly related to the explicit and observable impact it has on the consumer. Thus, these less tangible costs should have no impact on the relative ranking of design alternatives, as long as the costs that are associated with failure take into account the number of people impacted and the duration of the event.

8.6 CONCLUSIONS AND FUTURE WORK

In this chapter, we investigated the decision-making processes associated with the risk assessment and management of bulk power transmission systems. First, we presented a partitioned multiobjective risk method aimed at finding tradeoffs between various levels of resiliency ranging from low to high damage severity while minimizing the cost of the design. It is stated that while the ELOL index currently in use in power system planning captures failures of low or moderate damage severity via an N-1 security analysis, it does not account for the risk of rare but extreme events, the survivability analysis. As a result, the expansion that stems from such an index may not be resilient to cascading failures leading to blackouts. In addition, because hidden failures in the protection systems are instrumental in speading out failures across the networked system, their probabilities ought to be incorporated in the risk-based analysis. This is precisely what our proposed method does in that it assesses the risk of cascading events using a probabilistic algorithm that pinpoints the weak links of the network by simulating the propagation of the failures throughout the transmission network due to hidden failures in relays. These weak branches are the system components in which special controllers are to be installed.

Because the risk of a system failure, viewed as an event, is defined as the probability times the consequence of this event, an assessment of the consequence of this event must be carried out, for instance by means of its costs. In this chapter, these costs are defined as consisting of technical, business, and social costs, and a method for estimating them was proposed. Specifically, we found that technical costs are likely to be small compared to business costs, and that business costs scale linearly with the total lost load, which depends on the scope of the failure and its duration. In order to examine social costs, we used a novel form of text analysis, Centering Resonance Analysis (Corman et al. [7]), to examine media coverage of two high-profile power failures—the California crises and of the 2003 U.S. blackout. Over 5000 media articles were collected and examined in terms of their content and tone (e.g. how positive or negative the language was within the text), and then time series analyses were used to determine the dynamic characteristics of the news stream, thus inferring whether the failures were perceived by the public as escalating crises with interconnected events (e.g. a nonlinear social response) or discrete events connected only in theme (a linear social response). We found that in both cases power system failure stories do not "have legs," i.e., they do not generate sustained public interest regardless of their magnitude, thus we can assume that social costs scale linearly with business and technical costs and thus do not have to be explicitly taken into account when doing risk assessments.

As a future research work, the following problems need to be addressed:

- Define the characteristics of the design and specify the alternatives that can be implemented to improve the resiliency of the networks to catastrophic failures. Special attention has to be paid to the statistics of extreme events (Haimes [18], Gumbel [17]) since they require the modeling of the tails of the probability distributions.

- Formulate the design problem as an optimization problem under uncertainty. Game theory and the Pareto optimization method may be considered to solve conflicting multiple objective functions based on the partitioned multi-objective risk method. These methods should account for cascading failures in large-scale networked systems while realizing tradeoffs between resiliency to natural and man-made hazards and efficiency under normal operating conditions.
- Build simulation models that estimate the business costs associated with large-scale failures, taking into account the integrated nature of power systems and the underlying economic grid.
- Use social measurement techniques to more deeply examine public attitudes about power system failures and understand what are the factors that drive a negative versus neutral public response.
- Propose various monitoring and control schemes aimed at steering the system away from emergency states. One such a scheme would be a robust decentralized control system whose performance has to be assessed from a cost/effectiveness viewpoint.
- Develop a graceful degradation of the power system under emergency conditions via power system control separation. Indeed, once a critical infrastructure has experienced a catastrophic failure, it is of paramount importance to speed up its restoration so that the duration of the interruption of service is reduced. Unfortunately, due to the complexity of the operation of some critical infrastructures, this task is not an easy one. For instance, the restoration of a power system may take hours or even, in some extreme cases, days to complete (Adibi [1]). This is due to the fact that the reconnection of separated pieces of the system requires the careful initiation of a sequence of actions that should meet a host of operational constraints. Indeed, at every instant of time, a balance between the power generation and the load demand ought to be achieved without overloading the equipment, without violating the voltage lower and upper limits, and without inducing dynamic instabilities or voltage collapse. Consequently, decreasing the time of recovery in a graceful manner using advanced control systems and computer-aided software programs is an important issue that needs to be addressed adequately.

BIBLIOGRAPHY

[1] M. M. Adibi. *Power System Restoration: Methodologies and Implementation Strategies*. IEEE Press, 2002.

[2] R. J. Adler, R. E. Feldman, and M. S. Taqqu (Eds.). *A Practical Guide to Heavy Tails-Statistical Techniques and Applications*. Birkhauser, Boston, 1998.

[3] M. Amin, "Toward Secure and Resilient Interdependent Infrastructures," *Journal of Infrastructure Systems*, Sept. 2002.

[4] R. Billinton and W. Li. *Reliability Assessment of Electric Power Systems Using Monte Carlo Methods*. Plenum Press, NY, 1994.

[5] A. Charnes, W. Cooper, and E. Rhodes, "Short Communication: Measuring the Efficiency of Decision Making Units," *European Journal of Operations Research*, Vol. 3, No. 4, p. 339, 1978.

[6] W. G. Cochran. *Sampling Techniques*. John Wiley & Sons, 1977.

[7] S. Corman, T. Kuhn, R. McPhee, and K. Dooley, "Studying Complex Discursive Systems: Centering Resonance Analysis of Organizational Communication," *Human Communication Research*, Vol. 28, No. 2, pp. 157–206, 2002.

[8] T. D. Crocker, J. F. Shogren, and P. R. Turner, "Beliefs and Nonmarket Valuation," *Resource and Energy Economics*, Vol. 20, No. 2, pp.139–162, June, 1998.

[9] K. Dooley and S. Corman, "The Dynamics of Electronic Media Coverage," in B. Greenberg (Ed.). *Communication and Terrorism*. Creskill, NJ: Hampton Press, pp. 121–136, 2002.

[10] K. Dooley and S. Corman, "The Dynamics of Political Messages," *Patterns*, Vol. 2, No. 8, Aug. 13, 2004.

[11] K. Dooley and A. Van de Ven, "Explaining Complex Organizational Dynamics," *Organization Science*, Vol. 10, No. 3, pp. 358–372, 1999.

[12] EPRI. *The SPID Final Report of the EPRI/DoD Project*. University of Washington, Virginia Tech, Arizona State University, Iowa State University. EPRI Report, Dec. 2001.

[13] J. H. Gans. *Deciding What Is News: A Study of CBS News, NBC Nightly News, Newsweek, and Time*. Vintage Books Ed: New York, 1980.

[14] W. Gantz, "The Diffusion of News About the Attempted Reagan Assassination," *Journal of Communication*, Vol. 33, pp. 55–56, 1983.

[15] J. Goldstein, "Emergence as a Construct: History and Issues," *Emergence*, Vol. 1, pp. 49–72, 1999.

[16] B. Greenberg, "Diffusion of News of the Kennedy Assassination," *Public Opinion Quarterly*, Vol. 28, pp. 225–232, 1964.

[17] E. J. Gumbel. *Statistics of Extremes*. Columbia University Press, New York, 1958.

[18] Y. Haimes. *Risk Modeling, Assessment, and Management*. John Wiley & Sons, 1998.

[19] J. Hauer and J. E. Dagle. *Consortium for Electric Reliability Technology Solutions Grid of the Future White Paper on Review of Recent Reliability Issues and System Events*. Office of Power Technologies, Assistant Secretary for Energy Efficiency and Renewable Energy, U.S. Department of Energy, Dec. 1999.

[20] S. Hunt and S. Shuttleworth, "Unlocking the Grid," *IEEE Spectrum*, pp. 20–25, July 1996.

[21] ICF Consulting (2005), "The Economic Cost of the Blackout," in www.icfconsulting.com/energy.

[22] M. Ilic, F. Galiana, and L. Fink (Eds.). Power Systems Restructuring: Engineering and Economics. *Kluwer Academic Publishers*, 1998.

[23] R. L. Keeney and H. Raifa. *Decisions with Multiple Objectives: Preferences and Value Tradeoffs*. John Wiley & Sons, 1976.

[24] M. E. Khan and R. Billinton, "A Hybrid Model for Quantifying Different Operating States of Composite Power Systems," *IEEE Trans. on Power Systems*, Vol. 7, No. 1, pp. 187–193, Feb. 1992.

[25] S. Kotz and S. Nadarajah. *Extreme Value Distributions: Theory and Applications*. Imperial College Press, London, 2000.

[26] H. Kumamoto and E. J. Henley. *Probabilistic Risk Assessment and Management for Engineers and Scientists*. 2nd Ed., IEEE Press, 1996.

[27] C. Marnay and T. Strauss, "Effectiveness of Antithetic Sampling and Stratified Sampling in Monte Carlo Chronological Production Cost Modeling", *IEEE Trans. on Power Systems*, Vol. 6, No. 2, pp. 669–675, May 1991.

[28] G. McClure, "Electric Power Transmission Reliability Not Keeping Pace With Conservation Efforts," *IEEE Engineer*, Feb. 2005.

[29] G. McKoon and R. Ratcliff, "Memory-based Language Models: Psycholinguistic Research in the 1990s," *Annual Review of Psychology*, Vol. 49, pp. 25–42, 1998.

[30] R. McPhee, S. Corman, and K. Dooley, "Organizational Knowledge Expression and Management: Centering Resonance Analysis of Organizational Discourse," *Management Communication Quarterly*, Vol. 16, No. 2, pp. 130–136, 2002.

[31] L. Mili, F. Krimgold, J. Alwang, and J. E. Bigger, "Integrating Engineering, Economic, and Social Modeling in Risks of Cascading Failures Across Interdependent Complex Networks", *Proceedings of the 8th Internat. Conf. on Probabilistic Method Applied to Power Systems*, Iowa State University, Ames, Iowa, Sept. 12–16, 2004.

[32] L. Mili, Q. Qiu, and A. G. Phadke, "Risk Assessment of Catastrophic Failures in Electric Power Systems," *International Journal of Critical Infrastructures*, Vol. 1, No. 1, pp. 38–63, Feb. 2004.

[33] NERC. *NERC Disturbance Reports, 1984–1988.* North American Electric Reliability Council, New Jersey, 1988. http://www.nerc.com/dawg/database.html.

[34] G. C. Oliveira, M. V. F. Pereira, and S. H. F. Cunha, "A Technique for Reducing Computational Effort in Monte-Carlo Based Composite Reliability Evaluation," *IEEE Trans. on Power Systems*, Vol. 4, No. 4, pp. 1309–1315, Oct. 1989.

[35] C. Perrow. *Normal Accidents: Living with High-Risk Technologies.* Basic Books, NY, 1984.

[36] M. S. Poole, A. Van de Ven, K. Dooley, and M. Holmes. *Studying Processes of Organizational Change and Development: Theory and Methods.* Oxford University Press, 2000.

[37] E. Rogers. *The Diffusion of Innovations.* New York: Free Press, 1995.

[38] R. Y. Rubinstein. *Simulation and the Monte Carlo Method.* John Wiley & Sons, 1981.

[39] S. Rosenbush, "Satellite's Death Puts Millions Out of Touch," *USA Today*, May 21, 1998.

[40] S. Sagan. *The Limits of Safety: Organizations, Accidents, and Nuclear Weapons*, Princeton University Press, Princeton, New Jersey, 1993.

[41] A. Sankarakrishan and R. Billinton, "Sequential Monte Carlo Simulation for Composite Power System Reliability Analysis with Time Varying Loads," *IEEE Trans. on Power Systems*, Vol. 10, No. 3, pp. 1540–1545, Aug. 1995.

[42] L. Savaderi and R. Billinton, "A Comparison Between Two Fundamentally Different Approaches to Composite System Reliability Evaluation," *IEEE Trans. on Power Apparatus and Systems*, Vol. PAS-104, No. 12, pp. 3486–3492, Dec. 1985.

[43] K. Stahlkopf, "Technology Offers Solutions to the Current Power Crisis," *IEEE Spectrum*, pp. 14–15, June, 2001.

[44] W. Sweet and E. A. Bertz, "Energy Woes," *IEEE Spectrum*, pp. 48–53, July 2001.

[45] J. Taylor and P. Van Doren, "California's Electricity Crisis," *Policy Analysis*, July 3, 2001.

[46] J. S. Thorp, J. S., A. G. Phadke, S. H. Horowitz, and S. Tamronglak. "Anatomy of Power System Disturbances: Importance Sampling," *Electrical Power & Energy Systems*, Vol. 20, No. 2, pp. 147–152, 1998.

[47] U.S.-Canada Power System Outage Task Force. *Final Report on the August 14th, 2003 Blackout in the United States and Canada: Causes and Recommendations.* April 2004. https://reports.energy.gov/BlackoutFinal-Web.pdf.

[48] US Nuclear Regulatory Commission. Fault Tree Handbook. NUREG-81/0492, 1981.

[49] J. Von Neuman and O. Morgenstern. *Theory of Games and Economic Behavior.* Princeton University Press, 1944.

CHAPTER 9

MODELS FOR TRANSMISSION EXPANSION PLANNING BASED ON RECONFIGURABLE CAPACITOR SWITCHING

James McCalley, Ratnesh Kumar, Venkataramana Ajjarapu, Oscar Volij, Haifeng Liu, Licheng Jin, and Wenzhuo Shang

Iowa State University

EDITORS' SUMMARY: This chapter introduces models for power transmission system enhancement by integrating economic analysis of the transmission cost to accommodate an informed business decision. Continuous and discrete control schemes are proposed as an alternative to transmission expansion to optimize cost effectiveness. Furthermore, this chapter investigates complex planning issues such as transmission limits and general models of series and shunt capacitor switching that allow the planner to carry out an analysis of electricity market efficiency using alternate transmission expansion scenarios. An interesting conclusion is provided that advocates the design of an electricity market based on control expansion efficiency.

9.1 INTRODUCTION

Transmission expansion planning is the process of deciding how and when to invest in additional transmission facilities. It is complicated under any electric industry structure because resulting decisions can affect any stakeholder owning or operating interconnected facilities and are necessarily driven by predictions of uncertain futures characterized by changes in load and generation, and by potential of component unavailability from forced or scheduled outage. These decisions have significant consequences on the reliability and economy of the future interconnected power system; in addition, they usually involve large capital expenditures and

Economic Market Design and Planning for Electric Power Systems, Edited by James Momoh and Lamine Mili

complex regulatory processes, especially if they require obtaining right-of-way, and so represent high financial commitment to investors. Previous to deregulation when electric utilities were vertically integrated, overseeing generation, transmission, and distribution under one management structure, the necessary coordination between the highly interdependent functions was carried out in an intentionally integrated fashion, often involving the same people, targeting the objectives of the organization's management to whom the analysts and decision-makers reported. Transmission enhancements that affected multiple utilities were handled through bilateral coordination or through well-structured coordinating bodies. The utility paid for transmission upgrades and recovered regulatory-approved costs through customer rates. The most significant uncertainties faced by planners were load growth and component forced outage (due to a fault or failure), uncertainties for which historical data can be used in deriving associated probability distributions.

Under deregulation, the number of organizations involved in generation planning and transmission planning is significantly increased, each with their own objectives. Generation is planned by a multiplicity of companies seeking to maximize their individual profits through energy sales, while transmission is planned by transmission owners seeking to maximize their profits through transmission services, all overseen and coordinated by a centralized authority seeking to ensure grid reliability and market efficiency. The increased number of stakeholders requires procedures for coordinating among them the necessary analyses, decisions, and financial implications; in addition, it motivates the need for incentives so that organizations perceive transmission investment and ownership to be attractive. The number and nature of uncertainties have increased as well [1]. In addition to load uncertainty and component forced outages, planners must account for uncertainty in generation and transmission installation, in generation commitment and dispatch schedules, in wheeling (point-to-point power transactions), and in component economic outages due to financially-motivated decision on the part of the component owner.

Although electricity markets have been operating in the U.S. since the early 1990s, it is only recently that planning procedures and investment incentives have *begun* to mature. As a result, transmission investment has been inhibited during the early deregulation years, as indicated in Figure 9.1 [2], which compares U.S. annual average growth rates of transmission and load during three periods of time from 1982 to 2012, and Figure 9.2 [3], which compares U.S. investment trends in distribution, transmission, and generation from 1925 to 2020. The figures show transmission growth and investment at its lowest point during the period 1992–2002.

From an engineering perspective, there are four options for expanding transmission:

(1) build new transmission circuits,

(2) upgrade old ones,

(3) build new generation at strategic locations, and

(4) introduce additional control capability.

Although all of these continue to exist as options, options one through three are more capital-intensive than option four; right-of-way acquisition can sometimes

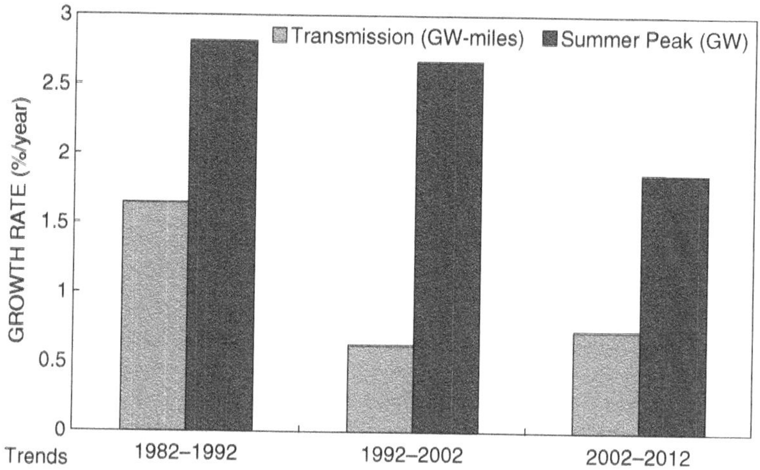

Figure 9.1. Annual avg. growth rates of transmission, load [2].

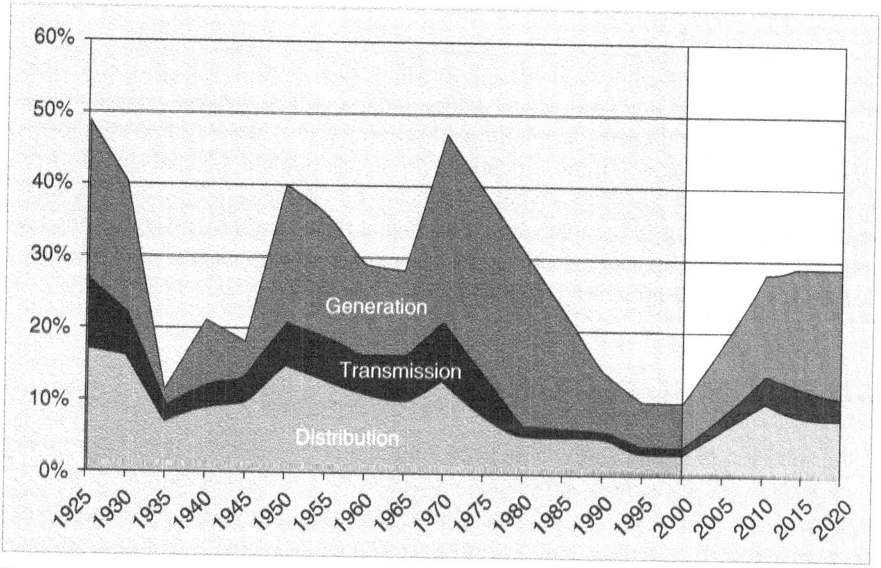

Figure 9.2. Capital investment as percentage of revenues [3].

prohibit option one, and option three as a transmission solution is almost always considered secondary to energy market profitability. Option four, control, although not always viable, is attractive when it is viable since it is relatively inexpensive, requires no right-of-way, and when not part of generation facilities, affects energy market operation only through the intended transmission expansion.

Although considerable work has been done in planning transmission in the sense of options one through three, there has been relatively little effort towards planning transmission control options in the sense of option four, yet the ability to consider these devices in the planning process is a clear need to the industry [4, 5, 6, 7]. Our interest therefore focuses on designing systematic control system planning algorithms. There are four types of control technologies that exist today: generation controls, power-electronic based transmission control, system protection schemes (SPS), and mechanically switched shunt and series devices (capacitors, reactors, and phase-shifters). Of these, the first two exert continuous feedback control action; the third and fourth exert discrete open-loop control action. Thus, power system control is hybrid [8, 9] in that it consists of continuous and discrete control. Since power systems are already hybrid, and since good solutions may also be hybrid, assessment of control alternatives for expanding transmission must include procedures for gauging cost and effectiveness of hybrid control schemes. Our emphasis is on the most promising and least expensive of the discrete control options, series and shunt capacitor switching; the aim is to provide flexible and inexpensive transmission expansion via reconfigurable switching of these controls in response to network disturbances that can occur.

In this chapter, we target planning methods and investment implications for enhancing transmission via discrete control. In Section 9.2, we summarize current market-based planning procedures because, owing to their recent development, the literature is relatively sparse on this topic; in addition, this summary illuminates the environment in which the methods described in this paper are intended for use. Section 9.3 describes and clarifies one particularly complex planning issue that is at the heart of our work: transmission limits. Section 9.4 provides engineering models capable of identifying solutions to planning problems. Section 9.5 analyzes electricity market efficiency under two types of transmission expansion options, new lines and control, resulting in the interesting conclusion that electricity markets allowing only control-based expansion are efficient, whereas markets that allow new transmission lines are not. Section 9.6 summarizes over conclusions.

9.2 PLANNING PROCESSES

A transmission planning study is an economic and engineering analysis of a transmission network to identify problems associated with expected future conditions together with solutions to those problems. Such a study may be motivated by the likely prospect of a single significant network change, e.g., the proposal of a large generation facility. However, it is essential to conduct planning studies periodically to account for normal load growth, retirement of old facilities, and changes in maintenance and operating policies. As a result, minimum planning frequency has generally been yearly, projecting conditions five to ten years ahead.

Order 2000 of the Federal Energy Regulatory Commission (FERC) stipulated that regional transmission organizations (RTOs) have "ultimate responsibility for both transmission planning and expansion within its region [10]." An RTO is an organization, independent of all generation or transmission owners and load-serving entities, which facilitates electricity transmission on a regional basis with responsibilities for grid reliability and transmission operation. Organizations approved or under consideration by FERC for approval as an RTO are shown in Figure 9.3 [11]

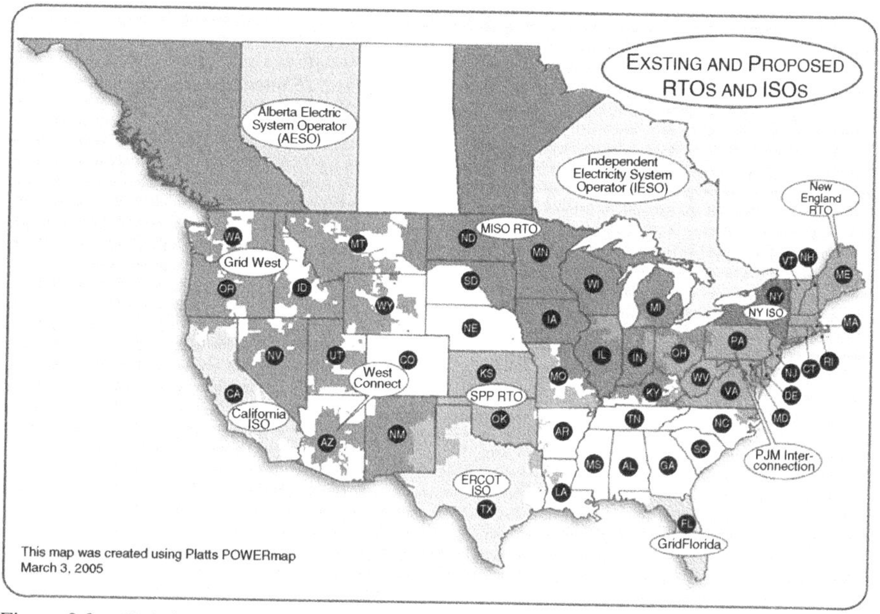

Figure 9.3. Existing and proposed RTOs [11].

as the white ovals. Two primary issues for RTO-based planning are coordinating plans of multiple stakeholders and provision of investment incentives including articulation of a cost-recovery path for transmission investors.

FERC also issued an important ruling in 2003, called Order 2003 [12], which required public utilities to "file revised open access transmission tariffs containing standard generator interconnection procedures and a standard agreement that the Commission is adopting in this order ..." These procedures, described in Order 2003, were encapsulated in a diagram contained in an appendix of Order 2003 [13]. Figure 9.4 provides a simplified version of this diagram.

In the remainder of this section, we describe some aspects of a planning process and cost-recovery approach used by one RTO, PJM Interconnection, based largely on [14, 15].

9.2.1 Engineering Analyses and Cost Responsibilities

Each planning cycle begins with an information gathering stage during which RTO engineers solicit information from a full range of stakeholders, including independent power producers (IPPs), interconnected transmission owners (ITOs) and transmission developers (TDs) proposing development plans, load serving entities (LSEs), and all regional reliability councils, independent system operators (ISOs), and transmission owners and operators within and adjoining the RTO network. Project queues are developed of proposed generation and transmission projects based on receipt of an interconnection request. Queued projects are assigned one of the following status indicators, in order of study sequence: feasibility study, impact study, facility study, interconnection service agreement (ISA), being built, or built.

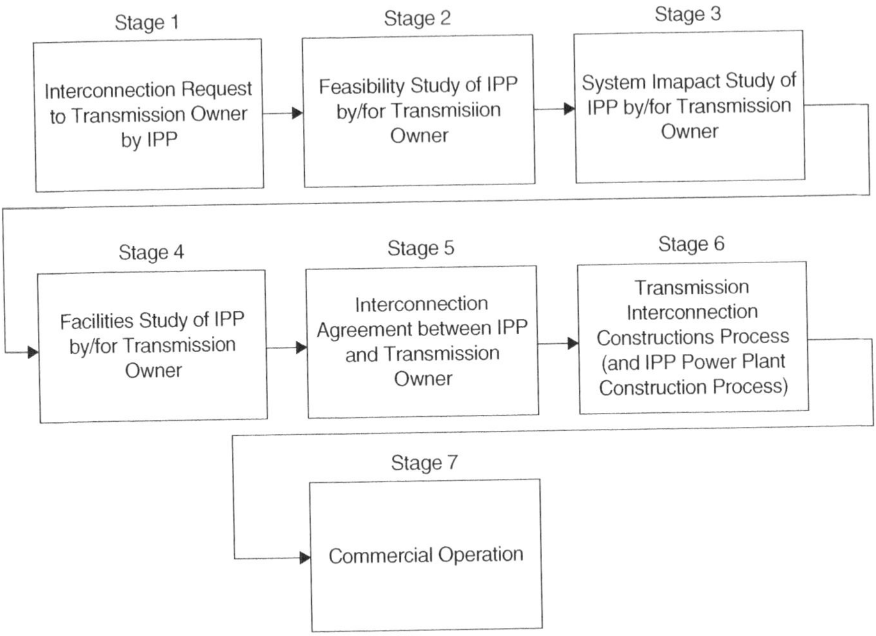

Figure 9.4. Simplified Illustration of FERC's interconnection procedures.

A baseline analysis of system reliability is performed by the RTO; this analysis models expected load growth and known transmission and generation projects, but it models development projects in the queue depending on their status. If a project has proceeded to the stage of "facility study," its associated system upgrades are modeled. If a project has proceeded to the stage of "ISA," it could be turned on in the basecase, even it has not been built. Power flow, voltage, time-domain (stability), and short-circuit studies are conducted to evaluate the reliability according to applicable criteria and to identify baseline expansion projects necessary to satisfy violated criteria that cause unhedgeable congestion (unhedgeable congestion is described in Section 9.2.3 below).

An initial feasibility study is performed for each interconnection request to provide a rough approximation of the transmission-related costs necessary to accommodate the interconnection in order to enable the developer to make an informed business decision, at which point the developer either drops out of the queue or signs a system impact study agreement. System impact studies are performed for each interconnection request remaining in the queue. System impact studies provide a more detailed assessment of interconnection requirements, revealing necessary enhancements. Such enhancements may include direct connection attachment facilities (required for new generation to "get to the bus") and/or network reinforcements to mitigate "network impact" effects that the proposed transmission development may have on the power system. Each interconnection project bears the cost responsibility for its own direct connection attachment facilities. The cost responsibility for network reinforcements is allocated among parties based on the percent impact which a given project has on a system element requiring upgrade. In the power flow cost allocation method, upgrade costs are allocated based on each party's MW impact on the need for the system upgrade, as determined by "distributed slack" power transfer distribution factors [16], defined as the MW impact on a line when transferring 1 MW of power from the new generating bus to all the rest of the generating buses. Such an approach is appropriate for cost allocation for new or re-conductored lines, for example. The short-circuit cost allocation method, applicable to upgraded circuit breakers, allocates costs in proportion to the fault level contribution of each proposed IPP. Identified network reinforcement costs, for a given capacity, are highly dependent on location, and developers have strong incentives to identify development locations that minimize these costs.

9.2.2 Cost Recovery for Transmission Owners

In addition to the investment or capital costs, transmission owners also incur ongoing costs due to operations and maintenance, administration, debt amortization, depreciation, and taxes. Transmission cost-recovery of all of these costs is accomplished in three primary ways.

- Network integration transmission service charges [17]: Network customers are so designated because they pay a transmission charge computed as the summation of their daily peak load multiplied by the annual network integration transmission service rate (in the zone in which the load is located) divided by 365. Typical service charges at the time of this writing range from 11,020–32,114 8/MW-year in the PJM area. Each transmission owner computes these service rates based on their annual transmission revenue requirements, which range from 812 million to 81.6 billion in the PJM area.
- Point-to-point transmission service charges [17]: Point-to-point customers obtain transmission service between a point of delivery to a point of receipt. Service may be firm (curtailed last) or non-firm (curtailed first); the calculation procedure for service charges, which is the same in both cases (but non-firm rates are less), is to multiply the capacity reserved by the rate. The published yearly firm rate at the time of this writing is 818.88/kw-year. Total firm charges are allocated to the transmission owners in proportion to their annual revenue requirements. Total non-firm charges are allocated to the firm point-to-point and network transmission customers based on percentage shares of their firm and network demand charges, respectively.
- Auction revenue rights (ARRs) [18]: ARRs are entitlements allocated annually to firm transmission service customers (which can include transmission owners) that entitle the holder to receive an allocation of the revenues from the annual FTR auction. FTRs are financial instruments that entitle the holder to rebates of congestion charges paid by firm transmission service customers. So transmission owners can purchase ARRs that give them the right to receive compensation from the proceeds of FTR sales. FTRs are sold to market participants to hedge against the possibility of paying congestion charges when flows on a transmission path exceed the path limit, and generation must be

uneconomically dispatched to avoid overload. That is, whenever congestion exists on the transmission system between sink and source points specified in a particular FTR, such that the locational marginal price (LMP) at the sink point (point of delivery) is higher than the LMP at the source point (point of receipt), the holder of that FTR receives a credit equal to the MW reservation specified in the FTR and the difference between the LMPs at the two specified points. (We assume that readers are familiar with LMPs, which are fundamental to understanding electricity markets. Basic treatment of LMPs may be found in [19, 20, 21].)

9.2.3 Economically Motivated Expansion

As described in Section 9.2.1, interconnection requests are placed in a study queue and motivate analysis to identify network expansion requirements and associated costs and cost responsibilities. Allowance is also made that *unhedgeable congestion* be identified and placed in the analysis queue by RTO engineers, and any transmission expansion resulting from this is referred to as economically motivated expansion. *Congestion* refers to the power flowing on a constrained circuit, i.e., a circuit for which the power flowing on it equals the transmission limit (transmission limits are addressed in Section 9.3). *Hedgeable congestion* is power flow on a constrained circuit for which FTRs have been purchased or for which economic local generation (defined in the footnote below[1] per Schedule 6 of [22]) is available. Therefore, *unhedgeable congestion* is power flow on a constrained circuit for which FTRs have not been purchased and economic local generation is unavailable.

Key to whether or not a constraint driven by unhedgeable congestion should be queued as a project is the cost-benefit analysis; i.e., the cost of the congestion to be relieved in comparison to the cost of the transmission solution that relieves it. Because the cost of the transmission solution can not be determined until a study is completed to identify that solution, proxies to this cost, called thresholds, are provided. To facilitate comparison to the cost of congestion, these thresholds are given in units of dollars/month. For example, at PJM, the identified thresholds are based on voltage levels and are 8,100k/month for facilities operating at voltages greater than 345kV, 850k/month for voltages operating at voltages of 100kV–345kV, and 825k/month for facilities operating at voltages less than 100kV [15].

The "congestion cost" to use in the comparison is the monthly unhedgeable congestion cost of a particular constraint. This cost is the sum of the hourly unhedgeable congestion costs for each hour during the month that the constraint is binding. The hourly unhedgeable congestion costs are the hourly gross congestion costs

[1]"Economic local generation" as defined shall mean the amount of generation capacity (in MW) (other than units that are running out of merit order at an offer-capped price pursuant to Section 6 of Schedule 1 of the Operating Agreement) that is on-line and available to affected load at each bus subject to the constraint, excluding generation at each bus that has a powerflow distribution factor on the constraint of less than 3%, at prices (as determined from generators' day-ahead price bids into the PJM Energy Market, provided that a price bid of zero shall be attributed to self-scheduled units) no greater than the PJM system marginal price.

(hedgeable plus unhedgeable congestion costs) that were not hedged. The hourly gross congestion costs are computed as the product of the shadow price (Lagrange multiplier) of the constraint, which represents the incremental reduction in congestion costs achieved by relieving the constraint by one MW, and the total affected load during each hour. The total affected load in each hour for a constraint is computed as the sum of the loads at each bus multiplied by the appropriate distributed slack power transfer distribution factor. In theory, every load bus in the network should be considered, but in practice, there is very little loss of accuracy if load buses are included that have distribution factors above a certain percentage, e.g., 3%.

Thresholds are the first step to identifying recommended economically motivated expansion. The identification of a path for which unhedgeable congestion exceeds threshold means that a market window has opened and market participants have the ability to propose projects through the queues for relieving the constraint. If no market solution to the congestion is present after one year, then a more detailed cost-benefit analysis is performed.

9.2.4 Further Reading

This section has provided a highly condensed view of existing planning processes for electric transmission systems as reported by PJM. Another reference useful in study of the PJM implementation includes [23]. Although other implementations of RTO-based planning processes share some similarities with that of PJM's, significant differences exist. Some other implementations at the time of this writing include that of the New York ISO [24], ISO-New England [25], Cal-ISO [26, 27], the Electric Reliability Council of Texas (ERCOT) [28, 29], and the Midwest ISO [30]. A different but equally important view on transmission expansion, from a pure transmission company, is provided in Section I of [31] and [32]. Some additional recommended reading includes [33], which provides historical context and reviews some of the other implementations and [34], which also surveys some of the other implementations. A book on related policy and strategy was also recently published [35].

9.3 TRANSMISSION LIMITS

The North American Electric Reliability Council (NERC), maintains an extensive set of planning standards [36] that address system reliability, system modeling data requirements, system protection and control, and system restoration. These standards require that under normal operating conditions, also called pre-contingency conditions, Level A performance requirements be met such as circuit loadings are within continuous ratings and voltage magnitudes lie within a specified range, e.g., 0.97–1.05 pu. In addition, reliability standards require that under contingency conditions, specified disturbance-performance criteria are met. A fundamental part of the reliability standards is the disturbance-performance table. This table is based on the planning philosophy that a higher level of performance (or lower level of severity) is required for disturbances having a higher occurrence likelihood. Typical disturbance-performance criteria are shown in Table 9.1. This table is similar in principle to NERC's table [37], where, for example, performance Level B requires that loss of a single element (an N-1 contingency) result in performance where: (a) transient criteria require that voltage dips may not exceed 25% of pre-contingency levels for any time, they may not exceed 20% for more than 20 cycles (0.333 sec), and frequency transients may not exceed 59.6 hz for more than 6 cycles, and (b) post-transient criteria require that voltage deviations remain within 5% of pre-contingency voltages, and all circuit loadings within their applicable ratings. Level C criteria applies to the less likely loss of two components (an N-2 contingency), but its performance criteria is less restrictive. Level D applies to very rare events with no explicit performance criteria specified, leaving the engineer to make a judgment. A voltage instability criterion is usually applied in planning studies, but to maintain simplicity, such a criterion is not indicated in Table 9.1.

Key to understanding power system flow limitations is the fact that limits on operating conditions (such as flows) can be imposed by violation of either Level A

TABLE 9.1. Example of Typical Disturbance-Performance Criteria.

Disturbance	Perf. Level	Performance Requirements			
		Transient Criteria		Post-transient criteria	
		Transient voltage dip criteria, ΔV_1	Minimum transient frequency	Post transient voltage dev, ΔV_2	Loading within emergency ratings
SLG fault or 3F fault w/loss of 1 generator or 1 circuit or DC monopole	B	—max *V* Dip—25% —max duration of *V* dip exceeding 20% is 20 cycles	max duration of freq <59.6 hz is 6 cycles	5%	Yes
SLG w/ or w/o delayed clearing or 3F fault w/loss of 2 generators or 2 circuits or DC bipole	C	—max *V* Dip—30% —max duration of *V* dip exceeding 20% is 40 cycles	max duration of freq <59.0 hz is 6 cycles	10%	Yes
Extreme events such as 3F fault with delayed clearing w/loss of 2 or more components	D	Evaluate for risks and consequences			

criteria or the contingency-driven Levels B or C criteria. In the case of Level A violation, transmission enhancements identified to relieve the violation must operate under normal conditions. In the case of Level B or C violation, transmission enhancements identified to relieve the violation (which is a post-contingency violation) need operate only following the contingency; thus, they may be active before the contingency as well, or they may not.

Yet regardless of whether the constraint is in the normal or in the contingency condition, and regardless of whether the relieving transmission enhancement is active in the pre- and post-contingency state, or only in the post-contingency state, the effect of the transmission enhancement is to relieve the limitation in the pre-contingency state. The significance of this observation lies in the fact that nodal-priced-based electricity markets operate almost all the time under the normal condition. Enhancements to raise transmission limits associated with Level A violations also affect the electrical characteristics of the network seen and thus the flows seen by the market. On the other hand, it is possible to raise transmission limits associated with Levels B or C violations so that the electrical characteristics of the network seen by the market do not change. This is done through the provision of a control that actuates only following the occurrence of a contingency with intention to eliminate the violation; most system protection schemes (SPS) [38] (in contrast to local protection which functions to isolate faults) are of this nature as are switched capacitors [38, 39, 40]. Many types of SPS, and all switched capacitors, are discrete-event controls. Switched capacitors are most common as switched shunt devices, in which case they alleviate mainly voltage violations, but they may also be switched as series devices, in which case they may alleviate both voltage and flow-related violations[2]. In this chapter, we explore the engineering and economic considerations for expanding transmission capability using switched shunt and series capacitors.

9.4 DECISION SUPPORT MODELS

The transmission planning process unavoidably includes a great deal of stakeholder input, human interaction, and subjective decision, and it is impractical to look for a single software application to provide the transmission planning solution. Yet software applications can and must be used in the process at appropriate times to guide and support human analysis, understanding, judgment, and decision, and suites of commercial tools are available today for this purpose. Good texts covering basic concepts used in developing many of these tools include [41, 42]. A more recent

[2]Series capacitor compensation has two effects that are not of concern for shunt capacitor compensation. First, series capacitors can expose generator units to risk of sub-synchronous resonance (SSR), and such risk must be investigated. Second, series capacitors also have significant effect on real power flows. In our work, we intend that both shunt and series capacitors be used as contingency-actuated controls (and therefore temporary) rather than continuously operating compensators. As a result, the significance of how they affect real power flows may decrease. However, the SSR risk is still a significant concern. To address this issue, the planner must identify *a-priori* lines where series compensation would create SSR risk and eliminate those lines from the list of candidates.

and quite comprehensive review of transmission planning models is given in [43]. Most of these tools endeavor to identify transmission enhancements that optimize the tradeoff between economy and reliability of electric energy delivery for given generation and load growth futures over a specified planning period. Almost all of these tools are therefore built upon optimization models.

In Section 9.4.1, we provide what we consider to be a comprehensive problem statement for the transmission planning problem, and in Sections 9.4.2 and 9.4.3, we describe and illustrate solution approaches to two sub-problems; in one case, transmission enhancements are limited to transmission circuits only, and in the other case, transmission enhancements are limited to switched shunt or series capacitors only.

9.4.1 Optimization Formulation

This section provides a comprehensive statement of the transmission expansion planning problem via an optimization model. The problem is to determine the time, type and location of new transmission facility additions given the cost of investment and production, the benefit of consumption, and constraints on reliability and equipment capabilities. The optimization model is a mixed-integer nonlinear programming problem that identifies the optimum among tradeoffs between surplus (consumption benefits less production costs) and transmission investments. From the perspective of a central system operator, the problem is posed as follows:

$$
\begin{aligned}
\min\ & \underbrace{\sum_{t_y\in T_y}\sum_{i\in\Omega_1}\beta_i(t_y)C_i(B_i(t_y),q_i(t_y))}_{\text{investment cost of shunt compensation}}+\underbrace{\sum_{t_y\in T_y}\sum_{j\in\Omega_2}\beta_j(t_y)C_j(X_j(t_y),q_j(t_y))}_{\text{investment cost of series compensation}} \\
&+\underbrace{\sum_{t_y\in T_y}\sum_{m\in N_m}\beta_m(t_y)C_m(n_m(t_y))}_{\text{investment cost of transmission line}}+\underbrace{\sum_{t_h\in T_h}\sum_{z\in N_g}\beta_z(t_y)C_{gz}(P_{gz}(t_y,t_h))}_{\text{production cost of real power generation}} \qquad (9.1) \\
&-\underbrace{\sum_{t_h\in T_h}\sum_{l\in N_l}\beta_l(t_y)R_l(P_{dl}(t_y,t_h))}_{\text{consumer benefit}}
\end{aligned}
$$

Subject to the following constraints:

- Transmission line expansion limit:

$$0\le\sum_{t_y\in T_y}n_m(t_y)\le n_{m,\max} \qquad (9.2)$$

- Capacity limit of switched shunt compensations:

$$0\le\sum_{t_y\in T_y}B_i(t_y)\le B_{i,\max} \qquad (9.3)$$

$$0\le B_i(t_y)\le q_i(t_y)B_{i,\max} \qquad (9.4)$$

$$q_i(t_y)=0,1 \qquad (9.5)$$

$$0 \le B_i^{(k)}(t_y, t_h) \le \sum_{t_y=1}^{t_y} B_i(t_y) \tag{9.6}$$

- Capacity limit of switched series compensations:

$$0 \le \sum_{t_y \in T_y} X_j(t_y) \le X_{j.\max} \tag{9.7}$$

$$0 \le X_j(t_y) \le q_j(t_y) X_{j.\max} \tag{9.8}$$

$$q_j(t_y) = 0, 1 \tag{9.9}$$

$$0 \le X_j^{(k)}(t_y, t_h) \le \sum_{t_y=1}^{t_y} X_j(t_y) \tag{9.10}$$

- Power flow equations under normal operating condition and contingencies:

$$\begin{aligned} &-P_i(t_y, t_h) + V_i^{(k)}(t_y, t_h) \sum_j V_j^{(k)}(t_y, t_h) \Big[\overline{G}_{ij}^{(k)}(t_y, t_h) \cos\theta_{ij}^{(k)}(t_y, t_h) \\ &+ \overline{B}_{ij}^{(k)}(t_y, t_h) \sin\theta_{ij}^{(k)}(t_y, t_h) \Big] = 0 \end{aligned} \tag{9.11}$$

$$\begin{aligned} &-Q_i(t_y, t_h) + V_i^{(k)}(t_y, t_h) \sum_j V_j^{(k)}(t_y, t_h) \Big[\overline{G}_{ij}^{(k)}(t_y, t_h) \sin\theta_{ij}^{(k)}(t_y, t_h) \\ &- \overline{B}_{ij}^{(k)}(t_y, t_h) \cos\theta_{ij}^{(k)}(t_y, t_h) \Big] = 0 \end{aligned} \tag{9.12}$$

- Voltage stability margin limit under normal operating condition and contingencies:

$$M^{(k)}(t_y, t_h) \ge M_{\min}^{(k)} \tag{9.13}$$

- Voltage magnitude limit under normal operating condition and contingencies:

$$V_{i.\min}^{(k)} \le V_i^{(k)}(t_y, t_h) \le V_{i.\max}^{(k)} \tag{9.14}$$

- Line-flow limit under normal operating condition and contingencies:

$$\left| S_{ij}^{(k)}(t_y, t_h) \right| \le S_{ij.\max}^{(k)} \tag{9.15}$$

- Generator output limit under normal operating condition and contingencies:

$$P_{gz.\min}(t_y) \le P_{gz}(t_y, t_h) \le P_{gz.\max}(t_y) \tag{9.16}$$

$$Q_{gz.\min}(t_y) \le Q_{gz}(t_y, t_h) \le Q_{gz.\max}(t_y) \tag{9.17}$$

- Consumer demand limit under normal operating condition and contingencies:

$$P_{dl.\min}(t_y) \le P_{dl}(t_y, t_h) \le P_{dl.\max}(t_y) \tag{9.18}$$

$$Q_{dl.\min}(t_y) \le Q_{dl}(t_y, t_h) \le Q_{dl.\max}(t_y) \tag{9.19}$$

- Generation/load growth with rate $a \ge 1$ and constant power factor:

$$P_{gz.\max}(t_y) = \alpha P_{gz.\max}(t_y - 1) \tag{9.20}$$

$$Q_{gz.\max}(t_y) = P_{gz.\max}(t_y) \tan(\text{acos}(\text{pf}_{gz})) \tag{9.21}$$

$$P_{dl,\max}(t_y) = \alpha P_{dl,\max}(t_y - 1) \quad (9.22)$$

$$Q_{dl,\max}(t_y) = P_{dl,\max}(t_y)\tan(\text{acos}(\text{pf}_{\text{dl}})) \quad (9.23)$$

Some clarifying remarks about this formulation follow:

- The objective function (1) is to minimize production and investment costs over the planning period. $\beta(t_y)$ is the discount factor for year t_y (we provide for different discount factors for different terms, reflecting the fact that different organizations may borrow at different interest rates); $q_i(t_y)$ and $q_j(t_y)$ are binary decision variables for switched shunt and series compensation at bus i and branch j, respectively, in year t_y; $C_i(B_i(t_y), q_i(t_y))$ is the cost of installing switched shunt compensation at bus i; $B_i(t_y)$ is the amount of switched shunt compensation under the installation decision $q_i(t_y)$; $C_j(X_j(t_y), q_j(t_y))$ is the cost of installing switched series compensation at branch j, $X_j(t_y)$ is the amount of switched series compensation under the installation decision $q_j(t_y)$; $C_m(n_m(t_y))$ is the cost of installing $n_m(t_y)$ number of circuits for branch m which could be between any pre-selected feasible pair of buses; $C_{gz}(P_{gz}(t_h))$ is the generator z's real power production cost function, $R_l(P_{dl}(t_h))$ is the consumer l's benefit function.
- The decision variables are $B_i(t_y)$, $q_i(t_y)$, $B_i^{(k)}(t_y, t_h)$, $X_j(t_y)$, $q_j(t_y)$, $X_j^{(k)}(t_y, t_h)$, $n_m(t_y)$, $P_{gz}(t_h)$, $P_{dl}(t_h)$. We assume V_i is known for each generator bus, and power factor is known for each load bus.
- T_h is set of all hours within a planning period.
- T_y is set of all years within a planning period.
- $n_m(t_y)$ is the number of circuits added for branch m in year t_y, $n_m(t_y)$ is a non-negative integer.
- Superscript $k = 0$ corresponds to no contingency, $k = 1$ to first contingency, $k = 2$ to second contingency, etc.
- $B_i^{(k)}(t_y, t_h)$, $X_j^{(k)}(t_y, t_h)$ are the amount of shunt/series compensations switched on under contingency k during year t_y and hour t_h.
- Ω_1, Ω_2 are candidate locations for shunt and series compensations respectively.
- N_m is the set of candidate locations for new transmission lines.
- N_g is the set of adjustable generators.
- N_l is the set of load buses.
- $P_{gz}(t_y, t_h)$ is the real power output of generator z during year t_y and hour t_h.
- $P_{dl}(t_y, t_h)$ is the real power consumption of load l during year t_y and hour t_h.
- $P_i(t_y, t_h)$ is the real power injection at bus i during year t_y and hour t_h.
- $Q_i(t_y, t_h)$ is the reactive power injection at bus i during year t_y and hour t_h.
- $\overline{G}_{ij}^{(k)}(t_y, t_h)$, $\overline{B}_{ij}^{(k)}(t_y, t_h)$ are functions of switched shunt/series compensations $B_i^{(k)}(t_y, t_h)$, $X_j^{(k)}(t_y, t_h)$, and newly added transmission lines $n_m(t_y)$.
- M, $M^{(k)}$ are voltage stability margin under normal condition and contingencies respectively and they are dependent on decision variables. Voltage stability

margin is defined as the distance between the nose point (the saddle node bifurcation point) of the system power-voltage (PV) curve and the total system real power load at a given operating condition. It can not be expressed with a closed-form function.

- V, $V^{(k)}$ are bus voltage magnitude under normal condition and contingencies respectively and they are dependent on decision variables.
- S, $S^{(k)}$ are the power flow through transmission lines under normal condition and contingencies respectively and they are dependent on decision variables.

The above formulation requires an optimized network solution together with a full contingency assessment for every hour of the planning period. Although rigorous, computational requirements render such a formulation impractical for large-scale networks. As a result, approximations are typically necessary and can include one or more of the following:

- *Hours:* Analysis in each year may be limited to only representative hours, e.g., typical hours in a day (peak, off-peak), for typical days (weekdays, weekend days), within a few seasons (summer, winter) to estimate the required attributes over the year.
- *Years:* Analysis may be limited to only certain years within the planning period; the simplest approximation would study include only the final year.
- *Decision variables and objective function:* Decision variables may be limited to only those associated with transmission circuits or to only those associated with switched reactive elements.

We consider two cases in the following sections where the hours and years are limited to only one, the peak load hour during the final year. In the first case, described in Section 9.4.2, the decision variables are limited to only those associated with transmission circuits. In the second case, described in Section 9.4.3, the decision variables are limited to only those associated with switched reactive elements.

9.4.2 Planning Transmission Circuits

The formulation of 9.4.1 reduces to the formulation presented in this section if we restrict our decision variables to just transmission circuits and make the following additional assumptions:

- The planning horizon is over T_y periods with the variable t representing a single period so that $t_y = 1, \ldots, T_y$. A period could be a single year, but it may be more appropriate to cover the range of loading conditions that it be quarters (i.e., fall, winter, spring, summer).
- Peak loading conditions are modeled for each period, and it is assumed that these conditions are constant throughout the period.
- Costs of planning and building a new transmission circuit are incurred during the period that it goes into service.

- The consumer utility is assumed to be a constant during each period (i.e., the consumer demand is fixed).
- We do not consider contingencies.
- The DC power flow model is adopted.

The formulation given in this section is adapted from that given in Section 6.3 of [1]. The objective function of our optimization problem can be formulated as the sum of the aggregate production costs C_E and the aggregate transmission circuit investment costs C_I in future periods, according to:

$$C = C_E + C_I = \sum_{t_y=1}^{T_y} \sum_{z=1}^{N_g} \beta_z(t_y) C_{gz}(t_y) P_{gz}(t_y) + \sum_{t_y=1}^{T_y} \sum_{b \in N_b} \beta_b(t_y) C_b(t_y) q_b(t_y) \qquad (9.24)$$

- $\beta_z(t_y)$ is the discount factor of real power production cost for period t_y, $\beta_b(t_y)$ is the discount factor of transmission circuit investment cost for period t_y.
- $C_{gz}(t_y)$ is average cost of producing 1 per-unit power at node z during period t_y.
- $P_{gz}(t_y)$ is the generation level for unit z at period t_y loading conditions.
- N_b is the set of candidate circuits.
- $C_b(t_y)$ is the investment cost of a circuit in branch m during period t_y.
- $q_b(t_y)$ is an integer 0 or 1. It is 1 if circuit $b \in N_b$ is put in service during period t_y, and 0 otherwise. In other words, each candidate transmission circuit is associated with a binary decision variable.

The equality constraints that we need are those which will force the solution to satisfy electrical laws associated with how power flows in the network. This is accomplished by enforcing the DC power flow equations:

$$A^T P_B = P \qquad (9.25)$$

$$P_B = (D \times A) \times \theta \qquad (9.26)$$

where A is the network node-arc incidence matrix and D is a diagonal matrix of negative branch susceptances. First set corresponding to (9.25) is as follows:

$$\sum_{b:B[b]=k} (-P_b) + \sum_{b:E[b]=k} P_b = \begin{cases} P_{di} - P_{gi}, i = 1, \ldots N_g \\ P_{di}, i = N_g + 1, \ldots N \end{cases} \qquad (9.27)$$

The second set corresponding to (9.26) is as follows. For existing branches ($b \in N_e$):

$$\theta(B[b]) - \theta(E[b]) = X_b P_b \qquad (9.28)$$

For candidate branches ($b \in N_b$):

$$\theta(B[b]) - \theta(E[b]) = X_b P_b + (z_b(t_y) - 1)G + U_b \qquad (9.29)$$

$$U_b \leq 2(1 - z_b(t_y))G \qquad (9.30)$$

$$U_b > 0 \qquad (9.31)$$

Here,

- P_b is the flow on branch b if that flow is in the defined direction.
- $B[b]$: This is the node from which branch b begins.
- $E[b]$: This is the node at which branch b ends.
- $\theta(B[b])$ is the angle variable at the begin node of branch b.
- $\theta(E[b])$ is the angle variable at the end node of branch b.
- P_{di} is the demand at node i.
- P_{gi} is the generation at bus i (previously defined).
- N_g is the total number of generator buses.
- N is the total number of buses.
- X_b: The branch reactance associated with branch b.
- N_e: The set of existing branches.
- N_b: The set of candidate branches (previously defined).
- U_b is a continuous fictitious variable included in the decision vector.
- G is a large constant.

$z_b(t_y)$ is an integer 0 or 1. It is 1 if circuit $b \in N_b$ is put in service before or during period t_y, and 0 otherwise. Therefore $z_b(t_y) = \sum_{t=1}^{t_y} q_b(t)$, and $\sum_{t=1}^{T_y} q_b(t) \leq 1$ $(b \in N_b)$.

Equations (9.29) through (9.31) need some explanation. Before we give that, we introduce inequality constraints.

The inequality constraints are for existing branches ($b \in N_e$):

$$-P_{b.\max} \leq P_b \leq P_{b.\max} \tag{9.32}$$

and for candidate branches ($b \in N_b$):

$$-z_b(t_y)P_{b.\max} \leq P_b \leq z_b(t_y)P_{b.\max} \tag{9.33}$$

To constrain generation levels, we have:

$$P_{gi} \leq P_{gi.\max} \tag{9.34}$$

And finally we constrain the following variables to be non-negative:

$$P_{gi}, \theta_i \geq 0 \tag{9.35}$$

When $z_b(t_y) = 1$ (branch b is in), then the equations (9.29), (9.30) and (9.31) reduce to:

$$\theta(B[b]) - \theta(E[b]) = X_b P_b \tag{9.29a}$$

$$U_b \leq 0 \tag{9.30a}$$

$$U_b > 0 \tag{9.31a}$$

Equation (9.29a) is just the line flow equation for branch b, and equations (9.30a) and (9.31a) constrain U_b to be exactly zero. When $z_b(t_y) = 0$ (branch b is out), then these equations reduce to:

$$\theta(B[b])-\theta(E[b])=-G+U_b \tag{9.29b}$$

This is because equation (9.33) forces P_b to be zero when $z_b(t_y) = 0$.

$$U_b \le 2G \tag{9.30b}$$

$$U_b > 0 \tag{9.31b}$$

Equations (9.29b), (9.30b) and (9.31b) indicate that when the angular difference $\theta(B[b]) - \theta(E[b])$ lies in a closed interval $[-G, G]$, there always exists a variable U_b such that equations (9.30b) and (9.31b) hold. That is to say, if the value of G is large enough, Equations (9.29b), (9.30b) and (9.31b) put no restriction on the angular variables. The above equality and inequality constraints are held for each t_y period. In addition, generation and load are assumed to be increased with rate $a \ge 1$:

$$P_{gz.\max}(t_y)=\alpha P_{gz.\max}(t_y-1) \tag{9.36}$$

$$P_{di}(t_y)=\alpha P_{di}(t_y-1) \tag{9.37}$$

The above mathematic model of transmission circuit planning is a mixed integer programming problem. It can be solved by the branch-and-bound method [61].

Example 1: Optimal transmission expansion by transmission circuits

The proposed transmission circuit planning model has been applied to a three bus power system shown in Figure 9.5. All the parameter values are in p.u. in the figure. For the simplicity of illustration, in this example, we only consider transmission circuit planning for one horizon year. The candidate transmission circuits are pre-selected to be Line 1-3B and Line 2-3B represented as the dashed lines in Figure 9.5. The parameter values adopted in the transmission circuit planning are given in Table 9.2.

The result indicates that Line 2-3B should to be built. The optimal output of the generators and the power flow results are shown in Figure 9.6.

TABLE 9.2. Parameter Values Adopted in Example 1.

C_{g1}	C_{g2}	C_{g3}	$C_{line2\text{-}3B}$	$C_{line1\text{-}3B}$	$P_{line1\text{-}2,max}$	$P_{line1\text{-}3,max}$	$P_{line2\text{-}3,max}$	$P_{line1\text{-}3B,max}$	$P_{line2\text{-}3B,max}$
1.0	1.5	3.0	1.0	1.0	1.5	1.0	1.0	1.0	1.0

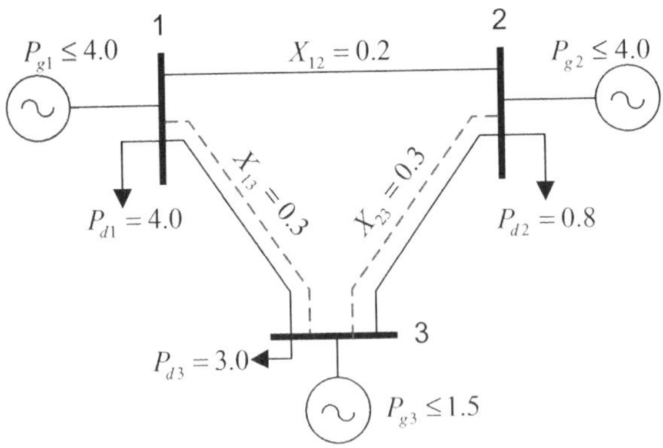

Figure 9.5. A three bus power system.

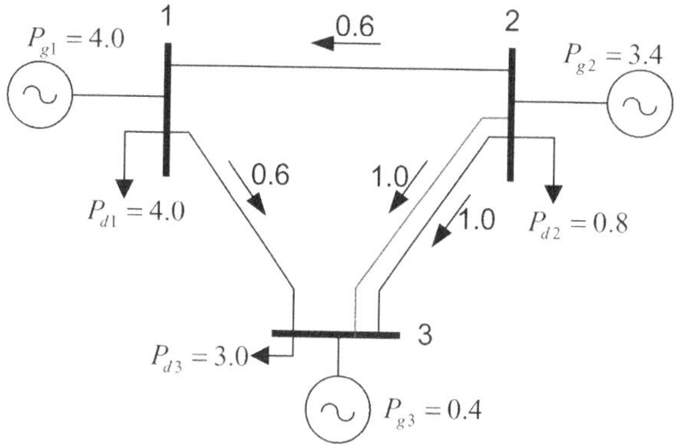

Figure 9.6. Simulation result for the transmission circuit planning.

9.4.3 Planning Transmission Control

As indicated in the introduction, additional control capability can be an attractive option for transmission expansion as it requires no new right-of-way and is generally less costly. In this section, we focus on planning reconfigurable reactive power control to increase the voltage stability limit and thus enhance transmission capability in voltage stability limited systems. In other words, we address the optimization formulation of Section 9.4.1 based on the following assumptions:

- No new transmission circuits may be installed, and generation expansion occurs only at existing generation facilities. This assumption represents the extreme form of relying on control to strengthen/expand transmission capability without building new transmission lines or strategically siting new generation.
- Decision variables are restricted to include only mechanically switched shunt/series capacitors.
- Expansion facilities are installed at the end of a particular year, and all costs of planning and building facilities are incurred in the period that they go into service.
- The consumer utility is fixed during each year (i.e., the consumer demand is constant).
- We represent only the effect of capacitive compensation on voltage stability margin, i.e., voltage and power flow magnitude constraints are excluded.
- The effects of production costs and consumer benefit on the planning decisions are not considered, and so the resulting objective is to identify the most cost-effective means of deploying switched capacitive compensation in order to satisfy voltage instability constraints.

These assumptions may be relieved at the cost of additional computational complexity.

In planning reconfigurable reactive power control, there are three problems to address:

(1) when is system enhancement needed;

(2) where to implement the enhancement;

(3) how much reactive power control is needed.

The first question is addressed using the techniques of continuation power flow (CPF) [44, 45, 46] and fast contingency screening [47, 48]. The last two questions are answered under an optimization framework, as has been done in a number of reactive power planning formulations [49, 50, 51, 52, 53, 54, 55]. Generally, the reactive power planning problem is formulated as a mixed integer nonlinear programming problem with objective to minimize the installation cost of reactive power devices subject to a set of equality and inequality constraints. Our efforts extend those mentioned in [49–55] by including contingency conditions so that identified controls have the capability of being reconfigured to secure the system given occurrence of a contingency. There have been relatively fewer reported efforts along these lines, with the exceptions summarized in what follows.

Yorino et al. in [56] proposed a mixed integer nonlinear programming formulation for reactive power control planning which takes into account the expected cost for voltage collapse and corrective controls. The Benders decomposition technique was applied to get the solution. As the authors indicated, they experienced poor convergence for some situations. Feng et al. in [57] used linear optimization with the objective of minimizing the control cost to derive reactive power controls based on voltage stability margin sensitivity [58, 59, 60], with formulation suitable to operational decision making, and therefore, without regard to modeling investment costs.

In the remainder of this section, a comprehensive methodology is described to address long-term reactive power control planning under the previous stated assumptions. Basic background on contingency screening and continuation power flow techniques are described in Section 9.4.3.1. Then the main steps of the proposed planning procedure, illustrated in Figure 9.7, are summarized as follows.

- Step 1: Identify the generation/load growth future (See Section 9.4.3.2A).
- Step 2: Assess voltage stability by fast contingency screening and the CPF techniques for each horizon year. The year when the voltage stability margin becomes less than the required value is the time to enhance the transmission system by adding reactive power controls (See Section 9.4.3.2B).
- Step 3: Select candidate control locations using a graph-search method (See Section 9.4.3.2C).
- Step 4: Refine location and amount of controls based on mixed integer programming and linear programming. The optimization formulation is to minimize the total installation cost including fixed cost and variable cost of new controls while satisfying the voltage stability margin requirement under normal and contingency conditions. The branch-and-bound and primal-dual interior-point methods [61] are used to solve the optimization problem (See Section 9.4.3.2D).

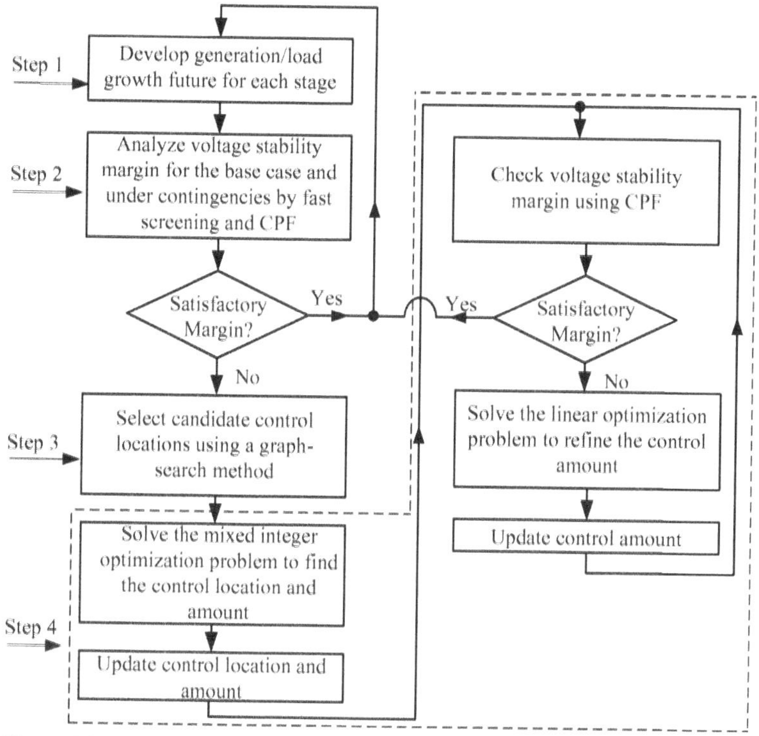

Figure 9.7. Flowchart for the reactive power control planning.

9.4.3.1 Voltage Stability Margin and Margin Sensitivity In this section, the notion of voltage stability margin and its sensitivity to parameters are defined, for such sensitivities are used in the planning procedure. Voltage stability margin is defined as the distance between the nose point (the saddle node bifurcation point) of the system power-voltage (PV) curve and the forecasted total system real power load as shown in Figure 9.8. The potential for contingencies such as unexpected component (generator, transformer, transmission line) outages often reduces the voltage stability margin [62, 63, 64]. Our objective is to find effective and economic reactive power controls to satisfy margin requirements under a set of specified contingencies. Reactive controls can be adopted to increase the voltage stability margin. Generally, series and shunt capacitors improve voltage stability margin [64]. Figure 9.8 shows the voltage stability margin under different operating conditions and controls.

One indicator that we will find very useful in planning reactive power controls is how much control is needed for the requirement of a given amount of margin increase. Margin sensitivities [58, 59, 60] are used to address this issue. Margin sensitivities provide the variation of the voltage stability margin with respect to any small change of power system parameter or control variable. The margin sensitivity may be used to estimate voltage stability margin if the variation of the control variable is small [59]. A typical voltage stability margin requirement is 5% under normal and "N-1" contingency conditions [65]. In addition, margin sensitivity is

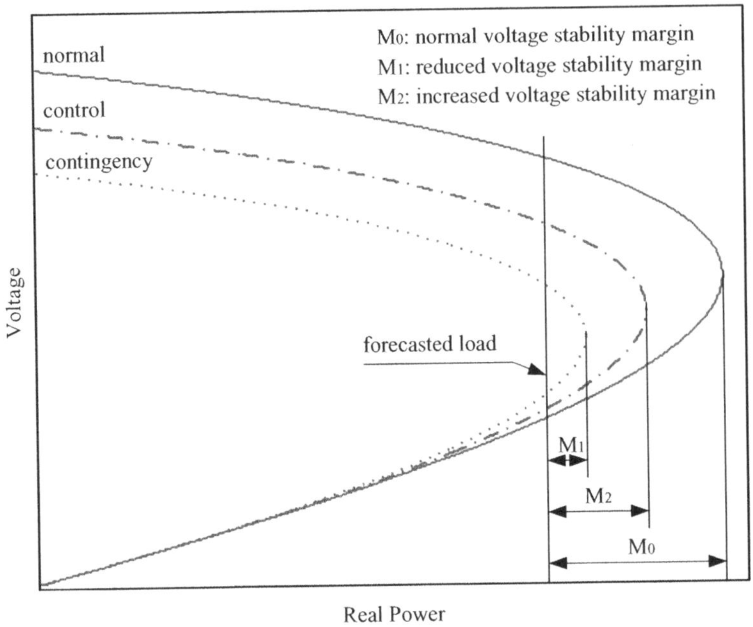

Figure 9.8. Voltage stability margin for different operating conditions.

useful in selecting candidate control locations [56, 57]. In the following, the analytical expression of the margin sensitivity is given. The details of the margin sensitivity can be found in [58, 59, 60].

Suppose that the steady state of the power system satisfies a set of equations expressed in the vector form:

$$F(x, p, \lambda) = 0 \tag{9.38}$$

where x is the vector of state variables, p is any parameter in the power system steady state equations such as the susceptance of shunt capacitors or the reactance of series capacitors, λ is the bifurcation parameter which is a scalar. At the nose point of the system PV curve, the value of the bifurcation parameter is denoted λ^*.

A specified system scenario can be parameterized by λ as:

$$P_{li} = (1 + K_{lpi}\lambda)P_{li0} \tag{9.39}$$

$$Q_{li} = (1 + K_{lqi}\lambda)Q_{li0} \tag{9.40}$$

$$P_{gj} = (1 + K_{gj}\lambda)P_{gj0} \tag{9.41}$$

where P_{li0} and Q_{li0} are the initial loading conditions at the base case where λ is assumed to be zero, and $Q_{li0} = P_{li0}\tan(\psi_i)$ (where ψ_i is the power factor angle of the ith load). K_{lpi} and K_{lqi} are factors characterizing the load increase pattern. P_{gj0} is the real power generation at bus j at the base case. K_{gj} represents the generator load pick-up factor. The voltage stability margin can be expressed as:

$$M = \sum_{i=1}^{n} P_{li} - \sum_{i=1}^{n} P_{li0} = \lambda^* \sum_{i=1}^{n} K_{lpi} P_{li0} \tag{9.42}$$

The sensitivity of the voltage stability margin with respect to the control variable at location i, S_i, is:

$$S_i = \frac{\partial M}{\partial p_i} = \frac{\partial \lambda^*}{\partial p_i} \sum_{i=1}^{n} K_{lpi} P_{li0} \tag{9.43}$$

In (43), the bifurcation parameter sensitivity with respect to the control variable p_i evaluated at the nose point of the system PV curve is:

$$\frac{\partial \lambda^*}{\partial p_i} = -\frac{w^* F_{pi}^*}{w^* F_{\lambda}^*} \tag{9.44}$$

where w is the left eigenvector corresponding to the zero eigenvalue of the system Jacobian F_x, $F\lambda$ is the derivative of F with respect to the bifurcation parameter λ, and F_{pi} is the derivative of F with respect to control variables such as shunt capacitor susceptance or series capacitor reactance.

9.4.3.2 *Reactive Control Planning Algorithm* The proposed reactive power control planning approach requires four steps:

1. development of generation/load growth future,
2. contingency selection,
3. selection of candidate control locations, and
4. refinement of locations and amounts of capacitive controls via a mixed integer programming and linear programming problems.

These steps are described in the remainder of this section. It is assumed that this algorithm is applied to a power system model representing a specific future year. This is a simplifying assumption that removes from the problem the issue of when different enhancements should be implemented.

9.4.3.2.1 Development of Generation and Load Growth Futures In this step, the generation/load growth future is identified, where the future is characterized by a load growth percentage for each load bus and a generation allocation for each generation bus. For example, one future may assume uniformly increasing load at 5% per year and allocation of that load increase to existing generation (with associated increase in unit reactive capability) based on percentage of total installed capacity. Such generation/load growth future can be easily implemented in the CPF program [44] by parameterization as shown in (9.39), (9.40) and (9.41).

9.4.3.2.2 Voltage Stability Assessment by Fast Contingency Screening We use the CPF program to calculate the voltage stability margin of the system under each credible contingency. However, the CPF algorithm is computation-intensive. Margin sensitivities can be used to reduce computation in the screening analysis, using a standard screening approach [48]. First, the CPF program is used to calculate the voltage stability margin for the base case, and second, margin sensitivities are computed with respect to line admittances S_l and bus power injections S_{pq}. For circuit outages, the voltage stability margin is estimated as

$$M^{(k)} = M^{(0)} + S_l \Delta l \tag{9.45}$$

where $M^{(k)}$ is the voltage stability margin under contingency k, $M^{(0)}$ is the voltage stability margin under base case conditions, and Δl is the negative of the admittance vector for the outaged circuits. For generator outages, the voltage stability margin is estimated as

$$M^{(k)} = M^{(0)} + S_{pq} \Delta pq \tag{9.46}$$

where Δpq is the negative of the output power of the outaged generators. Then contingencies are ranked from most to least severe according to the value of the estimated voltage stability margin. After the ordered contingency list is obtained, we evaluate each contingency using the CPF program and stop testing after encountering N sequential contingencies that have the voltage stability margin greater than or equal to the required value, where N depends on the size of the contingency list.

9.4.3.2.3 Selection of Candidate Control Locations To select appropriate candidate reactive power control locations [56, 57] the following procedure is applied:

1) Choose an initial set of switch locations using the bisection approach for each identified contingency possessing unsatisfactory voltage stability margin according to the following two steps:
   a) Rank the feasible control locations according to the numerical value of margin sensitivity in descending order with location one having the largest margin sensitivity and location n having the smallest margin sensitivity.
   b) Estimate the voltage stability margin with top half of the switches closed as

   $$M_{est}^{(k)} = \sum_{i=1}^{\lfloor n/2 \rfloor} X_{i\max}^{(k)} S_i^{(k)} + M^{(k)} \tag{9.47}$$

   where $M_{est}^{(k)}$ is the estimated voltage stability margin and $\lfloor n/2 \rfloor$ is the largest integer less than or equal to $n/2$. If the estimated voltage stability margin is greater than the required value, then the number of control locations is halved, otherwise the number of control locations is increased by adding the remaining half.
   c) Continue in this manner until the set of control locations that satisfies the voltage stability margin requirement are identified.
2) Refine candidate control locations for each identified contingency possessing unsatisfactory voltage stability margin using the backward/forward search algorithm (described below). The final candidate control locations are the union of the locations identified for all contingencies.

The backward/forward search algorithm is described as follows. Consider a graph where each node represents a configuration of discrete switches, and two nodes are connected if and only if they are different in one switch configuration.

The graph has 2^n nodes where n is the number of switches. We pictorially conceive of this graph as consisting of layered groups of nodes, where each successive layer (moving from left to right) has one more switch "on" (or "closed") than the layer before it, and the tth layer (where $t = 0, \ldots, n$) consists of a number of nodes equal to $n!/t!(n - t)!$. Figure 9.9 illustrates such a graph for the case of four switches, referred to as an automaton. The backward/forward search algorithm operates on this graph by beginning at an initial node and searching from that node in a prescribed direction, either backwards or forwards. The two extreme cases are either searching backward from the node corresponding to all switches closed (the strongest node) or searching forward from the node corresponding to all switches open (the weakest node). We give only the backward algorithm here since the forward algorithm is similar. The algorithm has four steps.

1. Select the node corresponding to all switches in the initial set that are closed.
2. For the selected node, check if voltage stability margin requirement is satisfied for the concerned contingency on the list. If not, then stop, the solution corresponds to the previous node (if there is a previous node, otherwise no solution exists).
3. For the selected node, eliminate (open) the switch that has the smallest margin sensitivity. We denote this as switch i^*:

$$i^* = \arg\left\{\min_{i\in\Omega_c} S_i^{(k)}\right\} \tag{9.48}$$

   where Ω_c = {set of closed switches for the selected node}, $S_i^{(k)}$ is the margin sensitivity with respect to the susceptance of shunt capacitors or the reactance of series capacitors under contingency k, at location i.
4. Choose the neighboring node corresponding to the switch i^* being off. If there is more than one switch identified in step three, i.e. $|i^*| > 1$, then choose any one of the switches in i^* to eliminate (open). Return to step two.

If step two of the above procedure results in no solution in the first iteration, then no previous node exists. In this case, we extend the graph in the forward direction by adding a new switch j^* that has the largest margin sensitivity, expressed by:

$$j^* = \arg\left\{\max_{i\in\Omega_c} S_i^{(k)}\right\} \tag{9.49}$$

9.4.3.2.4 Refinement of Location and Amount of Capacitive Controls This step is formulated as a mixed integer program (MIP) which minimizes control installation cost while increasing voltage stability margin to an arbitrarily specified percentage x:

Minimize:

$$F = \sum_{i\in\Omega_1}(C_{vi}B_i + C_{fi}q_i) + \sum_{j\in\Omega_2}(C_{vj}X_j + C_{fj}q_j) \tag{9.50}$$

subject to:

$$\sum_{i\in\Omega_1} S_i^{(k)}B_i^{(k)} + \sum_{j\in\Omega_2} S_j^{(k)}X_j^{(k)} + M^{(k)} \ge xP_{l0} \tag{9.51}$$

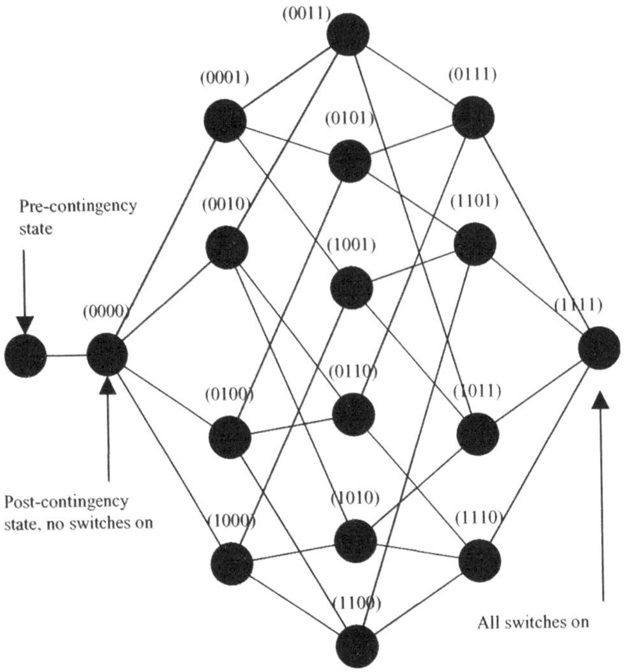

Figure 9.9. Automaton for four-switch problem.

$$B_{i\min}q_i \le B_i \le B_{i\max}q_i \tag{9.52}$$

$$X_{j\min}q_j \le X_j \le X_{j\max}q_j \tag{9.53}$$

$$0 \le B_i^{(k)} \le B_i \tag{9.54}$$

$$0 \le X_j^{(k)} \le X_j \tag{9.55}$$

$$q_{i,j} = 0, 1 \tag{9.56}$$

Here,

- C_f is fixed installation cost and C_v is variable cost of shunt or series capacitor switches;
- B_i is the size (susceptance) of the shunt capacitor at location i;
- X_j is the size (reactance) of the series capacitor at location j;
- $q_i = 1$ if the location i is selected for reactive power control expansion, otherwise, $q_i = 0$;
- the superscript k represents the contingency that leads the voltage stability margin to be less than the required value;
- O_1 is the set of candidate locations to install shunt capacitor switches;
- O_2 is the set of candidate locations to install series capacitor switches;
- $B_i^{(k)}$ is the size of the shunt capacitor to be switched on at location i under the contingency k;

- $X_j^{(k)}$ is the size of the series capacitor to be switched on at location j under the contingency k;
- $S_i^{(k)}$ is the sensitivity of the voltage stability margin with respect to the susceptance of the shunt capacitor at location i under contingency k;
- $S_j^{(k)}$ is the sensitivity of the voltage stability margin with respect to the reactance of the series capacitor at location j under contingency k;
- x is an arbitrarily specified voltage stability margin in percentage;
- P_{l0} is the forecasted system load;
- $M^{(k)}$ is the voltage stability margin under contingency k and without controls;
- $B_{i\min}$ is the minimal size of the shunt capacitor at location i;
- $B_{i\max}$ is the maximal size of the shunt capacitor at location i;
- $X_{j\min}$ is the minimal size of the series capacitor at location j, and
- $X_{j\max}$ is the maximal size of the series capacitor at location j.

For k contingencies that have the voltage stability margin less than the required value and n pre-selected candidate control locations, there are $n \cdot (k + 2)$ decision variables and $k + 3n + 2kn$ constraints. Fortunately, the number of candidate control locations can be limited to a relative small number even for problems of the size associated with practical power systems by assessing the combined effective index. Therefore, computational burden for solving the above MIP is not excessive even for large power systems. We solve this MIP using a branch and bound solution algorithm.

The output of the MIP is the control locations and amounts for all k contingencies and the combined control location and amount. For each contingency, the identified controls are switched in, and the voltage stability margin is recalculated to check if sufficient margin is achieved. However, because we use linear margin sensitivities to estimate the effect of the variations of control variables on the voltage stability margin, there may be contingencies that have voltage stability margin less than the required value after the network configuration is updated according to the results of the MIP. The control amount can be further refined by recomputing the margin sensitivity after the controls are updated under each contingency and adjusting the control amount via a second-stage linear program (LP) with control locations fixed at the locations found in the MIP. This LP is therefore formulated to minimize the adjusted installation cost subject to the constraint of the voltage stability margin requirement, as follows:

minimize:

$$F = \sum_{i \in \Omega_1'} C_{vi} \Delta B_i + \sum_{j \in \Omega_2'} C_{vj} \Delta X_j \tag{9.57}$$

subject to:

$$\sum_{i \in \Omega_1'} \overline{S}_i^{(k)} \Delta B_i^{(k)} + \sum_{j \in \Omega_2'} \overline{S}_j^{(k)} \Delta X_j^{(k)} + \overline{\overline{M}}^{(k)} \geq x P_{l0} \tag{9.58}$$

$$0 \leq B_i + \Delta B_i \leq B_{i\max} \tag{9.59}$$

$$0 \leq X_j + \Delta X_j \leq X_{j\max} \tag{9.60}$$

$$0 \le B_i^{(k)} + \Delta B_i^{(k)} \le B_i + \Delta B_i \tag{9.61}$$

$$0 \le X_j^{(k)} + \Delta X_j^{(k)} \le X_j + \Delta X_j \tag{9.62}$$

Here,

- ΔB_i is the adjusted size of the shunt capacitor at location i;
- ΔX_j is the adjusted size of the series capacitor at location j;
- Ω_1' is the set of identified locations to install shunt capacitors by solving the mixed integer programming problem;
- Ω_2' is the set of identified locations to install series capacitors by solving the mixed integer programming problem;
- $\overline{S}_i^{(k)}$ is the updated sensitivity of the voltage stability margin with respect to the susceptance of the shunt capacitor at location i under contingency k;
- $\overline{S}_j^{(k)}$ is the updated sensitivity of the voltage stability margin with respect to the reactance of the series capacitor at location j under contingency k;
- $\Delta B_i^{(k)}$ is the adjusted size of the shunt capacitor at location i under contingency k;
- $\Delta X_j^{(k)}$ is the adjusted size of the series capacitor at location j under contingency k;
- $\overline{M}^{(k)}$ is the updated voltage stability margin under the contingency k.

For k contingencies and n' computed control locations, there are $n' \times (k + 1)$ decision variables and $k + 2n' + 2kn'$ constraints. Again, by limiting the number of candidate control locations, computational requirements for this problem are not excessive, even for large systems. The above LP will provide good solutions because the voltage stability margin sensitivity can precisely predict the control amount under small deviation requirement of the voltage stability margin. Usually the deviation requirement of the voltage stability margin is relatively small after solving the first stage MIP. Re-solving once, beginning from the first solution, can result in small improvements, but we have not found subsequent solutions to significantly change. We solve this LP using a primal-dual interior-point method.

Example 2: Optimal transmission expansion by control

The approach described in the previous section is illustrated in this section using a small 9-bus test system modified from [66] and shown in Figure 9.10. The forecasted system load at the base case is 372.2 MW, and generators are economically dispatched. Table 9.3 shows the system loading and generation for the base case.

In the simulations, loads are modeled as constant power, voltage margin is computed assuming constant power factor at the loads, with load and generation scaled proportionally, and contingencies are assumed to be equally likely. In addition, the required voltage stability margin is assumed to be 15% for selection of candidate control locations (Step C) and 10% for refinement (Step D). The less restrictive margin requirement in location selection provides for a larger set of candidate locations that are used as input to the refinement set. Parameter values adopted in the procedure are given in Table 9.4.

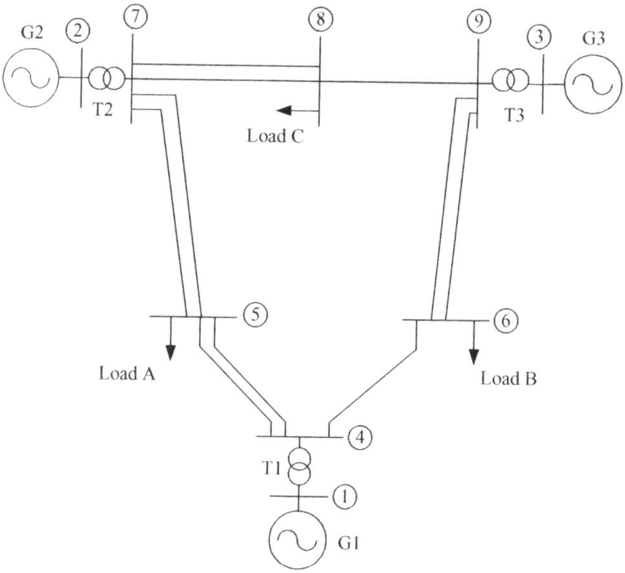

Figure 9.10. Modified WSCC nine-bus test system.

TABLE 9.3. Base Case System Loading and Generation.

	Load A	Load B	Load C	G1	G2	G3
MW	147.70	106.34	118.16	128.97	163.0	85.0
Mvar	59.08	35.45	41.36	41.39	16.72	−1.94

TABLE 9.4. Parameter Values Adopted in the Optimization Problem.

	Shunt Capacitor	Series Capacitor
Variable cost	$C_{vi} = 0.15$	$C_{vj} = 0.35$
Fixed cost	$C_{fi} = 0.13$	$C_{fj} = 0.25$
Maximum size	$B_{imax} = 0.16$	$X_{jmax} = 0.03$
Minimum size	$B_{imin} = 0.001$	$X_{jmin} = 0.001$

For each bus, consider the simultaneous outage of two components (generators, lines, transformers) connected to the bus. There exist two contingencies that reduce the post-contingency voltage stability margin to less than 10% as shown in Table 9.5.

We first plan candidate locations of shunt capacitors under the outage of lines 5-4A and 5-4B. Table 9.4 summarizes the steps taken by the backward search algorithm in terms of switch sensitivities, where we have assumed the susceptance of shunt capacitors to be installed at feasible buses $B_i^{(k)} = B_i = B_{i\max} = 0.16\ p.u.$ The initial network configuration has six shunt capacitors at buses 4, 5, 6, 7, 8, and 9 are switched on. The voltage stability margin with all six shunt capacitors switched on is 11.34%, which is greater than the required value of 10%. Therefore, the number of switches can be decreased to reduce the cost. At the first step of the backward search, we compute the margin sensitivity for all six controls as listed in the fourth column. From this column, we see that the row corresponding to the shunt capacitor at bus four has the minimal sensitivity. So in this step of backward search, this capacitor is excluded from the list of control locations indicated by the strikethrough. Continuing in this manner, in the next three steps of the backward search we exclude shunt capacitors at buses six, nine and eight, sequentially. However, as seen from the last column of Table 9.6, with only two controls at buses five and seven, the voltage stability margin is unacceptable at 9.51%. Therefore the final solution must also include the capacitor excluded at the last step, i.e., the shunt capacitor at bus eight. The location of these controls are intuitively pleasing given that, under the contingency, Load A, the largest load, must be fed radially by a long transmission line, a typical voltage stability problem.

Figure 9.11 shows the corresponding search via the graph. In the figure, node O represents the origin configuration of discrete switches from which the backward

TABLE 9.5. Voltage Stability Margin for Severe Contingencies.

Contingency	Voltage Stability Margin (%)
1. Outage of lines 5-4A and 5-4B	4.73
2. Outage of transformer T1 & line 4–6	4.67

TABLE 9.6. Steps Taken in the Backward Search Algorithm for Shunt Capacitor Planning.

No		no cntrl.	6 cntrls.	5 cntrls. (reject #6)	4 cntrls. (reject#5)	3 cntrls. (reject#4)	2 cntrls. (reject#3)
1	Sens. of shunt cap. at bus 5	0.738	0.809	0.808	0.807	0.804	0.756
2	Sens. of shunt cap. at bus 7	0.334	0.360	0.359	0.358	0.357	0.352
3	Sens. of shunt cap. at bus 8	0.240	0.263	0.262	0.261	~~0.260~~	
4	Sens. of shunt cap. at bus 9	0.089	0.098	0.097	~~0.096~~		
5	Sens. of shunt cap. at bus 6	0.046	0.051	~~0.051~~			
6	Sens. of shunt cap. at bus 4	0.019	~~0.021~~				
	loadability (MW)	389.8	414.4	414.0	413.2	411.7	407.6
	loading margin (%)	4.73	11.34	11.24	11.02	10.61	9.51

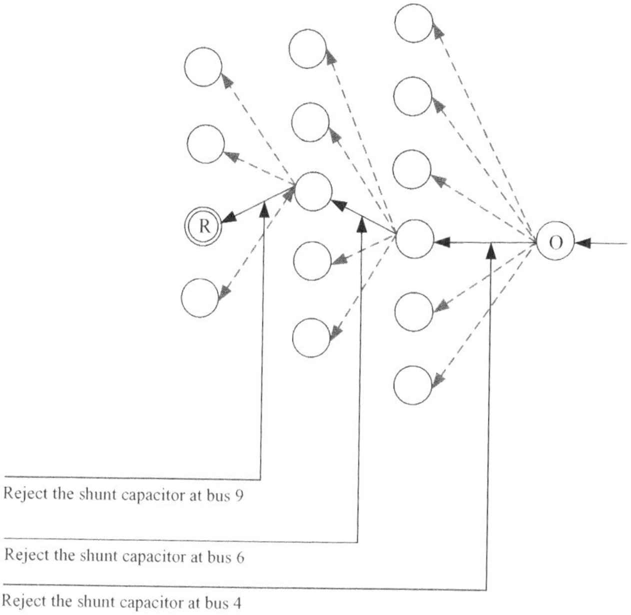

Figure 9.11. Graph for the backward search algorithm for shunt capacitor planning.

search originates, and node R represents the restore configuration associated with a minimal set of discrete switches which satisfies the voltage stability margin requirement (this is the node where the search ends).

For the outage of transformer T1 and line 4–6, the solution obtained by the forward search algorithm is: shunt capacitors at buses four and five. Therefore, the final candidate locations for shunt capacitors are buses four, five, seven and eight, which guarantee that the voltage stability margin under all prescribed N-2 contingencies is greater than the required value. In a similar way, we obtain the final candidate locations for series capacitors as lines 5-7A and 5-7B where we have assumed the reactance of series capacitor to be installed in feasible lines $X_i^{(k)} = X_i = X_{i\max} = 0.03$ $p.u.$. Therefore, the best six candidate locations are lines 5-7A, 5-7B to install series capacitor switches, buses four, five, seven and eight, to install shunt capacitor switches. We use these candidate locations to initialize the reactive power planning algorithm presented in Section III was carried out.

In order to demonstrate the efficacy of the proposed method, two cases are considered as follows. In case one, only shunt capacitor switches are chosen as candidate controls while both shunt and series capacitor switches are chosen as candidate controls in case two. Table 9.7 shows the results for case one where the optimal allocations for shunt capacitor switches are 0.16, 0.16, and 0.115 $p.u.$ at buses five, seven and eight, respectively, and these switches are fully used for the outage of transformer T1 and line 4–6. The total cost is 0.451 for the control allocations in Case one. On the other hand, the optimal control allocations for case two are shown in Table 9.8 indicating that a series capacitor switch of 0.03 $p.u.$ on line 5-7A and a shunt capacitor switch of 0.131 at bus five, and these switches are fully

TABLE 9.7. Control Allocations with Shunt Capacitors.

Candidate Locations for Shunt Capacitors	Maximum Size	Result for the Whole Problem	Result for Cont. 1	Result for Cont. 2
Bus 5	0.16	0.16	0.156	0.16
Bus 4	0.16	0.00	0.00	0.00
Bus 7	0.16	0.16	0.16	0.16
Bus 8	0.16	0.088	0.082	0.088

TABLE 9.8. Control Allocations with Shunt and Series Capacitors.

Candidate Locations for Shunt and Series Caps	Maximum Size for Shunt Caps	Result for the Whole Problem	Result for Cont. 1	Result for Cont. 2
Line 5-7A	0.03	0.03	0.03	0.03
Line 5-7B	0.03	0.00	0.00	0.00
Bus 5	0.16	0.131	0.105	0.131
Bus 4	0.16	0.00	0.00	0.00
Bus 7	0.16	0.00	0.00	0.00
Bus 8	0.16	0.00	0.00	0.00

TABLE 9.9. Voltage Stability Margin under Planned Controls.

Candidate Control	Iteration Number for LP	Voltage Stability Margin for Cont. 1	Voltage Stability Margin for Cont. 2
Shunt caps	1	9.98%	10.01%
Shunt and series caps	1	10.01%	9.99%

used for the outage of transformer T1 and lines four through six. For Case two, the total cost for control allocations is 0.41, which is 9.96% less than that of case one. This result shows that benefit can be obtained by pursuing a strategy of planning different types of discrete reactive power controls. Table 9.9 gives the verified results of the reactive power control planning with the continuation power flow program. The voltage stability margins of the concerned contingencies are approximately equal to the required value of 10% under the planned controls. The iteration number in the second column represents the number of times of solving the LP after solving the MIP.

This section has presented an optimization-based approach for planning reactive power control in electric power transmission systems to satisfy voltage stability margin requirements under normal and contingency conditions. The planned reactive power controls are capable of serving as control response for contingencies. Optimal locations and amounts of new switch controls are obtained by solving the MIP. The amount of control is further refined by solving the LP. The proposed algorithm can handle a large-scale power system because it significantly reduces computation burden by fully utilizing the information of the voltage stability margin sensitivity and overcomes the possible difficulty of convergence associated with nonlinear programming formulations. The effectiveness of the method is illustrated by using a test system. The results show that the method works satisfactorily to plan reactive power control.

9.4.4 Dynamic Analysis

In Section 9.4.3, we described procedures for planning control to expand transmission at minimum cost under constraints imposed by *post-transient* reliability criteria. These procedures resulted in optimal locations and amounts of shunt and series capacitors, but because analysis was based on purely static models, constraints associated with *transient* reliability criteria were not enforced. Although the resulting solutions provide very useful guides in regard to the investments necessary to appropriately expand transmission, there remain unanswered questions about control design, in particular, given the availability of the switchable shunt and series capacitors as determined from the static analysis, for each contingency: What sequence of switches and associated timing is necessary in order to satisfy the transient reliability criteria? Table 9.1 indicates two types of transient reliability criteria: one is on voltage deviations and the other is on frequency deviations. An implied necessary condition is that the system be stable.

The problem of designing the control for a specified contingency has three steps, as follows:

1. Identify the switches (each leading to series or shunt capacitor insertion) to be operated.
2. Identify the sequence of switch operations.
3. Determine the timing of each switch operation.

We assume that the solution for Step 1 is obtained from the static analysis described in Section 9.4.3. This assumption may not always be valid, i.e., the controls that result in acceptable post-transient conditions do not necessarily provide for acceptable transient conditions independent of their sequence or timing, and in such a case, one may need to augment the switches identified in Step 1[3]. For Step 2, there are $n!$ possible solutions, where n is the number of switches identified in Step 1. Each possible solution corresponds to a path through the automaton; for example, in the four-switch case of Figure 9.9, there are $4! = 24$ paths or switching sequences. There may be a number of acceptable sequences, depending on the range of switching times as determined in Step 3.

A commonly used method for the dynamic performance analysis of a control is time domain simulation. For each initial fault-on state, one performs simulation

[3]One efficient way to gain insight into the effectiveness of a group of switched capacitive compensators on transient performance for a transient is to simulate the transient under the conditions that all capacitors are switched at the earliest conceivable moment (simultaneous with fault clearing), and if transient criteria are not satisfied, then the group should be either be augmented, or faster controls should be employed such as static var compensators (SVCs) or thyristor-controlled-series-capacitors (TCSC).

for each control in order to find an effective one. For the planning problem, when there are multiple operating conditions, multiple contingencies, and multiple control possibilities, this is ad-hoc and labor-intensive method. Alternatively, methods based on determination of stability regions may be more efficient for control design, particularly when discrete control strategies are considered. We approach the problem by identifying the stability region of post-fault stable equilibria associated with different switching configurations. For a general nonlinear autonomous system, the stability region is defined as the set of all points from which the trajectories start and eventually converge to the stable equilibrium point (SEP) as time approaches infinity [67].

To motivate the potential for using stability regions to address the dynamic performance analysis, consider a one machine-infinite bus system equipped with shunt and series controls as shown in Figure 9.12.

The system model is given as follows:

$$\begin{cases} \dfrac{d\delta}{dt} = \omega \\ \dfrac{d\omega}{dt} = \dfrac{P_m - P_e^M \sin(\delta)}{M} \end{cases} \tag{9.63}$$

Here, δ is the machine rotor angle and ω is the relative angular velocity of the rotor. Suppose the inertial constant $M = T_J/\omega_0 = 0.026\,\text{sec}^2/\text{rad}$, the damping coefficient $D = 0.12$, the mechanical power $P_m = 1.0$ per unit, and the maximum electrical power transferred is $P_e^M = EU/X^{(i)}$, where E is the voltage of the generator, U is the voltage of the infinite bus, and $X^{(i)}$ is reactance between the source and infinite bus at Mode i. The mode is the particular combination of switch positions that lead to a specific control approach. For example, Mode 1 is no control, and Mode 2 is control using the series capacitor only.

Consider the following scenario: a fault occurs at the middle of the transmission line at time $t = 0$ and is cleared after 0.1 second. We use this scenario to illustrate the effectiveness of using the stability region to guide the design of stabilizing controls via comparison of Mode 1 to Mode 2. For our two-state system, the stability region can be displayed in the two-dimensional state plane, which in this case is the plane of generator speed and angle. In Figure 9.13, the stability region of the Mode 1 post-control equilibrium is given by the solid line, so that application of the Mode 1 control within this region is guaranteed to result in stable performance.

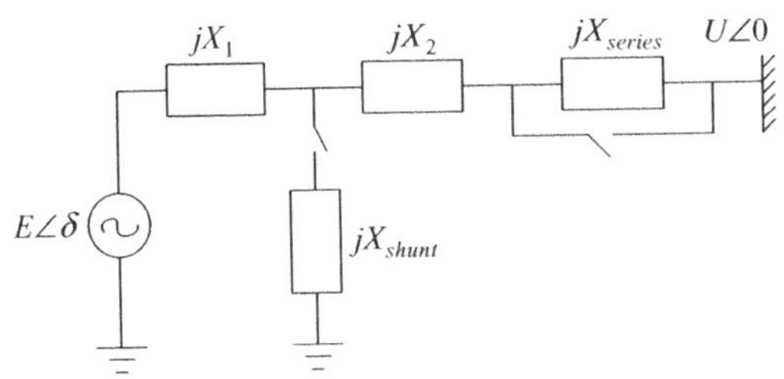

Figure 9.12. System model with shunt and series control strategies.

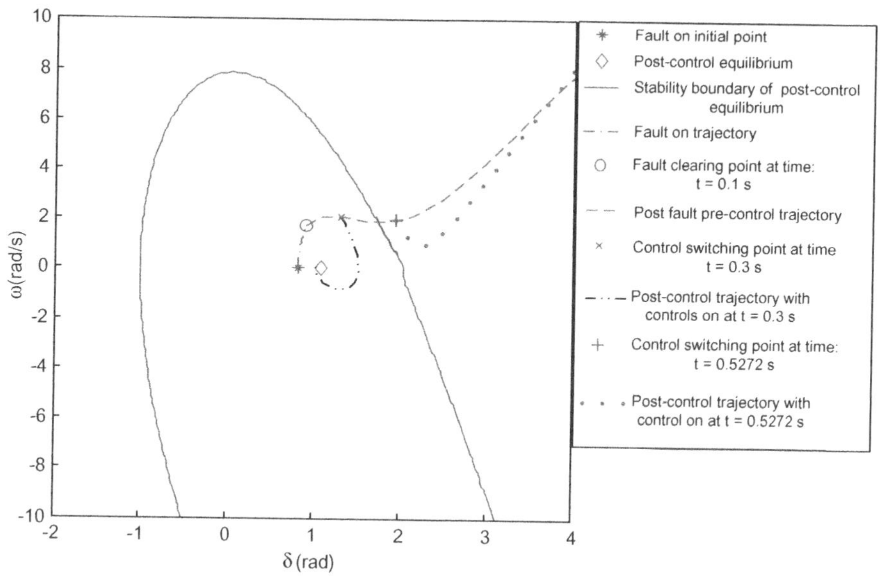

Figure 9.13. Control based region of attraction.

Study of the legend and corresponding points and curves is illuminating. The dashed-dotted curve emanating from the fault-on initial point represents the fault-on trajectory. After the fault is cleared, the post-fault pre-control trajectory represented by the dashed line results in continuously increasing speed, indicating that the system is unstable if no control is applied. If the control is switched on early enough (prior to system trajectory leaving the stability region of post-control equilibrium), the system stabilizes, as indicated by the dashed-dot-dot curve that converges to the post-control equilibrium. On the other hand, if the control is switched too late, outside of the stability regions, the trajectory diverges, as indicated by the dotted curve. We observe from this illustration that a control mode's stability region provides the ability to assess the effectiveness of the mode and to determine maximum switching times for stability.

In the last three decades numerous efforts have been undertaken to determine the stability region with the goal of power system transient stability analysis. The studies [68, 69, 70, 71] provided the theoretical foundations for the geometric structure of the stability region. Reference [70] proved that the stability boundary of a stable equilibrium point (SEP) is the union of the stable manifolds of the type one unstable equilibrium points. It also proposed a numerical algorithm to determine the stability region. This method, however, is not always applicable, and even when it is applicable, the computation of stable manifold of a type one unstable equilibrium point is not easy for a large system, and even when the computation is feasible, the method can only provide local approximation of the stable manifold. Recently, some algorithms have been developed to approximate the stable manifold of an unstable equilibrium point (UEP). For example, in [72, 73] the Taylor expansion is used to obtain a quadratic approximation, whereas in [74, 75] the stable manifolds around an UEP are approximated by the normal form technique and the energy function methods [76]. A well-known alternative method called the closest unstable equilibrium point method [77] finds a subset of the true stability region and thereby need not obtain the stable manifold of an UEP. It is shown in [78] that the stability region estimated by the closest UEP method is optimal in the sense that it is the largest region within the stability region, which can be characterized by the corresponding energy function. These energy function/Lyapunov based methods however can only provide a conservative estimate of a stability region. Furthermore, these methods can not compute a stability region for a hybrid system.

Our method for computing the stability region of an SEP of a power system is based on backward reachability analysis as reported in [79]. Reachability analysis focuses on finding reachable sets of a target set. Reachable sets are a way of capturing all at once the behavior of entire groups of trajectories. The backwards reachable set is the set of states which give rise to trajectories leading to the target set. Given a post-fault stable operating point (an SEP), there must exist an open neighborhood of it that is contained in its stability region. This means that if we choose a sufficiently small ball of radius ε around the SEP as the target set, any trajectory entering that target set is guaranteed to converge to the associated SEP. Thus, as time goes to infinity, the backward reachable set of the target set approaches the stability region of the system.

One way of describing a subset of states is via an implicit surface function representation as shown in Figure 9.14. Consider a closed set $S \subseteq R^n$. An implicit surface representation of S would define a function $\phi(x): R^n \to R$ such that $\phi(x) \le 0$ if $x \in S$ and $\phi(x) > 0$ if $x \notin S$.

In [80], the author formulates the backward reachable set in terms of a Hamilton-Jacobi-Isaacs (HJI) Partial Differential Equation (PDE) and proves that the viscosity solution of this PDE is an implicit surface representation of the backward reachable set. This HJI PDE can be solved with the very accurate numerical methods drawn from the level set literature. In [81] we applied the stability region computation to determine the minimal amount of load shedding in voltage stability control.

The following algorithm summarizes the procedure to determine the stability region of a post-fault power system.

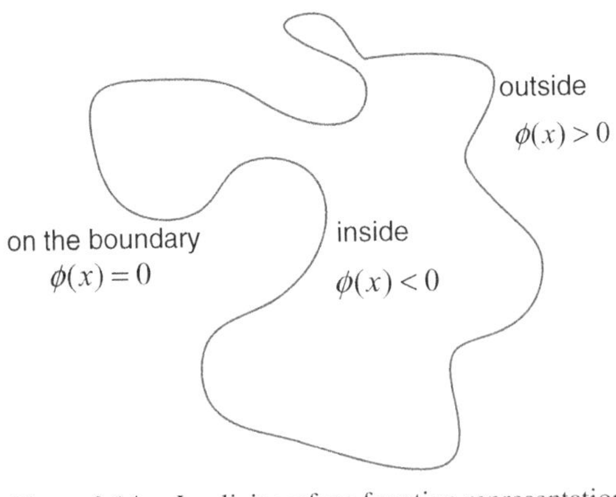

Figure 9.14. Implicit surface function representation.

(1) Form the state space equations of the post-fault power system, $\frac{dx}{dt} = f(x)$

(2) Find the stable equilibrium point of this autonomous nonlinear system, by solving $f(x) = 0$ and let $x^* \in R^n$ be a SEP.

(3) Specify a ε ball centered at the stable equilibrium point with sufficiently small radius ε.

(4) Define an implicit surface function at $t = 0$ as:

$$\phi(x, 0) = \|x - x^*\| - \varepsilon \tag{9.64}$$

Then the target set is the zero sublevel set of the function $\phi(x, 0)$, i.e, it is given by:

$$\{x \in R^n | \phi(x, 0) \leq 0)\} \tag{9.65}$$

Therefore, a point x is inside the target set if $\phi(x, 0)$ is negative, outside the target set if $\phi(x, 0)$ is positive, and on the boundary of the target set if $\phi(x, 0)$.

(5) Propagate in time the boundary of the backward reachable set of the target set by solving the following HJI PDE:

$$\phi_x^T f(x, t) + \phi_t = 0 \tag{9.66}$$

with terminal conditions (9.64). The zero sublevel set of the viscosity solution $\phi(x, t)$ to (9.64), (9.66) is the backward reachable set at time t is:

$$\{x \in R^n | \phi(x, t) \leq 0)\} \tag{9.67}$$

(6) The backward reachable set of the ε ball around the stable equilibrium point is computed using a software tool from [60]. It is always possible to find a certain epsilon-ball contained in the stability region of a stable equilibrium point. As t goes to infinity, the backward reachable set approaches the true stability region. If the stability region is bounded, the level set based numerical computation of the backward reachable set eventually converges to the stability region within a finite computation time.

We present an example to illustrate stability region computation and its application in dynamic analysis of a control strategy.

Example 3: Stability region identification for single-machine-infinite-bus

Consider the system in Figure 9.12. Define the system with no controls on as Mode 1, with series control on as Mode 2, with shunt control on as Mode 3, and with both series control and shunt control on as Mode 4. As the mode is changed, the transmission line reactance changes causing the P_e^M as well as the equilibrium point to change. Each of the four modes (corresponding to two different binary controls), the associated transmission line reactance, the P_e^M value, and the equilibrium point are summarized in Table 9.10.

The stability regions of all the four modes are shown in Figure 9.15. The stability region of Mode 1 is inside the dotted curve that of Mode 2 is inside the dashed-dot curve, that of Mode 3 is inside the dashed curve, and that of Mode 4 is inside the solid curve.

TABLE 9.10. Four Control Modes and Their Certain Parameters.

Mode	Series Capacitor	Shunt Capacitor	$X(^i)$	P_e^M Value	Equilibrium Points
1	Off	Off	$X_1 + X_2$	1.35	(0.8342, 0)
2	On	Off	$X_1 + X_2 - X_{series}$	2.25	(0.4603, 0)
3	Off	On	$X_1 + X_2 - \frac{X_1 X_2}{X_{shunt}}$	1.543	(0.7084, 0)
4	On	On	$X_1 + (X_2 - X_{series}) - \frac{X_1(X_2 - X_{series})}{X_{shunt}}$	2.3478	(0.4400, 0)

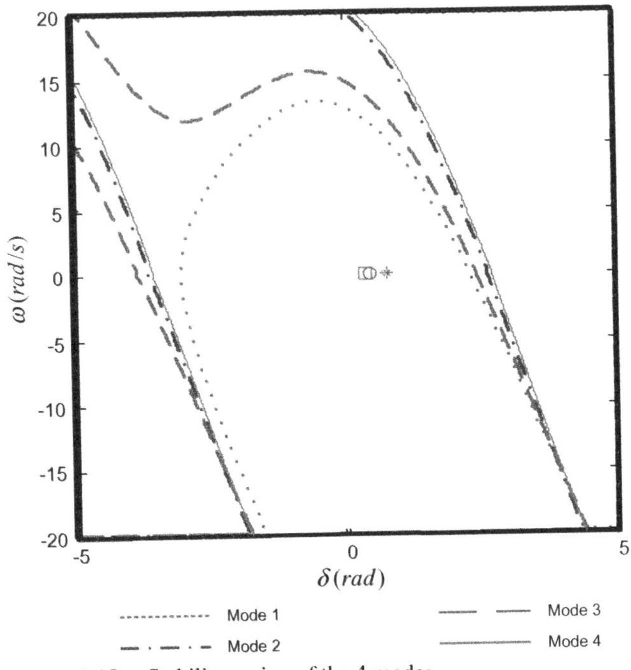

Figure 9.15. Stability region of the 4 modes.

Using the stability regions, we can verify the effectiveness of different control strategies and also provide an effective control strategy for a given post-fault initial state. When the post-fault state is inside the stability region of Mode 1, no control is needed because the state will eventually reach the stable equilibrium point. When the post-fault state is outside the stability region of Mode 1, some control needs to be switched-on to stabilize the post-fault state. For example, if the initial post-fault state is inside the stability region of Mode 2 and outside the stability region of Mode 3, we have two choices to stabilize the system: Switch on the series capacitor or switch on both the series and shunt capacitors. The system will then converge to the stable equilibrium point of Mode 2 or Mode 4 accordingly. In general, if the post-fault initial state is inside the union of stability regions of all such modes, the transient stability can be achieved by switching on one or more controls.

We identified the importance of dynamic analysis besides the static analysis. A dynamic analysis is needed to determine the domain over which a control strategy computed using a static analysis is effective. Stability region forms the basis of a dynamic analysis, and we presented an example to illustrate how stability region associated with various control modes can be used in devising a contingency control strategy. Our method uses backward reachability analysis involving propagation of level-sets for computing a stability region. A limitation of our method is that the computation complexity grows exponentially in the number of system dimension. This is because a computation of the backward reachable set is based on gridding of the state space. As part of future research we plan to explore faster and/or approximate techniques for reachability computation. This includes possibility of parallelization, of hierarchical computation, and Sum of Squares (SOS) based approach.

9.5 MARKET EFFICIENCY AND TRANSMISSION INVESTMENT

As competitive reforms are introduced into the electricity power industry, much attention has been focused on the potential market organization of the industry's transmission sector. Questions have naturally risen: Shall deregulation and competition be applied to transmission, as they are to generation and distribution? To what degree shall market play a role in transmission expansion and investment?

There is some literature on transmission and transmission investment. Hogan [82] proposes a contract network pricing model, using congestion payments as the rental fee for use of the capacity rights. Within this contract network regime, Bushnell and Stoft [83] analyze the incentives for grid investment. They show that under certain conditions this contract network approach can effectively deter detrimental investments, some of which are encouraged under other regimes. Chao and Peck [84] define a trading rule and property rights so that a competitive market could be established for transmission services to achieve a social optimum within a power pool. In Bushnell and Stoft [85], a process is outlined by which transmission planning and investment would be undertaken by competitive entities in a lightly regulated environment. More recently, Joskow and Tirole [86] examine the performance attributes of a merchant transmission investment framework that relies on "market driven" investment to increase transmission network capacity and conclude that inefficiencies may result from reliance on such a framework. In what follows, we will show our work that addresses something not explicitly identified in this literature.

It is well known that generation of power can be efficiently decentralized by means of a price system and competitive markets. Indeed, Chao and Peck [84] showed that for a given grid, the competitive equilibrium is efficient. That is, the competitive equilibrium nodal and transmission prices induce an efficient dispatch. It is also known that this result breaks down as soon as the grid itself is endogenous. The main example of this inefficiency is based on the fact that adding or removing a line has a dramatic change in the flow of power for any given set of injections. In the economist's jargon, there is a market failure in the power market once investment in transmission is allowed. The alleged reason for this market failure is the externalities created by loop flows[4]. That loop flows are responsible for the market failure is clear, since a power market with endogenous investment in a radial network can be efficiently decentralized by a market mechanism. However, the nature of the externalities created by loop-flows has, to the best of our knowledge, never been identified. Is the addition or removal of circuits necessary for markets to fail? In other words, if we only allow investment that results in an upgrade of the line capacities of a given grid, can a competitive equilibrium allocation fail to be efficient? Are the externalities created by loop flows due to the fact that changes in the line capacities affect the set of feasible injections into the grid? Are the loop-flow induced externalities related to the fact that the allowable injections in one bus depend on the injections in the other buses?

In this section we clarify the nature of the externalities introduced by the loop flows. The bottom line is that transmission investment introduces an externality only if it affects the flow of power along the lines *for any given set of injections*. For instance, the addition or removal of a new circuit will affect the flow of power for any given set of injections, unless of course we are adding or removing part of a radial network. But the increase of the operational capacity of a line will not introduce an externality, even if it does change the set of feasible injections, unless it affects the flow of power for any given set of injections.

As a result, we can answer the above questions as follows. The addition or removal of lines are not necessary for markets to fail: the competitive equilibrium will not be efficient even if the grid topology is restricted to remain the same but upgrades of line capacities that change the power flow are allowed. The change in the set of feasible injections itself is not responsible for the market failure: as long as the line capacities are changed in a flow-preserving way, there will be no externalities associated with the investment. Finally, the fact that injections in one bus affect the set of allowable injections in other buses is not the source of externalities. The truth of the last two statements can be seen by observing a two-bus network: in such a network, the flow structure is always the same; namely, each MW injected in one bus transits the only line, independently of its capacity.

In this section we present two examples. The first one shows that unless it leaves the flow of power for any given set of injections unaffected, transmission investment will induce externalities that cause the market to fail. The second example considers a type of investment, which consists of enhancing the operational capacity of a line by adding a capacitor that will be switched on only in case a contingency occurs. Since under normal circumstances the impedances are constant, whether the capacitor is installed or not, this type of capacity enhancement will not affect the

[4]There are externalities when the actions of one agent **directly** affect the payoff of the other agents associated with a given action.

flow structure of the network and as a result the competitive allocation will be efficient.

Example 4: Transmission-induced capacity enhancement

Consider Figure 9.16, where there are three interconnected buses, buses one, two and three ($n = 1, 2, 3$). Let lines one, two and three denote the lines connecting buses two and three, buses one and three and buses one and two, respectively. Originally, each line has some capacities. The capacities of line one and two are so large, that they are never congested. Let k_0 denote the initial capacity on line three. To make things interesting, suppose k_0 is less than the socially efficient capacity.

A generator is attached to buses one and two, respectively, denoted by G_1 and G_2. G_1's cost function is $C_1(P_{g1})$ and G_2's cost function is $C_2(P_{g2})$, where, for $n = 1,2$, P_{gn} is the amount of power generated and C_n is a strictly convex function that satisfies $C_n(0) = 0$. The only load, of a constant 1000 MW, is located at bus three. There is an investment firm that produces transmission capacity. It only chooses to build lines between buses one and two, since the other two lines already have enough capacity. The investment firm's cost function is $C(I)$, which is assumed to be strictly convex, and where I is the capacity of the new line it builds between buses one and two.

A dispatch is a pair of injections (P_{g1}, P_{g2}) that satisfy the load, i.e. $P_{g1} + P_{g2} = 1000$. Not all dispatches are feasible. In order for a dispatch to be *feasible*, the flow along each line should not exceed the line's capacity. In this example, since we assume that lines one and two have large enough capacity, we are only concerned with the flow along line three, from bus one towards bus two. Clearly, the flow of power along this line depends on the dispatch. But it may also depend on the capacity of the line. For the purpose of the analysis we will adopt a linear approximation and assume that the flow of power *from bus 1 to bus 2* is given by $P_{12} = \alpha(k)(P_{g1} - P_{g2})$, where $k = k_0 + I$ is the capacity of line 3, after an investment I has been made. As we will see, the dependence of the coefficient α on the capacity of the line is the source of the market failure in the transmission investment market.

With this formulation in hand, we can define a feasible allocation to consist of a dispatch (P_{g1}, P_{g2}) and a capacity investment I, such that—$(k_0 + I) \leq \alpha(k_0 + I)$

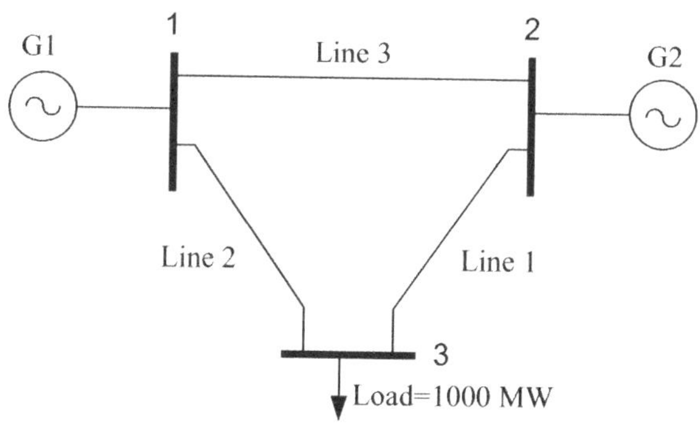

Figure 9.16. 3-bus example.

$(P_{g1} - P_{g2}) \leq (k_0 + I)$. With this notation, we are ready to solve for the optimal allocation. The optimal allocation is the transmission investment I and the dispatch (P_{g1}, P_{g2}) that satisfies the load in the least expensive way. Formally, it is the solution to the following "social planner's" problem:

$$\left.\begin{aligned} &\min_{P_{g1},P_{g2},I} C_1(P_{g1})+C_2(P_{g2})+C(I) \\ &\text{s.t. } P_{g1}+P_{g2}=1000 \\ &-(k_0+I)\leq \alpha(k_0+I)(P_{g1}-P_{g2})\leq (k_0+I) \end{aligned}\right\} \tag{9.68}$$

For the sake of the analysis, assume that $\alpha(\cdot)$ is such that the set of feasible allocations is convex. Also, for simplicity assume that the above problem has an interior solution and, without loss of generality, that at that solution $\alpha(k_0 + I)\,(P_{g1} - P_{g2}) \geq 0$.[5] Let λ and μ be the Lagrangian multipliers of the above constraints, respectively. Then the first order conditions for an interior solution are:

$$\begin{aligned} &\frac{\partial C_1(P_{g1}^*)}{\partial P_{g1}} = \lambda - \mu\alpha(k_0+I^*) \\ &\frac{\partial C_2(P_{g2}^*)}{\partial P_{g2}} = \lambda + \mu\alpha(k_0+I^*) \\ &\frac{\partial C(I^*)}{\partial I} = \mu - \mu\frac{\partial\alpha(k_0+I^*)}{\partial k}(P_{g1}^* - P_{g2}^*) \\ &P_{g1}^* + P_{g2}^* = 1000 \\ &\alpha(k_0+I^*)(P_{g1}^* - P_{g2}^*) = k_0 + I^* \end{aligned} \tag{9.69}$$

It follows that at an interior efficient allocation $(P_{g1}^*, P_{g2}^*, I^*) >> \mathbf{0}$,

$$\frac{\partial C_2(P_{g2}^*)}{\partial P_{g2}} - \frac{\partial C_1(P_{g1}^*)}{\partial P_{g1}} = 2\frac{\dfrac{\partial C(I^*)}{\partial I}}{\left(1-\dfrac{\partial\alpha(k_0+I^*)}{\partial k}(P_{g1}^* - P_{g2}^*)\right)}\alpha(k_0+I^*) \tag{9.70}$$

We now want to compare the efficient allocation with the outcome of decentralized trade through a price mechanism. For this we need to define the competitive equilibrium. In the following definition of economic equilibrium, there will be electricity prices associated to each bus (the nodal prices), and one transmission charge. The concept of nodal prices, is central in power system economics, and was introduced by Schweppe et al. [19].

A **competitive equilibrium** consists of an allocation $((P_{g1}^*, P_{g2}^*), I^*)$ and a price vector $(\pi_1, \pi_2, \pi_3, \tau)$ that satisfy the following conditions.

a. The transmission investing firm maximizes its profits, given the transmission price on the newly built line: it chooses I^* so as to solve

[5] Sufficient conditions for an interior solution would be that marginal costs of generation and investments are 0 when evaluated at 0.

$$\max_{I} \tau I - C(I) \tag{9.71a}$$

b. Each generator maximizes its profit, given its respective nodal prices: More specifically, given the nodal prices π_n, $n = 1,2$, generator G_n, for $n = 1,2$, solves:

$$\max_{P_{gn}} \pi_n P_{gn} - C_n(P_{gn}) \tag{9.71b}$$

c. Markets clear:

$$P_{g1}^* + P_{g2}^* = 1000 \tag{9.71c-1}$$

$$\alpha(k_0 + I^*)(P_{g1}^* - P_{g2}^*) = k_0 + I^* \tag{9.71c-2}$$

d. No arbitrage opportunity exists.

$$\pi_3 = \pi_1 + \alpha(k_0 + I^*)\tau \tag{9.71d-1}$$

$$\pi_3 = \pi_2 - \alpha(k_0 + I^*)\tau \tag{9.71d-2}$$

Conditions (9.71a) and (9.71b) can be replaced by the corresponding necessary and sufficient conditions for profit maximization as follows:

$$\frac{\partial C(I^*)}{\partial I} = \tau \tag{9.71a'}$$

$$\frac{\partial C_1(P_{g1}^*)}{\partial P_{g1}} = \pi_1 \tag{9.71b'-1}$$

$$\frac{\partial C_2(P_{g2}^*)}{\partial P_{g2}} = \pi_2 \tag{9.71b'-2}$$

Suppose that an allocation $((P_{g1}^*, P_{g2}^*), I^*)$ and a price vector $(\pi_1, \pi_2, \pi_3, \tau)$ constitute a competitive equilibrium. Then, from the no-arbitrage conditions we have:

$$\begin{aligned} \pi_3 &= \pi_1 + \alpha(k_0 + I^*)\tau \\ \pi_3 &= \pi_2 - \alpha(k_0 + I^*)\tau \end{aligned} \tag{9.72}$$

Substituting into the generators' first order conditions (9.71b'-1) and (9.71b'-2), we get:

$$\begin{aligned} \pi_3 &= \frac{\partial C_1(P_{g1}^*)}{\partial P_{g1}} + \alpha(k_0 + I^*)\tau \\ \pi_3 &= \frac{\partial C_2(P_{g2}^*)}{\partial P_{g2}} - \alpha(k_0 + I^*)\tau \end{aligned} \tag{9.73}$$

Replacing the transmission prices with the marginal costs, from equation (9.71a') and rearranging it, we get:

$$\frac{\partial C_2(P_{g2}^*)}{\partial P_{g2}} - \frac{\partial C_1(P_{g1}^*)}{\partial P_{g1}} = 2\frac{\partial C(I^*)}{\partial I}\alpha(k^*) \tag{9.74}$$

Comparing equations (9.70) and (9.74), we see that unless $\frac{\partial \alpha(k_0 + I^*)}{\partial k} = 0$, that is, unless investment does not affect the power distribution factor α, we cannot guarantee that the competitive equilibrium be efficient. This allows us to conclude that the source of the market failure lies on the fact that investment in transmission capacity affects the power flow through the lines for any given dispatch.

Can, nevertheless, some government intervention achieve an efficient allocation via a decentralized market mechanism? The answer is yes, if we have enough information to apply the optimal Pigouvian tax. Suppose that the government imposes an ad-valorem tax t on capacity enhancement. Then the investment firm's profit maximization problem becomes:

$$\max_{I} \tau(1-t)I - C(I) \tag{9.75}$$

The profit maximizing investment satisfies the first order condition:

$$\frac{\partial C(I^*)}{\partial I} = \tau(1-t) \tag{9.76}$$

and the conditions that a competitive equilibrium satisfies are:

$$\begin{aligned}
&\frac{\partial C(I^*)}{\partial I} = \tau(1-t) \\
&\frac{\partial C_1(P_{g1}^*)}{\partial P_{g1}} = \pi_1 \\
&\frac{\partial C_2(P_{g2}^*)}{\partial P_{g2}} = \pi_2 \\
&P_{g1}^* + P_{g2}^* = 1000 \\
&\alpha(k^*)(P_{g1}^* - P_{g2}^*) = k^* \\
&\pi_3 = \pi_1 + \alpha(k^*)\tau \\
&\pi_3 = \pi_2 - \alpha(k^*)\tau
\end{aligned} \tag{9.77}$$

By comparison, we can see that if allocation ((P_{g1}^*, P_{g2}^*), I^*) solves the social optimum problem (9.68) with associated Lagrangian multipliers (λ, μ), then the same allocation ((P_{g1}^*, P_{g2}^*), I^*) together with the price vector (π_1^*, π_2^*, π_3^*, τ^*) and ad-valorem tax rate t^* defined by:

$$\begin{aligned}
&P_{g1}^* = \lambda - \alpha(k^*)\mu \\
&P_{g2}^* = \lambda - \alpha(k^*)\mu \\
&\pi_3^* = \lambda \\
&\tau^* = \mu \\
&t^* = \frac{\partial \alpha(k_0 + I^*)}{\partial k}(P_{g1}^* - P_{g2}^*)
\end{aligned} \tag{9.78}$$

is a competitive equilibrium. Conversely, if allocation $((P_{g1}^*, P_{g2}^*), I^*)$ together with price vector $(\pi_1, \pi_2, \pi_3, \tau)$ and ad-valorem tax rate $t^* = \frac{\partial \alpha(k_0 + I^*)}{\partial k}(P_{g1}^* - P_{g2}^*)$ constitute a competitive equilibrium, then the same allocation $((P_{g1}^*, P_{g2}^*), I^*)$ together with the Lagrangian multiplier (λ, μ) defined by:

$$\lambda = \pi_3^* \tag{9.79}$$

$$\mu = \tau^* \tag{9.80}$$

solve the social optimum problem (9.68).

Example 5: Capacitor-induced capacity enhancement

Consider the three-bus network shown in Figure 9.17. Buses two and three are connected by line one, with impedance one; buses one and two are connected by line three, also with impedance one; finally, buses one and three are connected by two parallel lines, line 21 and line 22, each with impedance two, so that the impedance of the path two, from bus one to bus three, is one. For simplicity, assume that lines one and three have Large enough capacities so that they are never congested. Each of the two parallel lines that connect buses one and three, on the other hand, has a capacity of k_1. Figure 9.17 illustrates this three-node transmission network under normal conditions.

In a contingency, line 21, but not any other, can fail. When line 21 fails, the capacity on line 22 will be k_2, where $k_2 > k_1$, because the pre-reserved capacity for line 22 is released in the contingency. As a result, a higher flow is allowed to move along line 22 when line 21 breaks, but the pre-reserved margin is usually small. Suppose that $k_2 = 110\% \; k_1$. That is, there is a 10% margin reserved for capacity of line 22. Capacities k_1 and k_2 should not be interpreted as a "physical limit" on the flow transmitted through the lines but as "operational limit" that results from the satisfaction of disturbance performance criteria for the network. The network in case of a contingency is shown in Figure 9.18.

A generator is attached to nodes one and two, respectively, denoted by G_1 and G_2. Generator G_1 generates power with a technology whose associated cost function is denoted by $C_1(P_{g1})$. Similarly, generator G_2's cost function is $C_2(P_{g2})$. That is, for $n = 1, 2$, $C_n(P_{gn})$ is the minimum cost for G_n of generating P_{gn} MW in 1 hour. It is assumed that both cost functions are differentiable, strictly convex, and satisfy

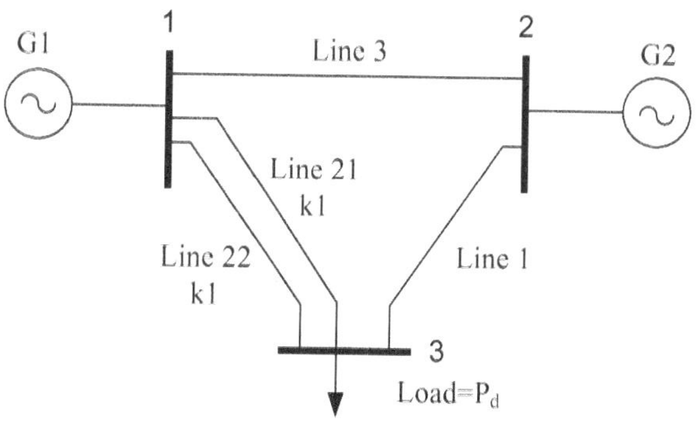

Figure 9.17. 3-node network under normal conditions.

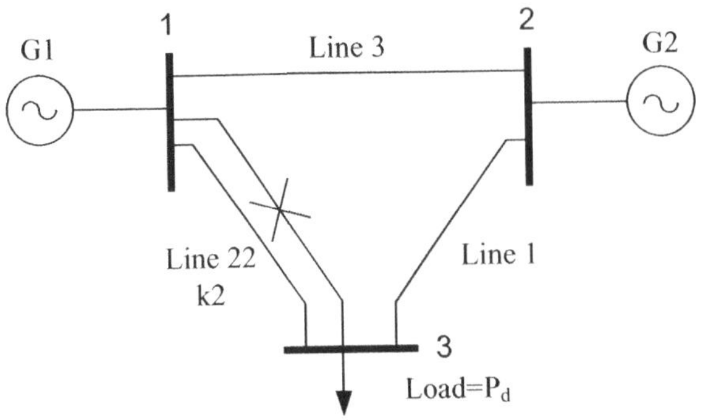

Figure 9.18. Contingency with no capacitor.

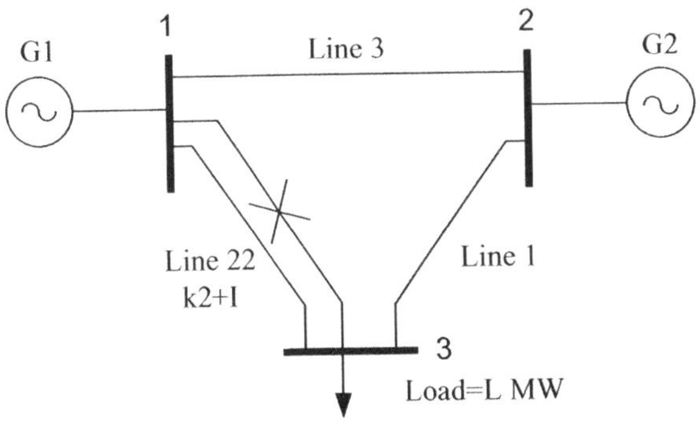

Figure 9.19. Contingency with capacitor switched on.

$C_n(0) = 0$, for $n = 1, 2$. At node three, there is a constant load of P_d MW. Apart from generators and consumers, there is an investment firm that can increase the capacity of the network by installing capacitors. When the capacitor is switched on the maximum acceptable flow on a given line is enhanced by some I units, assuming that the path from bus 1 to bus 3 is limited only by voltage constraints. Specifically for our example, when the capacitor is switched on the capacity of line 22 becomes $k_2 + I$. The magnitude of I is a decision variable of the investment firm. The cost of increasing the contingent capacity by I is given by, $C(I)$, where again, C is a differentiable and strictly convex cost function, with $C(0) = 0$. Figure 9.19 illustrates the network under the contingency when a capacitor is installed and switched on.

In order to satisfy the load in bus three, the total generation of the system must satisfy $P_{g1} + P_{g2} = P_d$. However, not every pair of injections (P_{g1}, P_{g2}) is allowable. Only those pairs that induce flows on the lines 21 and 22 that respect their capacity constraints are allowed. Given the basic data of the network, under normal circumstances the flow through lines 21 and 22 will be given by $P_{21}(P_{g1}, P_{g2}) = P_{22}(P_{g1}, P_{g2}) = \frac{1}{3}P_{g1} + \frac{1}{6}P_{g2}$. This flow should not exceed the maximum acceptable flow of k_1. Similarly, if the contingency occurs and line 22 is the only line that remains connecting buses 1 and 3, the flow through that line will be $P_{22}^{c}(P_{g1}, P_{g2}) = \frac{1}{2}P_{g1} + \frac{1}{4}P_{g2}$, and in order for the injections (P_{g1}, P_{g2}) to be allowable, their associated contingent flow should not exceed $k_2 + I$. The forgoing discussion suggests the following.

Definition: A *feasible allocation* $((P_{g1}, P_{g2}), I)$ is a specification of a production plan P_{gn} of each generator $n = 1, 2$, and an investment plan I of the investment firm, such that:

$$P_{g1} + P_{g2} = P_d \tag{9.81}$$

$$\frac{1}{3}P_{g1} + \frac{1}{6}P_{g2} \leq k_1 \tag{9.82}$$

$$\frac{1}{2}P_{g1} + \frac{1}{4}P_{g2} \leq k_2 + I. \tag{9.83}$$

Condition (9.81) requires that the generation should satisfy the load. Condition (9.82) dictates that the flow through either line 21 or line 22 under normal conditions should not exceed its capacity. Condition (9.83) says that in a contingency where line 21 fails, the flow through the remaining line should not exceed the operating capacity of that line when a capacitor is switched on that provides additional transmission capacity of I.

Although all feasible allocations satisfy the load and respect the capacity and contingency constraints, not all of them are equally attractive. We are interested in those feasible allocations that minimize the cost of carrying them out. These allocations are called *efficient allocations*.

Definition: A feasible allocation $((P^*_{g1}, P^*_{g2}), I^*)$ is *efficient* if there is no alternative feasible allocation $((P_{g1}, P_{g2}), I)$ such that $C_1(P_{g1}) + C_2(P_{g2}) + C(I) < C_1(P^*_{g1}) + C_2(P^*_{g1}) + C(I^*)$. Efficient allocations are optimal because they satisfy the load and it is impossible to do so in a less expensive way.

By the definition, an efficient allocation $((P^*_{g1}, P^*_{g2}), I^*)$ solves:

$$\min_{P_{g1}, P_{g2}, I \in R^3_+} C_1(P_{g1}) + C_2(P_{g2}) + C(I) \tag{9.84}$$

$$\text{s.t. } P_{g1} + P_{g2} = P_d \tag{9.85}$$

$$\frac{1}{3}P_{g1} + \frac{1}{6}P_{g2} \leq k_1 \tag{9.86}$$

$$\frac{1}{2}P_{g1} + \frac{1}{4}P_{g2} \leq k_2 + I \tag{9.87}$$

Since the cost functions are assumed to be strictly convex, and the constraints are linear, this problem has a unique solution.

Before we solve this problem, let us note that for every pair of injections (P_{g1}, P_{g2}) that satisfy the load, the associated flow through line 22 under normal circumstances is lower than the flow in case of a contingency: $\frac{1}{3}P_{g1} + \frac{1}{6}P_{g2} < \frac{1}{2}P_{g1} + \frac{1}{4}P_{g2}$. This means that since $k_1 > k_2$, in the absence of a capacitor ($I = 0$) constraint (9.86) will not bind. In other words the capacity of the lines connecting buses one and three will be underutilized. The benefit of adding a capacitor consists precisely of allowing a more efficient use of the line capacities under normal circumstances. Obviously, this benefit should be compared to the cost of the capacitor and the incremental cost of the new dispatch.

Now let us solve problem (9.84) above. Let λ, μ and η be the Lagrangian multipliers of the constraints in that problem. Then the FOCs are:

$$\frac{\partial C_1(P_{g1}^*)}{\partial P_{g1}} \geq \lambda - \frac{1}{3}\mu - \frac{1}{2}\eta \quad \text{with equality if } P_{g1}^* > 0 \tag{9.88}$$

$$\frac{\partial C_2(P_{g2}^*)}{\partial P_{g2}} \geq \lambda - \frac{1}{6}\mu - \frac{1}{4}\eta \quad \text{with equality if } P_{g2}^* > 0 \tag{9.89}$$

$$\frac{\partial C(I^*)}{\partial I} \geq \eta \quad \text{with equality if } I^* > 0 \tag{9.90}$$

$$P_{g1}^* + P_{g2}^* = P_d \tag{9.91}$$

$$\frac{1}{3}P_{g1}^* + \frac{1}{6}P_{g2}^* \leq k_1 \quad \text{with equality if } \mu > 0 \tag{9.92}$$

$$\frac{1}{2}P_{g1}^* + \frac{1}{4}P_{g2}^* \leq k_2 + I^*, \quad \text{with equality if } \eta > 0 \tag{9.93}$$

In order to understand the above conditions, consider an interior efficient allocation (P_{g1}^*, P_{g2}^*, I^*) > 0. Since generation at both buses is positive, constraints (9.88) and (9.89) are satisfied with equality. By inspection, this implies that the marginal cost of a MWH at bus one is lower than the marginal cost of a MWH at bus two. If we could generate ΔP additional units at the cheaper bus one and ΔP less units at the costly bus two, we could save:

$$\frac{\partial C_2(P_{g2}^*)}{\partial P_{g2}}\Delta P - \frac{\partial C_1(P_{g1}^*)}{\partial P_{g1}}\Delta P$$

and still satisfy the load. The problem is that we cannot transfer ΔP units of generation from generator two to generator one without violating contingency constraint (9.93). Therefore, if we want to enjoy the above savings we have to relax contingency constraint (9.93) by means of an increase in the operational capacity of line 22 under the contingency. We should increase this operational capacity by a small unit as long as its cost is no bigger than the savings induced by the redispatch that this investment allows. At the optimum, the marginal cost of the capacity should be equal to its marginal benefit:

$$\frac{\partial C(I^*)}{\partial I} = \frac{\partial C_2(P_{g2}^*)}{\partial P_{g2}}\Delta P - \frac{\partial C_1(P_{g1}^*)}{\partial P_{g1}}\Delta P$$

And this is precisely one of the implications of the first order conditions (9.88–91) above. To see this, note that since $I^* > 0$, equation (9.90) is satisfied with equality, and hence $\frac{\partial C(I^*)}{\partial I} = \eta$. Since by assumption, the marginal cost of capacitor-induced capacity is positive, $\eta > 0$, and consequently constraint (9.93) is binding. It can be shown that in this case $\mu = 0$.[6] On the other hand, a unit of additional capacity in case of a contingency allows us to change the injections in buses one and two by ΔP_{g1} and ΔP_{g2}, respectively, where ΔP_{g1} and ΔP_{g2} satisfy:

[6]Note that $\frac{1}{2}P_{g1} + \frac{1}{4}P_{g2} = \frac{2}{3}(\frac{1}{3}P_{g1} + \frac{1}{6}P_{g2})$. Therefore, constraint (9.87) can be written as $\frac{1}{2}P_{g1} + \frac{1}{4}P_{g2} \leq \frac{2}{3}k_1$. If constraint (9.93) is binding then $k_2 + I = \frac{1}{2}P_{g1} + \frac{1}{4}P_{g2} \leq \frac{2}{3}k_1$ and it is satisfied. Consequently, $\mu = 0$.

$$\frac{1}{2}\Delta P_{g1} + \frac{1}{4}\Delta P_{g2} = 1$$
$$\Delta P_{g1} + \Delta P_{g2} = 0$$

This means that the unit of additional capacity allows us to redispatch in a way that $\Delta P_{g1} = 4$, and $\Delta P_{g2} = -4$. The savings in the generation cost that this new redispatch induces is:

$$\begin{aligned}
-\frac{\partial C_1(P_{g1}^*)}{\partial P_{g1}}\Delta P_{g1} - \frac{\partial C_2(P_{g2}^*)}{\partial P_{g2}}\Delta P_{g2} &= -\left(\lambda - \frac{1}{3}\mu - \frac{1}{2}\eta\right)\Delta P_{g1} - \left(\lambda - \frac{1}{6}\mu - \frac{1}{4}\eta\right)\Delta P_{g2} \\
&= -4\left(\left(\lambda - \frac{1}{3}\mu - \frac{1}{2}\eta\right) - \left(\lambda - \frac{1}{6}\mu - \frac{1}{4}\eta\right)\right) \\
&= \frac{2}{3}\mu + \eta \\
&= \eta
\end{aligned}$$

where the last equality follows from the fact that $\mu = 0$.

Now that we have a characterization of the efficient allocation, we can ask how to implement it. One alternative would be to impose it by a central planning committee. This entity knows what the efficient allocation is and can, in principle, dictate the optimal generation levels to the generators and the optimal capacitor-induced capacity to the investment firm. In reality, however, trying to impose an allocation to the different players may be an impossible task. One would have to know the cost structure of every generator and of the investment firms, and more importantly, one would have to have the power to impose on them the optimal generation and investment levels. Another alternative would be to decentralize the decisions by means of a price system and a competitive market. The idea of such a price system is to allow the generators and investment firms to decide for themselves the generation and investment levels, respectively, taking electricity prices and transmission charges as given. The objective is still the same, but the huge task of determining the optimal allocation is now subdivided into many small tasks, each performed by each economic agent. Nobody needs to know the technology and cost structure of all the firms. It is enough for each firm to know its own cost function. Similarly, it is not needed for any omniscient central planner to figure out the optimal allocation. Each economic agent will try to maximize its own profits given the market prices. Presumably, the players will decide what is best for themselves but if the prices are right, these prices will induce the players to choose the quantities that correspond to the efficient allocation.

In the following definition of economic equilibrium, there will be electricity prices associated to each bus (the nodal prices), and two different transmission charges. Both transmission charges are related to congestion on the 1–3 corridor. One charge can be associated to the transmission on the lines under normal circumstances, and the other to the transmission under the contingency. The generators and investment firm will take these prices as given and will choose their generation and investment decisions optimally.

Definition: An allocation ((P_{g1}^*, P_{g2}^*), I^*) and a price vector (π_1^*, π_2^*, π_3^*, τ^*, ω^*) constitute a *competitive equilibrium* if the following conditions are satisfied:

1. Generators' profit maximization: each generator, G_n, for $n = 1, 2$, chooses its generation level P_{gn}^* so as to maximize its profits given the nodal price π_n^*:

$$\pi_n^* P_{gn}^* - C_n(P_{gn}^*) \geq \pi_n^* P_{gn} - C_n(P_{gn}) \quad \forall P_{gn} \geq 0, n = 1, 2$$

2. Investment firm's profit maximization: the investment firm, chooses the capacitor-induced capacity I^* so as to maximize its profits given the contingency transmission charge ω^*:

$$\omega^* I^* - C(I^*) \geq \omega^* I - C(I) \quad \forall I \geq 0$$

3. Power market clears; power supply equals the load:

$$P_{g1}^* + P_{g2}^* = P_d$$

4. Transmission market clear: demand for transmission, both under normal circumstances and under the contingency, should not exceed the capacity. And the associated transmission charge is positive only if demand for transmission equals capacity:

$$\frac{1}{3} P_{g1}^* + \frac{1}{6} P_{g2}^* \leq k_1 \quad \text{with equality if } \tau^* > 0$$

$$\frac{1}{2} P_{g1}^* + \frac{1}{4} P_{g2}^* \leq k_2 + I^* \quad \text{with equality if } \omega^* > 0$$

5. No arbitrage conditions: it should not be possible to make a profit by buying power at one of the buses at its market price, transmitting it to another node and paying the corresponding transmission charge, and selling it there at that bus's market price:

$$\pi_3^* = \pi_1^* + \frac{1}{3} 2\tau^* + \frac{1}{2} \omega^* \tag{9.94}$$

$$\pi_3^* = \pi_2^* + \frac{1}{3} 2\tau^* + \frac{1}{4} \omega^* \tag{9.95}$$

In order to understand conditions (9.94) and (9.95), note that if we inject one MW at bus one and eject it at bus three, under normal circumstances, one-third of the MW will transit through line 21 and 1/3 of the MW will transit through line 22.[7] If the contingency occurs, then half of the MW will transit through the remaining line 22. Therefore, each MW injected at bus one and ejected at bus three must pay one-third of the price of transmission along line 21and 1/3 of the price of transmission along line 22, under normal circumstances, and half of the price of transmission along line 22 under the contingency. If we add the price/hour of the MW at bus one, we obtain that the cost of buying one MWH at bus one and transmitting it to bus three is $\pi_1^* + \frac{1}{3} 2\tau^* + \frac{1}{2}\omega^*$. The first no-arbitrage condition states that this cost should

[7] The other third will transit through lines 1 and 3.

equal the price that one would obtain by selling this MWH at the destination bus. A similar interpretation applies to no-arbitrage condition (9.95).

Let's characterize a competitive equilibrium. For this purpose assume that allocation $((P_{g1}^*, P_{g2}^*), I^*)$ and a price vector $(\pi_1^*, \pi_2^*, \pi_3^*, \tau^*, \omega^*)$ constitute a competitive equilibrium. Then the generation level P_{gn}^*, for $n = 1, 2$, satisfies the first order conditions of the generator's profit maximization problem:

$$\frac{\partial C_1(P_{g1}^*)}{\partial P_{g1}} \geq \pi_1^* \quad \text{with equality if } P_{g1}^* > 0 \tag{9.96}$$

$$\frac{\partial C_2(P_{g2}^*)}{\partial P_{g2}} \geq \pi_2^* \quad \text{with equality if } P_{g2}^* > 0 \tag{9.97}$$

Also, the investment firm capacitor-induced capacity satisfies the first order conditions of its profit maximization problem:

$$\frac{\partial C(I^*)}{\partial I} \geq \omega^* \quad \text{with equality if } I^* > 0 \tag{9.98}$$

As a result, a competitive equilibrium is characterized by conditions (9.96–98) and the market clearing and nor-arbitrage conditions above.

By comparison, we can see that if allocation $((P_{g1}^*, P_{g2}^*), I^*)$ solves the social optimum problem (9.84) with associated Lagrangian multipliers λ, μ and η, then the same allocation $((P_{g1}^*, P_{g2}^*), I^*)$ together with the price vector $(\pi_1^*, \pi_2^*, \pi_3^*, \tau^*, \omega^*)$ defined by:

$$\begin{aligned}
P_{g1}^* &= \lambda - \frac{1}{3}\mu - \frac{1}{2}\eta \\
P_{g2}^* &= \lambda - \frac{1}{6}\mu - \frac{1}{4}\eta \\
\pi_3^* &= \lambda \\
\tau^* &= \mu/2 \\
\omega^* &= \eta
\end{aligned}$$

is a competitive equilibrium. Conversely, if allocation $((P_{g1}^*, P_{g2}^*), I^*)$ together with price vector $(\pi_1^*, \pi_2^*, \pi_3^*, \tau^*, \omega^*)$ constitute a competitive equilibrium, then the same allocation $((P_{g1}^*, P_{g2}^*), I^*)$ together with the Lagrangian multipliers defined by:

$$\begin{aligned}
\lambda &= \pi_3 \\
\mu &= 2\tau^* \\
\eta &= \omega^*
\end{aligned}$$

solve the social optimum problem (9.84).

The above analysis shows that the competitive equilibrium leads to the efficient allocation. In particular, the competitive equilibrium induces the optimal amount of capacitor-induced capacity enhancement.

9.6 SUMMARY

The transmission planning process is receiving a great deal of attention today because it is arguably the most significant technical element of competitive electric markets for which consensus has not yet been reached in regards to its implementation. Impediments to achieving that consensus include difficulties in siting and obtaining right-of-ways, the significant investment cost, uncertainties associated with predicting future information affecting network operation, and difficulties in identifying beneficiaries, assignment of cost responsibilities, and how cost recovery takes place for investors. Some organizations have provided answers to these questions, and those answers for one such organization are summarized in this document. We also provide a general optimization framework for transmission planning, and we illustrate its use for two difference cases; when solutions are restricted to new circuits only, and when solutions are restricted to switchable shunt and series capacitors only. Although the latter case cannot always be implemented alone, it is economically attractive when it is a feasible solution. We provide a detailed description and illustration of practical optimization procedures for identifying optimal location and amount of switchable shunt and series capacitors to increase contingency-limited transmission capacity through network reconfiguration. This approach is attractive because it is conceptually based on the automaton, a key element in addressing dynamic performance for discrete-event systems. We describe our implementation of system design, in terms of sequence and timing of configurable switches, based on identification of stability regions corresponding to the considered switching mode. The last part of our chapter focuses on the efficiency of the electricity market with inclusion of the facility investment effects. The interesting conclusion is that, when transmission expansion is limited to contingency-driven switchable capacitors, the competitive equilibrium leads to the efficient allocation.

ACKNOWLEDGMENTS

The authors express appreciation to Qiming Chen and John Condren of PJM Interconnection for the helpful suggestions related to Section 9.2 of this material, and to John Paserba of Mitsubishi Electric Power Products, Inc. for his careful review and helpful comments throughout the paper.

BIBLIOGRAPHY

[1] M. Buygi, G. Balzer, H. Shanechi, and M. Shahidehpour, "Market-based transmission expansion planning," *IEEE Trans. on Power Systems*, Vol. 19, No. 4, pp. 2060–2067, November 2004.

[2] E. Hirst, "U.S. Transmission Capacity: Present Status and Future Prospects," 2004, at www.ee.iastate.edu/~jdm/ee458/TransmissionCapacityFinal_Herst.pdf.

[3] M. Shahidehpour, "Investing in expansion: the many issues that cloud transmission planning," *IEEE Power and Energy Magazine*, pp. 1–18, Jan/Feb. 2004.

[4] L. Dale, "Reactive compensation using static VAR compensators-a planning case study using security constrained techniques," *International Conf. on AC and DC Power Transmission*, pp. 50–54, Sept. 17–20, 1991,

[5] R. Koessler, J. Mountford, L. Lima, and J. Rosende, "The Brazilian interconnected system: a study on transfer limits, reactive compensation and voltage collapse," *Proc. of the IEEE Power Engineering Society Winter Meeting*, Vol. 3, pp. 1147–1153, Jan. 28-Feb. 1, 2001.

[6] P. Pourbeik, R. Koessler, and B. Ray, "Tools and techniques for analyzing voltage stability related reliability challenges," *Proc. on the IEEE Power Engineering Society Transmission and Distribution Conference and Exposition*, Vol. 1, pp. 417–421, Sept. 12, 2003.

[7] A. Hammad and M. El-Sadek, "Prevention of transient voltage instabilities due to induction motor loads by static VAr compensators," *IEEE Transactions on Power Systems*, Vol. 4, Issue 3, pp. 1182–1190, Aug. 1989.

[8] M. S. Branicky, "Modeling Analysis and Control of Hybrid Systems", PhD Thesis Massachusetts Institute of Technology, 1994.

[9] C. J. Tomlin, J. Lygeros, and S. Shankar Sastry, "A Game Theoretic Approach to Controller Design for Hybrid Systems," *Proceedings of the IEEE*, Vol. 88, No. 7, pp. 949–970, Jul. 2000.

[10] "Regional Transmission Organizations," Docket No. RM99-2-000; Order 2000 of the Federal Energy Regulatory Commission (FERC), issued Dec.20, 1999; www.ferc.gov/legal/ferc-regs/land-docs/RM99-2A.pdf.

[11] www.ferc.gov/industries/electric/indus-act/rto.asp

[12] Final Rule—Order 2003, "Standardization of Generator Interconnection Agreements and Procedures," Docket No. RM02-1-000, Issued July 24, 2003.

[13] "Flow Chart of the Large Generating Facility Interconnection Process," in Appendix A to the Final Rule—Order 2003 "Standardization of Generator Interconnection Agreements and Procedures."

[14] Schedule 6 of "Operating agreement of PJM Interconnection L.L.C.," June 22, 2005, at www.pjm.com/documents/agreements.html.

[15] "PJM Manual 14B: Generation and Transmission Interconnection Planning," Revision: 05, June 23, 2005, at www.pjm.com/contributions/pjm-manuals/manuals.html.

[16] T. Wu, Z. Alaywan, A. Papalexopoulos, "Locational Marginal Price Calculations Using the Distributed-Slack Power-Flow Formulation," *IEEE Transactions on Power Systems*, Vol. 20, No. 2, May 2005.

[17] "PJM Manual 27: Open Access Transmission Tariff Accounting Revision," Revision 46, May 1, 2005, at www.pjm.com/contributions/pjm-manuals/manuals.html.

[18] "PJM Manual 06: Financial Transmission Rights," Revision 7, April 15, 2005, at www.pjm.com/contributions/pjm-manuals/manuals.html.

[19] F. Schweppe, M. Caramanis, R. Tabors, and R. Bohm, "*Spot Pricing of Electricity*," Boston, Kluwer Academic Press, 1988.

[20] N. Rau, "*Optimization Principles: Practical Application to the Operation and Markets of the Electric Power Industry*," John Wiley, 2003.

[21] D. Kirschen and G. Strbac, "*Power system economics*," John Wiley, 2004.

[22] "Operating agreement of PJM Interconnection, L.L.C.," June 22, 2005, at http://www.pjm.com/documents/downloads/agreements/.

[23] P. Jaskow, "Transmission Policy In The United States," Utilities Policy, 13:95–115, 2005. http://econ-www.mit.edu/faculty/download_pdf.php?id=1003

[24] New York Independent System Operator, "Transmission Expansion and Interconnection Manual," Sept. 1999, at http://www.nyiso.com/public/documents/manuals/planning.jsp.

[25] ISO New England Planning Procedures, at http://www.iso-ne.com/rules_proceds/isone_plan/index.html

[26] Z. Alaywan, "Understanding the California Firm Transmission Rights," at www.caiso.com/aboutus/articles/.

[27] Z. Alaywan and T. Wu, "Effects of Firm Transmission Rights on Transmission Revenue Allocation in California," at www.caiso.com/aboutus/articles/.

[28] S. Jones, "ERCOT ISO Experience with Transmission Operating and Planning Criteria in a Competitive Environment," *Proc. of the 1999 IEEE Power Engineering Society Summer Meeting*, Vol. 1, July 18–22, 1999, pp. 140–142.

[29] Public Utilities Commission of Texas, P.U.C. Subst. R. 25.101, "Transmission Planning, Licensing and Cost-Recovery for Utilities within the Electric Reliability Council of Texas," Effective data, April 13, 2005.

[30] "Midwest ISO Transmission Expansion Plan (MTEP) 2005," June 2005, at www.midwestiso.org/plan_inter/expansion.shtml.

[31] "American Transmission Company 2004 10-Year Assessment Update," March 2005, at www.atcllc.com/IT2b.shtml.

[32] "American Transmission Company Planning Criteria," Document number PLG-CR-0001-v08, July 5, 2005.

[33] S. Harvey, W. Hogan, and S. Pope, "Electricity Market Design and Structure: Working Paper on Rate and Transition Issues in Standardized Transmission Service and Wholesale Electric Market Design," Submission to the Federal Regulatory Commision, Docket No. RM01-12-000, May, 2002.

[34] E. Hirst and B. Kirby, "Transmission Planning and the Need for New Capacity," at http://certs.lbl.gov/NTGS/ISSUE_4.PDF.

[35] F. Woolf. *Global Transmission Expansion—Recipes for Success*. Penn Well Corporation, Tulsa Oklahoma, 2003.

[36] Standards of the North American Reliability Council, located at www.nerc.com/~filez/pss-psg.html.

[37] "NERC Planning Standards," Section I System Adequacy and Security, Standard S1.

[38] "System Protection Schemes in Power Networks," by Task Force 38.02.19, chaired by Daniel Karlsson and Xavier Waymel, A technical brochure for International Conference on Large High Voltage Electric Systems (CIGRE), January, 2001.

[39] W. Mittelstadt, C. Taylor, M. Klinger, J. Luini, J. McCalley, and J. Mechenbier, "Voltage Instability Modeling and Solutions As Applied to the Pacific Intertie," *International Conference on Large High Voltage Electric Systems* (CIGRE), June 1990.

[40] N. Miller, R. D' Aquila, K. Jimma, M. Sheehan, G. Comegys, "Voltage stability of the Puget Sound system under abnormally cold weather conditions," *IEEE Trans. on Power Systems*, Vol. 8, No. 3, pp. 1133–1142, Aug. 1993.

[41] R. L. Sullivan, *Power System Planning*. New York: McGraw-Hill, 1977.

[42] X. Wang and J. R. McDonald, *Modern Power System Planning*. London: McGraw-Hill, 1994.

[43] G. Latorre, R. D. Cruz, J. M. Areiza, and Andres Villegas, "Classification of Publications and Models on Transmission Expansion Planning," *IEEE Trans. Power Systems*, Vol. 18, pp. 938–946, May 2003.

[44] V. Ajjarapu and C. Christy, "The continuation power flow: a tool for steady state voltage stability analysis," *IEEE Trans. on Power Systems*, Vol. 7, pp. 416–423, Feb. 1992.

[45] C. A. Canizares and F. L. Alvarado, "Point of collapse and continuation methods for large AC/DC systems," *IEEE Trans. Power Systems*, Vol. 8, pp. 1–8, Feb. 1993.

[46] H. D. Chiang, A. J. Flueck, K. S. Shah, and N. Balu, "CPFLOW: a practical tool for tracing power system steady-state stationary behavior due to load and generation vacations," *IEEE Trans. on Power Systems*, Vol. 10, pp. 623–634, May 1995.

[47] S. Greene, I. Dobson, F. L. Alvarado, "Contingency ranking for voltage collapse via sensitivities from a single nose curve," *IEEE Trans. on Power Systems*, Vol. 14, pp. 232–240, Feb. 1999.

[48] M. Ni, J. McCalley, V. Vittal, S. Greene, C. Ten, V. S. Ganugula, and T. Tayyib, "Software implementation of online risk-based security assessment," *IEEE Trans. on Power Systems*, Vol. 18, pp. 1165–1172, Aug. 2003.

[49] O. O. Obadina and G. J. Berg, "Var planning for power system security," *IEEE Trans. Power Systems*, Vol. 4, pp. 677–686, May 1989.

[50] K. H. Abdul-Rahman and S. M. Shahidehpour, "Application of fuzzy sets to optimal reactive power planning with security constraints," *IEEE Trans. Power Systems*, Vol. 9, pp. 589–597, May 1994.

[51] V. Ajjarapu, P. L. Lau, and S. Battula, "An optimal reactive power planning strategy against voltage collapse," *IEEE Trans. on Power Systems*, Vol. 9, pp. 906–917, May 1994.

[52] J. A. Momoh, S. X. Guo, E. C. Ogbuobiri, and R. Adapa, "The quadratic interior point method solving power system optimization problems," *IEEE Trans. Power Systems*, Vol. 9, pp. 1327–1336, Aug. 1994.

[53] K. Y. Lee, X. Bai, and Y. M. Park, "Optimization method for reactive power planning by using a modified simple genetic algorithm," *IEEE Trans. Power Systems*, Vol. 10, pp. 1843–1850, Nov. 1995.

[54] E. Vaahedi, Y. Mansour, C. Fuchs, S. Granville, M. L. Latore and H. Hamadanizadeh, "Dynamic security constrained optimal power flow/Var planning," *IEEE Trans. on Power Systems*, Vol. 16, pp. 38–43, Feb. 2001.

[55] W. D. Rosehart, C. A. Canizares, and V. H. Quintana, "Effect of detailed power system models in traditional and voltage-stability-constrained optimal power flow problems," *IEEE Trans. on Power Systems*, Vol. 18, pp. 27–35, Feb. 2003.

[56] N. Yorino, E. E. El-Araby, H. Sasaki, and S. Harada, "A new formulation for FACTS allocation for security enhancement against voltage collapse," *IEEE Trans. on Power Systems*, Vol. 18, pp. 3–10, Feb. 2003.

[57] Z. Feng, V. Ajjarapu, and D. Maratukulam, "A comprehensive approach for preventive and corrective control to mitigate voltage collapse," *IEEE Trans. Power Systems*, Vol. 15, pp. 791–797, May 2000.

[58] I. Dobson and L. Lu, "Computing an optimal direction in control space to avoid saddle node bifurcation and voltage collapse in electrical power systems," *IEEE Trans. on Automatic Control*, Vol. 37, pp. 1616–1620, Oct. 1992.

[59] S. Greene, I. Dobson, and F. L. Alvarado, "Sensitivity of the loading margin to voltage collapse with respect to arbitrary parameters," *IEEE Trans. on Power Systems*, Vol. 11, pp. 845–850, May 1996.

[60] B. Long and V. Ajjarapu, "The sparse formulation of ISPS and its application to voltage stability margin sensitivity and estimation," *IEEE Trans. on Power Systems*, Vol. 14, pp. 944–951, Aug. 1999.

[61] G. L. Nemhauser and L. A. Wolsey, *Integer and Combinatorial Optimization*. New York: Wiley, 1988.

[62] IEEE/CIGRE joint task force on stability terms and definitions, "Definition and classification of power system stability," *IEEE Trans. on Power Systems*, Vol. 19, pp. 1387–1401, Aug. 2004.

[63] C. W. Taylor, *Power System Voltage Stability*. EPRI Power System Engineering Series. New York: McGraw Hill, 1994.

[64] T. Van Cutsem and C. Vournas, *Voltage Stability of Electric Power Systems*. Boston: Kluwer Academic Publishers, 1998.

[65] A. M. Abed, "WSCC voltage stability criteria, undervoltage load shedding strategy, and reactive power reserve monitoring methodology," in *Proc. IEEE Power Eng. Soc. Summer Meeting*, Vol. 1, pp. 191–197, 1999.

[66] P. M. Anderson and A. A. Fouad, *Power System Control and Stability*, 2nd ed. Piscataway, N.J.: IEEE Press; Wiley-Interscience, 2003.

[67] H. K. Khalil, "*Nonlinear Systems*," Third Edition, Prentice Hall, New Jersey, 2002.

[68] P. Varaiya, F. Wu, and R. Chen, "Direct methods for transient stability analysis of power systems: recent results," *Proceedings of the IEEE*, Vol. 73, pp. 1703–1715, 1985.

[69] J. Zaborszky, J. Huang, B. Zheng, and T. Leung, "On the phase portraits of a class of large nonlinear dynamic systems such as the power systems," *IEEE Trans. on Automatic Control*, Vol. 33, pp. 4–15, 1988.

[70] H. Chiang, M. Hirch, and F. Wu, "Stability regions of nonlinear autonomous dynamic systems," *IEEE Trans. on Automatic Control*, Vol. 33, pp. 16–27, 1988.

[71] H. Chiang and F. Wu, "Foundations of the potential energy boundary surface method for power system transient stability analysis," *IEEE Trans. on Circuits and Systems*, Vol. 35, pp. 712–728, 1988.

[72] V. Venkatasubramanian and W. Ji, "Numerical approximation of (n-1)-dimensional stable manifolds in large systems such as power system," *Automatica*, Vol. 33, pp. 1877–1883, 1997.

[73] D. Cheng and J. Ma, "Calculation of stability region," *Proceedings of the 42nd IEEE Conference on Decision and Control*, Vol. 6, pp. 5615–5620, maui, Hawaii, USA, 2003.

[74] S. Saha, A. A. Fouad, W. H. Kliemamm, and V. Vittal, "Stability boundary approximation of a power system using the real normal form of vector fields," *IEEE Trans. on Power Systems*, Vol. 12, pp. 797–802, 1997.

[75] R. Qi, D. Cook, W. Kliemann, and V. Vittal, "Visualization of stable manifolds and multidimensional surfaces in the analysis of power system dynamics," *Journal of Nonlinear Science*, Vol. 10, pp. 175–195, 2000.

[76] N. Anthony, A. Fouad, and V. Vittal, "Power system transient stability using individual machine energy functions," *IEEE Trans. on Circuits and Systems*, Vol. 30, pp. 266–276, 1983.

[77] M. Ribbens-Pavella and F. Evans, "Direct methods for studying dynamics of large scale power system—a survey," *Automatica*, Vol. 32, pp. 1–21, 1985.

[78] H. Chiang and J. Thorp, "The closest unstable equilibrium point method for power system dynamic security assessment," *IEEE Trans. on Circuits and Systems*, Vol. 36, pp. 1187–1199, 1989.

[79] L. Jin, H. Liu, R. Kumar, V. Ajjarapu, J. McCalley, N. Elia, and V. Vittal, "An application of reachable set analysis in power system transient stabiltiy assessment," *Proceedings of the 2005 IEEE PES General Meeting*, San Francisco, June 2005.

[80] I. M. Mitchell, "Application of level set methods to control and reachability problems in continuous and hybrid systems," PhD thesis, Stanford University, Stanford, 2002.

[81] H. Liu, L. Jin, V. Ajjarapu, R. Kumar, J. McCalley, N. Elia, and V. Vittal, "Reachability analysis based minimal load shedding determination," *Proceedings of the 2005 IEEE PES General Meeting*, San Francisco, June 2005.

[82] W. W. Hogan, "Contract Network for Electric Power Transmission," *Journal of Regulatory Economics*, Vol. 4, pp. 211–242, 1996.

[83] J. B. Bushnell and S. E. Stoft, "Electric Grid Investment Under a Contract Network Regime," *Journal of Regulatory Economics*, Vol. 10, pp. 61–79, 1996.

[84] Chao, Hung-Po and Peck, Stephen, 1996. "A Market Mechanism for Electric Power Transmission," *Journal of Regulatory Economics*, Springer, Vol. 10, Issue 1, pp. 25–59.

[85] J. B. Bushnell and S. E. Stoft, "Improving Private Incentives For Electric Grid Investment," *Resource and Energy Economics*, Vol. 19, pp. 85–108, 1997.

[86] P. Joskow and J. Tirole, "Merchant Transmission Investment," *Journal of Industrial Economics*. Vol. 53, Issue 2, pp. 233–264, June 2005.

CHAPTER **10**

NEXT GENERATION OPTIMIZATION FOR ELECTRIC POWER SYSTEMS

James Momoh
Howard University

EDITORS' SUMMARY: This chapter presents a summary tutorial based on a course titled "Next Generation Optimization for Power Systems." It utilizes state-of-the-art research in optimization Systems Engineering, Operation Research, Intelligent Systems, AI communities to solve the grand challenge problems of electric power networks. The work is inspired by the initiative led by the author on interdisciplinary research and education at the National Science Foundation (NSF). The initiative aims to develop unification of knowledge through research and education. The scope of the course includes mathematical formulations, concepts, algorithms, and practical applications of advanced optimization methods to power system with illustrative examples and benchmark test beds. As part of new power system curriculum at Howard University, a new course titled, "Next Generation Optimization for Electric Power Systems" is proposed for graduate students. We summarize in this paper, the highlights of the topic and demonstration of the new optimization technique to power system problems.

10.1 INTRODUCTION

Over the past few yeas, the traditional systems engineering program has not been taught in a majority of the engineering schools' curriculum. We are graduating engineering students with minimal background in System Theory, Control Systems Optimization, and Computation Intelligence tools needed for solving large scale power systems problems. A recent National Science Foundation (NSF) initiative was developed, aimed at promoting broader unified knowledge of system engineering

Economic Market Design and Planning for Electric Power Systems, Edited by James Momoh and Lamine Mili

topics for addressing control coordination in real time, reliability issues, resource allocation, risk assessment, self healing ability, and optimum planning and operation of secured large power system networks. This paper serves as a preliminary work under preparation for addressing many of the computation challenges in addressing the above problems [1, 14, 29].

The process of analysis and synthesis of large scale systems utilizes optimization theory concepts that include the outcome of each possible action or feasible solution predicted based in analysis, evaluation of outcomes according to some scale of value or desirability, and a criterion for decision based objectives of the system being used to determine the most desirable action or optimal solution. The process of optimal decision-making is shown in Figure 10.1.

There are many methods of decision-making, which are useful tools for systems analysis and optimization [10, 11, 15–18]. Although many engineering problems are deterministic and ill-structured, several non-deterministic problems have led to a set of decisions that results in uncertainty and anticipatory in nature.

The mathematical tools for solving such problems have grown over the years and they range from classical optimization methods, critical path programming, dynamic programming, stochastic programming, decision support tools, to Intelligent System (IS) based tools [19, 22–24]. Intelligent Systems spanning a broad category including Artificial Neural Networks (ANN), Expert systems (ES), Genetic Algorithm (GA), and Evolutionary Programming (EP). Recently, Adaptive Dynamic Programming (ADP) [4–9], which handles both dynamic and uncertainty due to conventional probability and statistical interface technique, have emerged as a state-of-the-art problem solving technique for a wide range of optimization problems [21, 25, 26,]. This has led to new optimization tools capable of handling non-deterministic problems.

The reinforcement learning techniques has been accomplished by physiologists and Intelligent System (IS) research community as a tool for enhancing classical dynamic programming to handle optimization problems with stochastic and uncertainty [4]. It provides an optimal time saving search technique (storing only useful trajectory needed for the solution). Thus, overcoming the so-called "curse of dimensionality" [5, 13, 38]. The result of the techniques lead to savings in computation hence can easily be used in real-time problems. Also, there are many variants of ADP optimization techniques applicable for different applications [5].

Decision Analysis (DA) has been used for decision-making under uncertainty, a vital factor when there is a need to determine a course of action consistent with

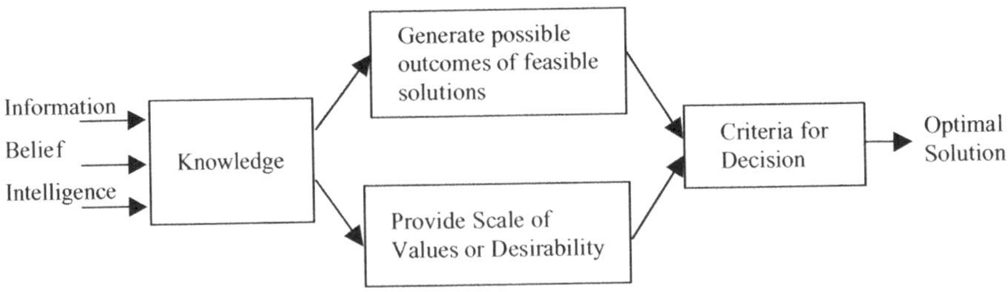

Figure 10.1. Optimal decision-making.

personal basic judgments and preferences. Over the decades, professionals including system engineers and engineering practitioners have demonstrated several real-time applications [28, 31]. While this method is mainly for deterministic problems, improvements to extend its applications to handle stochastic processes with risk factors are needed [25]. Furthermore, storage problem and computational complexity for handling real life problems such as power system planning and operation requires enhanced knowledge of reinforcement learning and other geometric theoretical techniques.

The Analytical Hierarchical Process (AHP) is another decision-making tool fundamental to multi-criteria decision of a constrained problem with multiple solutions in the decision space. It employs principles of hierarchy for a given assignment or allocation planning or operational task. A practical comparison process based on values and its priority is utilized for optimal decision. Systems engineers have demonstrated this technique applied to real-time or practical problems. Again, further knowledge and integration of these operational methods will enhance future generations of optimization methods and applications to power systems.

Other tools exist such as game theory, critical path-finding networks planning methods for scheduling and planning large-scale events. Optimization graphs were first introduced to power systems topology, resource allocation planning in the mid-sixties in order to mathematically formulate decision-making processes. This technique featured a myriad of objectives subject to technical and non-technical constraints and stochastic decisions as well as dynamic changes in data, topology, etc.

Most recently, there have been tremendous surge in use of optimization techniques, intelligent system and some variety of decision support tools for deterministic problems [21, 30]. As power system planning and operation include stochastic/dynamic (anticipating changes), significant research is needed for the development of next generation optimization to achieve highly efficient and autonomous power systems of the future. It is with this intention that a survey course summarizing overall optimization tools and reviewing the formulation Decision Analysis (DA) methods, selected classical Optimization techniques, and Adaptive Dynamic Programming (ADP) tools for developing the future Dynamic Stochastic Optimal Power Flow (DSOPF) are proposed. An ongoing research at the Center for Energy System and Controls (CESaC), Howard University employs these tools for research and enhancing education materials for solving optimization problems is discussed in this tutorial paper.

10.2 STRUCTURE OF THE NEXT GENERATION OPTIMIZATION

10.2.1 Overview of Modules

The organization of the topics for the Next Generation Optimization course being offered at Howard University under a research grant by the National Science Foundation (NSF) covers, but is not limited to, the following core topics:

a. Decision Analysis (DA), Analytical Hierarchical Processes (AHP)
b. Decision support tools such as Game Theory
c. Intelligent Systems (IS)
d. Classical Optimization methods and their extensions
e. Dynamic Programming (DP)
f. Adaptive Dynamic Programming (ADP) and its variants
g. Dynamic Stochastic Optimal Power Flow (DSOPF)
h. Benchmark Systems—Applications and solutions to challenge problems

The course material is organized in modules as follows:

- Module I: Review *of Decision Analysis Tools.* A review of decision analysis methods, concepts, tools, and modeling of decision-making under uncertainty will be provided. We will evaluate the use of decision support such as Analytical Hierarchical Processes (AHP), game theory, and their relationships to learning algorithms. The application of the methods to solve power system problems such as control coordination, optimal reconfiguration, etc., will be discussed.
- Module II: *Review of Classical Optimization Techniques.* A review of optimization techniques will be presented to include static optimization techniques such as linear and nonlinear programming, Interior Point method and its variants, etc. The concepts and algorithms as well as illustrative examples applicable to state estimation, control coordination, and extensions to stochastic optimal power flow will also be provided.
- Module III: *Dynamic Optimization Techniques.* The course will also present an overview of optimal control, dynamic programming, and underlying concepts such as the generalized Hamiltonian-Jacobi, Pontryagin's Principle, and Bellman's optimality conditions. Application of dynamic optimization to power systems (such as stability, fault analysis, unit commitment, etc.) will be described.
- Module IV: *Adaptive Dynamic Programming (ADP).* This module of the course provides an overview of Adaptive Dynamic Programming (ADP) principle, formulation, variants, and potential applications to power systems control, operation, and planning problems. ADP and its applications to power system, economics, and other areas will be discussed.
- Module V: *Dynamic Stochastic Optimal Power Flow (DSOPF).* This section of the course introduces generalized formulations of ADP for solving different classes of OPF problems with stochastic variables and input power system parameters. Examples of ADP to power system problems such as unit commitment, reconfiguration, reliability, restoration, fault studies and remedial control, dynamitic security assessment, and voltage security assessment will be discussed in the framework of Dynamic Stochastic Optimal Power Flow (DSOPF).

- Module VI: The development of the research and education material and bench mark systems will be built, tested and disseminated to schools for evaluating the tools against well known results from other researchers. The results will be presented in a book to be published by CRC Publishing Company.

10.2.2 Organization

An overview of the topics and possible applications to different topics in power system is presented herein. To date, in a deregulated power system environment, the following topics have been proven to be of current interest. Table 10.1 summarizes applications of classical and hybrid optimization methods or mathematical programming methods that are being applied to power systems. The sufficiency of these tools to solve each of the selected problems are discussed and demonstrated throughout the textbook.

TABLE 10.1. Summary of Applied Mathematical Programming Methods for Power Systems Applications.

Topic	Brief Description of Typical Electric Power System Applications	Computational Tools and Mathematical Methods
Risk Assessment	Involves assessing decisions under deterministic/uncertain conditions—e.g., technical implications of investment options.	▪ Game Theory ▪ Analytical Hierarchical Processes (AHP)
Reliability	Determination of power system adequacy and efficiency in Loss of Load Probability, Expected Unserved Energy, etc.	▪ Optimization theory ▪ Probability theory
Resource Allocation	Optimal siting/setting of controls such as VAr Planning and Unit Commitment.	▪ Dynamic Programming ▪ LP / NLP optimization ▪ AI techniques
Power system operational challenges	Optimal power flow and cost of generation dispatch (economic dispatch of different mix of generation types and units)	▪ Classical optimization ▪ AI techniques
Optimal Control Coordination	Optimizing the size of control cost effective location of controls devices and equipment.	▪ Classical optimization ▪ AHP
Power System Planning	Optimal mix of decisions under budget, resource, and time constraints.	▪ AHP ▪ Game Theory
Operation and Maintenance (O&M)	Optimization of maintenance schedules and reliability assessments of complex, interacting networks.	▪ Game theory ▪ Decision Analysis (DA)

10.3 FOUNDATIONS OF THE NEXT GENERATION OPTIMIZATION

10.3.1 Overview

This section of the course provides an overview to several formulations and algorithms for Next Generation Optimization methods as well as global optimization techniques that handle complexity, stochastic and dynamic changes in optimization process applicable to power systems. Future electric power systems needs certain criteria to satisfy the efficiency, reliability, reconfiguration, survivability and self-healing feature as defined in [39, 40] and the vision for the Electric Power Networks Efficiency and Security (EPNES) initiative.

A review of decision analysis methods, concepts, tools, and modeling for decision-making under uncertainty will be provided. We will evaluate the use of decision supports such as Analytical Hierarchical Processes (AHP), and game theory, and their relationships to learning algorithms. The application of these methods to solve power system problems such as control coordination, optimal reconfiguration, etc., will be discussed.

In this chapter, we evaluate the different optimization techniques that are candidates for enhancing and contributing to the next generation optimization methods for power system operation and planning. We will present the formulations of the optimization methods, procedures or algorithms for implementing the optimization process and subsequently provide solutions to the grand challenges. Solution strategies will be provided to overcome the drawbacks in computation burdens and **adequacy in handling the stochasticity and dynamics,** and subsequently foresight in dealing with challenges of developing a robust optimization methods/decision support systems that give derivative of existing system energy tools. The ultimate goal is to develop new optimization techniques or hybrids. This has been demonstrated by many workshops held at NSF/ECS by the author in Panel discussions aimed at promoting the NSF/ONR sponsored initiative on Electric Power Networks Efficiency and Security (EPNES) [39, 40]. EPNES aims at developing future power systems that is self-healing, reconfigurable, reliable, and efficient.

The optimization methods are summarized in this course with the hope that researchers can evaluate their potential and scope to include the concept of anticipatory events and decisions, dynamics (time-scale), or stochastic changes borrowed from the Adaptive Dynamic Programming (ADP) community to enhance the applications and capability for power systems.

A framework of applying next generation optimization methods to power systems is proposed using Adaptive Dynamic Programming (ADP) and Interior Point method for Optimal Power Flow (OPF). This will be termed the Dynamic Stochastic Optimal Power Flow (DSOPF), which handles anticipatory situations and solves new class of optimal power flow challenges. Table 10.2 lists some typical power system applications and the new trends in applied optimization methods and next generation optimization techniques for the future power systems.

TABLE 10.2. List of Typical Applications, Classical Optimization, and New Trends.

Optimization Problems	Currently Used Optimization Techniques	Next Generation Optimization Techniques
Unit Commitment / Hydro dispatch	Dynamic Programming (DP)	ADP & its variants
Control Coordination	Decomposition Optimization	ADP, AHP, classical optimization, Evolutionary Programming
Machine Controls and Stabilization	Optimal Control	ADP and Evolutionary Programming
Optimal Reconfiguration	Mixed Integer Programming	Dynamic Stochastic Optimal Power Flow (DSOPF) and its variants
Loss Minimization	Nonlinear programming (NLP) and Interior Point methods	
Economic dispatch	NLP, DP	
Locational Marginal Pricing	Linear Programming	ADP
Data Mining	State estimation (SE)	ADP and Evolutionary Programming
Optimal Sensor Placement	Intelligent Systems such as Artificial Neural Networks (ANN)	ADP and Decision Analysis (DA)

10.3.2 Decision Analysis Tools

10.3.2.1 Decision Analysis (DA) Decision Analysis [28] is a method of decision-making under uncertainty. The final decision is based on the expected monetary value calculated from probabilistic parameters and actual earnings dependant on the outcome of the decision process. Decision analysis is a powerful tool that makes a total uncertainty problem appear as a perfectly rational decision based on numerical values for comparing and yielding fast results. However, there is always a risk, even if the expected loss is reduced to its lowest, it cannot be cancelled.

A decision has to be made and the result of this decision will yield a profit or a loss. There is a probability for the result to occur in one-way or another at the beginning. However the decision maker can spend or take more or less risk by sampling or buying some accurate information. Therefore, decision analysis (DA) is the discipline comprising the philosophy, theory, methodology, and professional practice necessary to address important decisions in a formal manner.

Decision analysis includes many procedures, methods, and tools for identifying, clearly representing, and formally assessing the important aspects of a decision situation. Decision analysis is for computing the recommended course of action by applying the maximum expected utility action axiom to a well-formed representation of the decision, and for translating the formal representation of a decision and its corresponding recommendation into insight for the decision-maker and other decision participants. Multi-Criteria Decision Analysis (MCDA) is a form of DA and is a procedure aimed at supporting decision maker(s) whose problem involves numerous and conflicting evaluations. MCDA aims at highlighting these conflicts and deriving a way to come to a compromise in a transparent process. Analytical Hierarchical Processing (AHP) is also a form of this MCDA.

Real decisions are complex; the purpose of analysis is to not capture decisions in all its complexity but to simplify the decision enough to meet the decision maker's needs. An important challenge then is to determine how to simplify an analysis without diminishing its usefulness and accuracy. A useful simplification is to ignore some uncertainties, so the value of an action is assumed to be more "certain" than in reality. In other words, the chance of an event is either near zero or one. For instance, in deciding which departments need additional funds, the decision maker might choose to assess current levels of needs and ignore the uncertainty about future needs. Of course, such simplifications are only appropriate when using them will make little difference in the results of the analysis.

Alternatively, the analyst may assume that uncertainty is the only issue and that the other values and actions can be addressed without the help of analysis. For example, the principal challenge in strategic planning may be diagnosing what would our target customers need. Presumably, after knowing unmet customers needs, the decision maker's action would be relatively clear and the analyst would not need to examine the decision maker's preferences over different outcomes.

In developing a Decision Analysis support, the following two-stage operation must be done:

- Stage 1: Evaluate the EMV (expected monetary value) from the profit and loss data and the probability associated with them. Draw the first decision flow tree. This should yield a best decision based on the highest EMV and/or the lowest expected loss.
- Stage 2: Consider the possibilities of sampling and accurate information and reevaluate the new EMV. Draw the new decision flow diagram with the one in step 1 included. This should yield a best decision based on the highest EMV and/or the lowest expected loss.

DA must be implement with care; if available data is inadequate to support the analysis, it is difficult to evaluate the effectiveness, and leading to oversimplification of the problem. The outcomes of decision analyses are not amenable to traditional statistical analysis. Strictly, by the tenets of decision analysis, the preferred strategy or treatment is the one that yields the greatest utility (or maximizes the occurrence of favorable outcomes) no matter how narrow the margin of improvement.

10.3.2.2 Analytical Hierarchical Programming (AHP) The Analytic Hierarchy Process (AHP) is a decision-making approach that presents the alternatives and criteria, evaluates the trade-off, and performs a synthesis to arrive at a final decision. AHP is especially appropriate for cases that involve both qualitative and quantitative analysis. It is a general theory of measurement that takes into consideration several factors simultaneously, in order to arrive at a conclusion. This synthesis can be a decision-making or planning and resource allocation, or conflict resolution. AHP has a special concern relating to departure form consistency and its measurement, as well as the dependence within and between the groups of elements of its structure.

In order to make a decision, several criteria have to be examined before an absolute or relative measurement can be made. This measurement depends on preferences developed from experience for the first case, for relative comparisons, alternatives compared in pairs according to a common attribute. From these measurements ratio scales are derived and priorities set for the criteria. Finally, alternatives scored and ranked by checking their ratings under each criterion and summing for all the criteria.

AHP has found its widest application in multi-criteria decision making, in planning and resource allocation, and in conflict resolution [18]. In its general form, the AHP is a nonlinear framework for carrying out both deductive and inductive thinking without use of the syllogism by taking several factors into consideration simultaneously and allowing for dependence and for feedback, and making numerical tradeoffs to arrive at a synthesis or conclusion.

The composite priorities of each alternative at the bottom level of a hierarchy may be represented as a multi-linear form:

$$\sum_{i_1,\ldots,i_p} x_1^{i_1} x_2^{i_2} \ldots x_p^{i_p} \tag{10.1}$$

Consider a single term of this sum and for simplicity denote it by $x_1, x_2, \ldots, x_p$. We have a product integral given by:

$$x_1 x_2 \cdots x_p = e^{\log x_1 x_2 \cdots x_p} = \prod_{i=1}^{n} e^{\log x_i} = e^{\sum_{i=1}^{n} \log x_i} \rightarrow e^{\int \log x(\alpha) d\alpha} \tag{10.2}$$

Typical steps in Analytic Hierarchy Process (AHP) include:

1. Determine the overall goal to reflect the expected accomplishment or goal-oriented target.
2. Select sub-goals of overall goal. If relevant, identify time horizons that affect the decision.
3. Identify criteria that must be satisfied to fulfill sub-goals of the overall goal.
4. Identify sub-criteria under each criterion. Note that criteria or sub-criteria may be specified in terms of ranges of values of parameters or in terms of verbal intensities such as high, medium, low.
5. Identify actor, goal, and policies involved and identify actor option or outcomes.
6. For yes-no decisions take the most preferred outcome and then compare benefits and costs of making the decision with not making it.
7. Do benefit/cost analysis using marginal values. Because we are dealing with dominance hierarchies, ask which alternative yields the greatest benefit; for costs, which alternative costs the most. Proceed similarly if a risks hierarchy is included.

The main features of this algorithm are:

- *Problem Decomposition*—It consists of making a decomposition of the problem into a hierarchy. At the top of the analytical hierarchy is the overall goal. Then, the criteria that contributes to the goal. On the bottom or third level are the possible candidates for the outcome or decisions to be made.

- *Development of Criteria Matrix and develop Priority Vectors*—Here, we make a comparative judgment, by arranging the criteria according to their importance with respect to the overall goal. This will yield a matrix that performs a one by one comparison between the criteria. The elements of the matrix will be the ratio of importance between one criterion and the other, for example 1/5. The first row and the first column will contain the criteria. The right end column will contain the priority vector, which is obtained by summing the ratios on the same rows.

  Then is the comparison of the possible candidates for each criteria. The same type of matrix is built as many times as the number of criteria and the priority vectors are derived too.

- *Synthesize Priorities*—A new matrix is may be constructed by using the operation $C = A * B$

Where A represents the priority vector for the candidate, B represents the priority vector for the criteria, n is the number of candidates, m is the number of criterion, and C is the synthesized matrix of priority.

A is an $n \times n$ matrix with weight vectors, w to be determined. For each row in A, geometric mean methods are used to obtain the weights given by:

$$v_i = \sqrt[n]{\prod_{j=1}^{n} a_{ij}} \qquad (10.3)$$

Then normalize the v_i's using:

$$w_i = \frac{v_i}{\sum_{j=1}^{n} v_j} \qquad (10.4)$$

The rows are summed up to yield an additional column at the right end with the composite or global priority vector of the candidates. We can then deduct the winning candidate that will have the highest score. An ideal matrix can be built out of C by dividing each element in a column of the matrix by the highest number in the column.

Overall, AHP has been found to be very useful in a wide range of applications where decision-making based on criteria and comparison is to be made. Its limitations reside in the fact that it requires expert judgment to create the scales for rating alternatives.

10.3.2.3 Analytical Network Process (ANP) ANP provides a way to input judgments and measurements to derive ratio scale priorities for the distribution of influence among the factors and groups of factors in the decision. The process is based on deriving ratio scale measurements, so it can therefore be used to allocate resources according to their ratio-scale priorities [18]. It is a more general form of AHP, incorporating feedback and interdependent relationships among decision attributes and alternatives. This provides a more accurate approach for modeling complex decision environment.

The ANP consists of coupling of two phases. The first phase consists of a control hierarchy of network of criteria and sub-criteria that control the interactions. The second phase is a network of influences among the elements and clusters. The network varies from criteria to criteria and thus different super-matrices of limiting influence are computed for each control criteria. Finally, each one of these super-matrices is weighted by the priority of its control criteria and results are synthesized through addition for the entire control criterion.

Advantages of ANP include: ability to handle multiple decision criteria, integrate subjective judgments with numerical data, and incorporate participation and encourages a process of learning, debate and revision [18]. Some limitations include curse of dimensionality and requires expertise to create scales for rating.

AHP Algorithm—The steps include:

1. Determine the control hierarchies and their criteria and sub-criteria for comparing the elements and components of the lower system according to influence. The will be a control hierarchy for each process (benefits, opportunities, costs risk, etc).
2. For each terminal or covering control criterion or sub-criterion, determine the clusters of the lower level system and their elements.
3. Number and arrange the clusters and their elements for each control criterion.
4. Determine the approach you want to follow in the analysis of cluster or element.
5. For each control criterion, construct a table with the labels of all the clusters of the lower models, clusters that are influenced by the lower models, and clusters that are influenced by that cluster.
6. For each table above, perform paired comparisons on the cluster as they influence each other or are influenced by it, with respect to that control criterion. Use the derived weights later to weight the elements of the corresponding column blocks of the super-matrix corresponding to the control criterion.
7. Perform paired comparisons on the elements within the clusters using a criterion of the control hierarchy or compare the elements in a cluster according to their influence or impact on each interconnected element in another cluster.
8. Construct the super-matrix by laying out the clusters in the order they are numbered and elements in each cluster and compute the limiting priorities of the super-matrix.
9. Include the alternatives in the super-matrix if they influence other clusters. Otherwise, their priorities can be derived by keeping them out and after computing the limiting super-matrix.
10. Multiply the priorities of the alternatives by the priority of the governing control criterion.

11. Synthesize the weights of the alternative for all the control criteria in each of the four control hierarchies. This yields four sets of weights for the alternatives, one each for benefits, opportunity, costs and risks.

Finally, given the final priority of each alternative, calculate the decision criteria, such as (benefits x opportunities)/(costs x risk), and select the option with the largest value.

The next section presents a brief review of selected classical optimization methods such as Linear Programming (LP), Nonlinear Programming (NLP), and Interior Point (IP).

10.3.3 Selected Methods in Classical Optimization

The classical optimization [10, 13, 37] for given scalar objective functions with or without constraints—equality and/or inequality—is mathematically stated as:

$$\text{Minimize} \quad f(\mathbf{x}, \mathbf{u}) \tag{10.5}$$

$$\text{s.t.} \quad g(\boldsymbol{x}, \boldsymbol{u}) = 0 \quad m \text{ equality constrains} \tag{10.6}$$

$$C_i \le h_i(x, u) \le D_i \quad m+1 \text{ to } n \text{ inequality constraints} \tag{10.7}$$

This class of problem is solvable using Linear Programming (LP), Nonlinear Programming (NLP) methods for continuous variables. There are additional constraints that include discrete and stochastic variables. This class can be solved using LP and NLP extensions and its variants, and Integer Programming methods such as the commonly used branch and bound method [2, 3]. Commonly used Linear Programming method includes the Simplex method, revised Simplex methods, Interior Point optimization, and Barrier method. These are extended to include stochastic features. An adequacy summary of these techniques for this class of optimization is summarized [13] for further reading.

10.3.3.1 Linear Programming (LP) Linear Programming is one of the most important scientific advances of the mid-twentieth century. It was first developed by Dantzig in 1948 and has been significantly used since then. General problems solved by linear programming include allocation of limited resources among competing activities. Linear programming uses a mathematical model to describe the problem with linear objectives and linear constraints [10, 13, 12]. In this context, programming does not necessarily mean computer programming. It involves planning of activities to obtain an optimal result, i.e., a result that reaches the specified goal best (according to the mathematical model) among the feasible alternatives. Mathematically, the linear programming problem involves complete linearization of the classical optimization model presented in Equations (10.5)~(10.7), and it is commonly stated as:

$$\text{Maximize } c^T x \tag{10.8}$$

$$\text{s.t. } Ax \le b \tag{10.9}$$

$$\text{and } x_i \ge 0 \quad \forall i \in \{1, n\} \tag{10.10}$$

with:

$$\text{Decision matrix:} \quad x = [x_1, \quad x_2, \quad \ldots \quad x_n]^T \tag{10.11}$$

$$\text{Cost coefficient array:} \quad c^T = [c_1, \quad c_2, \quad \ldots \quad c_n] \tag{10.12}$$

$$\text{Constant array:} \quad b = [b_1, \quad b_2, \quad \ldots \quad b_m]^T \tag{10.13}$$

The process to achieve the global optimum is done using Simplex like techniques or the Interior Point method. In summary, the procedure for solving this class of problems involves:

10.3.3.1.1 The Simplex Method

1. Initialization Step: introduce slack variables (if needed) and determine initial point as a corner point solution of the equality constraints.
2. At each iteration, move from the current basic feasible solution to a better adjacent basic feasible solution.
3. Determine the entering basic variable: Select the non-basic variable that, when increased, would increase the objective at the fastest rate. Determine the leaving basic variable: select the basic variable that reaches zero first as the entering basic variable is increased.
4. Determine the new basic feasible solution.
5. Optimality Test and Termination Criteria: check if the objective can be increased by increasing any non-basic variable by rewriting the objective function in terms of the non-basic variables only and then checking the sign of the coefficient of each non-basic variable. If all these coefficients are non-positive, then this solution is optimal, so stop. Otherwise, go to the iterative step.

10.3.3.1.2 Interior Point Optimization Method [13]

1. Determinate a feasible point within the inner space of the constrain boundaries.
2. Compute the corresponding objectives (cost) for the initial feasible points.
3. For the situation in which the objective is not optimum, compute the new increase in cost by computing the new trajectory or projection to achieve an improvement in the objective, without exiting the space.
4. Optimality and Termination Criteria: A feasible direction, along with the objective function increases, is found and then an approximate step length is determined to guarantee the new feasible solution which is strictly better then the previous one. The stopping criteria are determined from the relative changes in the objective function at iteration on the changes in iterations. The optimality condition is computed until the maximum (or minimum) is satisfied.

Interior Point has several variant such as the primal, affine, dual affine, etc. [12, 13]. And, the IP technology has been used to solve a special class of Quadratic Programming (QP), which has quadratic objective function and linear constraints of continuous variables. This has lead to innovations such as Quadratic and Extended Quadratic IP (QUIP/EQUIP [41, 42]) for power system applications such as VAr

Planning, Loss Minimization, Phase Shifter optimization, Generation Dispatch, etc. [41, 42]

10.3.3.2 Nonlinear Programming (NLP) Briefly stated, the following steps in the Nonlinear Programming (NLP) method involve the following steps:

1. Determine the initial feasible set based on investigation of extrema of the functions with or without constraints.
2. Check the optimality conditions.
3. Determine candidate solution for local or global optimum.
4. Perform further optimization and evaluate the optimal value to the objective function that satisfies the constraints.

This process involves application of Kuhn-Tucker (KT) and Extended Kuhn-Tucker first and second order necessary and sufficient conditions [13, 33]. This can be applied to functions as well as functional.

10.3.4 Optimal Control

Optimal Control theory is a mathematical field that is concerned with control policies that can be deduced using optimization algorithms. The objective of optimal control is to determine the control signals that will cause a process to satisfy the physical constraints and at the same time minimize (or maximize) some performance criterion [11, 33].

The state vector $\boldsymbol{x}(t)$ and the control vector $\boldsymbol{u}(t)$ are related by:

$$\dot{x}(t) = a(x(t), u(t), t) \tag{10.14}$$

The performance of a system is evaluated by:

$$J = h(x(t_f), t_f) + \int_{t_0}^{t_f} g(x(t), u(t), t)\, dt \tag{10.15}$$

where t_0 is the initial time and t_f is the final time.

10.3.4.1 Type 1—Minimum Time Problem The goal is to transfer a system from an arbitrary initial state $x(t_o) = x_o$ to a specified target set {S} in minimum time. The performance measure to be minimized is:

$$J = t_f - t_0 = \int_{t_o}^{t_f} dt \tag{10.16}$$

With t_f the first instant of time when $x(t)$ and {S} intersect. This problem is applicable to space missions, missile interception, and rescue mission.

10.3.4.2 Type 2—Terminal Control Problem The goal is to minimize the deviation of the final state of a system from its desired value $r(t_f)$. A possible performance measure is:

$$J = \sum_{i=1}^{n} [x_i(t_f) - r_i(t_f)]^2 \tag{10.17}$$

Since positive and negative deviations are equally undesirable, the error is squared. Absolute values could also be used, but the quadratic form in the above equation is easier to handle mathematically. Using matrix notation, we have:

$$J = \sum_{i=1}^{n} [x_i(t_f) - r_i(t_f)]^T [x_i(t_f) - r_i(t_f)] \tag{10.18}$$

$$= \| x_i(t_f) - r_i(t_f) \|^2 \tag{10.19}$$

where $\| x_i(t_f) - r_i(t_f) \|$ is the vector norm of $[x_i(t_f) - r_i(t_f)]$.

To allow greater generality, we can insert a real symmetric positive semi-definite $n \times n$ weighting matrix H to obtain the closed form solution in quadratic form as:

$$J = [x(t_f) - r(t_f)]^T H [x(t_f) - r(t_f)] \tag{10.20}$$

10.3.4.3 Further Insights to Optimal Control The methods to solve optimal control problem are dynamic programming, the calculus of variations, and iterative numerical techniques:

- Dynamic Programming: The dynamic programming leads to a functional recurrence relation when a continuous process is approximated by a discrete system. The primary limitation is the "curse of dimensionality".
- Calculus of Variations: The calculus of variations generally leads to a non-linear two-point boundary value problem that requires the use of iterative numerical techniques for solution.

A statement of a typical optimal control problem can be expressed as obtain the state equation and its initial condition of a system to be controlled, provide defined objective set, and determine a feasible control such that the system starting from the given initial condition transfers its state to the objective set, and minimizes a performance index.

The control that minimizes a cost functional is called the optimal control. The performance of the control system is measured by the criteria of optimality: steady state error, gain margin and phase margin. In optimal control problem, the system measure of performance or performance index is not fixed and the system is only considered as an optimum control system when the system parameters are adjusted so that the index is either maximized or minimized. The performance index is a function of error between the actual and ideal responses. The best system is then defined as the system that minimized this index.

Control systems are optimized mainly by applying the Bellman's Optimality Principle which states: "*An optimal policy (or a set of decisions) has the property that whatever the initial state and initial decisions are, the remaining decisions must constitute an optimal policy with regard to the state resulting from the first decision*" [7, 12].

The general framework of optimal control is given a system with input $\boldsymbol{u}(t)$, output $y(t)$ and state $\boldsymbol{x}(t)$, $y(t) = f(\boldsymbol{x}(t), u(t))$

The cost functional, which is a measure that the control designer is to minimize, can be defined as:

$$J = \int_0^\infty x^T(t)Qx(t) + u^T(t)Ru(t)dt \tag{10.21}$$

Where the matrix Q is positive semi-definite and R is positive-definite.

This cost function is in terms of penalizing the control energy (measured as a quadratic form) and the time it takes the system to reach zero-state. This function could seem rather useless since it assumes that the operator is driving the system to zero-state, and hence driving the output of the system to zero. This is indeed right, however the problem of driving the output to the desired level can be solved after the zero output one is. In fact, it can be proved that this secondary problem can be solved in a very straightforward manner.

The optimal control problem defined with the previous functional is usually called State Regulator Problem and its solution the Linear Quadratic Regulator (LQR) which is no more that a feedback matrix gain with Gain K. This is typically solved using Continuous Time Dynamic Riccati Equation [10, 12].

10.3.5 Dynamic Programming (DP)

Optimization over time in a single or multi-stage decision process is generally formulated as Dynamic Programming involving large number of variables under different stages [4, 5, 6, 13]. DP can be defined as an operational research technique to facilitate the solution of sequential problems. It is a method of solving multi-stage problems in which the decisions at one stage become the conditions governing the succeeding stages. The advantage of DP is that each stage can be optimized; on the other hand, the advantage lies in the complexity of its solution for large system, the so-called "curse of Dimensionality." With this in mind, applications of DP have been limited. Of course, new advances and approximations are in place to enhance its usefulness to large-scale systems. Recent work to enhance DP method involves work in approximate dynamic programming, Genetic Algorithm (GA), and annealing methods [19].

In the formulation of a DP problem, Any decision process is characterized by certain input parameters, X (or data), certain decision variables (U) and certain output parameters (T) representing the outcome obtained as a result of making the decision. For any physical system, that is represented as a single stage decision process shown in Figure 10.2.

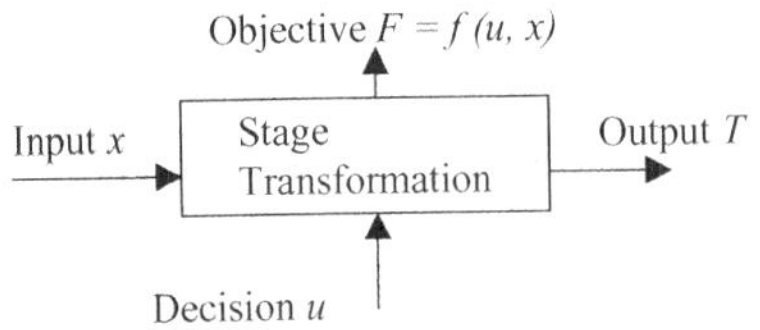

Figure 10.2. Single Stage Decision problem.

The output of this single stage is $T(x, F)$ given by:

$$x_i = t_i(x_{i+1}, u_i) \quad \forall i \in \{1, n\} \tag{10.22}$$

$$F_i = f_i(x_{i+1}, u_i) \quad \forall i \in \{1, n\} \tag{10.23}$$

Where u_i denotes the vector of decision variables at stage i.

The objective of a multistage decision process is to find $u_1, u_2, \ldots, u_n$ so as to optimize some function of the individual stage returns, say, $F(f_1, f_2 \ldots f_n)$ and satisfy Equations (10.22) and (10.23). In general, an additive objective function in DP optimization is:

$$F = \sum_{i=0}^{\infty} f_i(u_i, x_{i+1}) \tag{10.24}$$

where f_i is the individual stage i return. This is for either addictive or multiplicative objectives that employ a multistage decision process. The multiplicative objective takes the form:

$$F = \prod_{i=1}^{n} f_i(x_{i+1}, u_i) \tag{10.25}$$

These objectives are generally subject to:

$$u_i = t_i(x_i, u_{i+1}) \tag{10.26}$$

The solution to this problem results in a multistage process can be classified into:

- *Initial Value Problem*

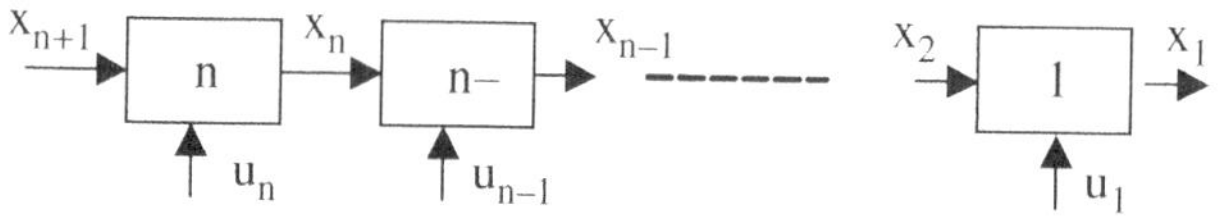

- *Final Value Problem*

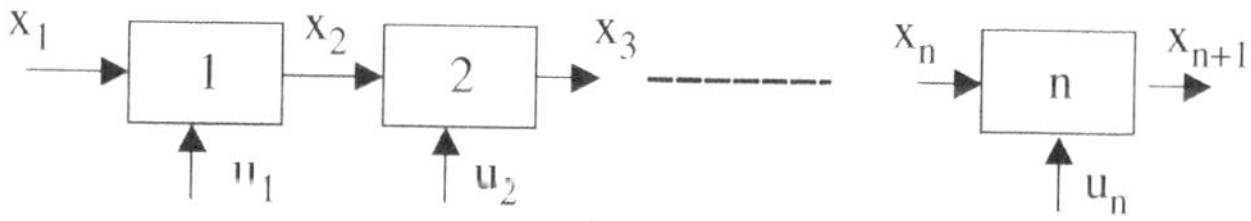

- *Boundary Value Problem*

The boundary value problem is a combination of both the initial value and final value problem. Here, the values of both the input and output variables are specified, the problem is called a boundary value problem.

10.3.6 Adaptive Dynamic Programming (ADP)

Nomenclature:

$u(t)$: Action vectors

$U(t)$: The utility which the system is to maximize

$X(t)$: Senor inputs

r: Usual discount rate or interest rate that is needed only in infinite-time-horizon problems (or only sometimes)

$< >$: Denote the expectation value

J: Secondary or strategic utility function

$R(t)$: Complete state description of the plant to be controlled at time t

A: Action network

F_W_{ij}: Derivatives of error with respect to all weights W_{ij}

Adaptive Dynamic Programming (ADP) is a computational intelligence technique that incorporates time dependency of deterministic or stochastic data required for the future. Also called "reinforcement learning," "adaptive critics," "neural-dynamic programming," and "approximate dynamic programming [5, 6]." ADP consider the optimization over time by using learning approximation to handle problems that severally challenge conventional methods due to their very large scale and lack of sufficient prior knowledge. ADP overcomes the "curse" of dimensionality in Dynamic Programming (DP). Traditionally, there is only one exact and efficient way to solve problems in optimization over time, in general case where noise and nonlinearity are present: dynamic programming.

ADP determines optimal control laws for a system by successively adapting two Neural Networks. One is action neural network (which dispenses the control signals) and the other is critic network (which learns the desired performance index for some function associated with the performance index) [35, 36]. Figure 10.3 shows the structure of the coupled neural networks used in adaptive dynamic programming where $X(t)$ is the system state, $u(t)$ is the action, and $J(t)$ is the secondary or strategic utility function.

In dynamic programming, the user supplies both a utility function- the function to be maximized and a stochastic model F of the external plant or environment [8]. ADP is designed to maximize the expected value of the sum of future utility over all future time periods:

$$\text{Maximize} \quad \sum_{k=0}^{\infty}(1+r)^{-k}U(t+k) \tag{10.27}$$

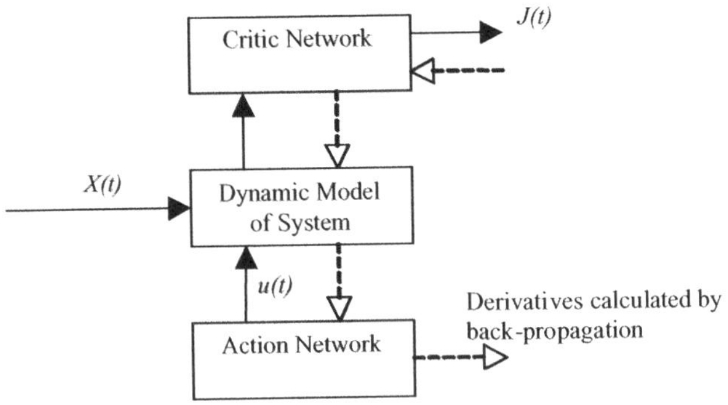

Figure 10.3. Structure adaptive dynamic programming system.

ADP may be defined as design that attempt to approximate dynamic programming in the general case. The cost of running true dynamic programming is proportional to the number of possible states in the plant or environment; that number, in turn, grows exponentially with the number of variables in the environment [7, 35]. Therefore, approximate methods are needed even with many small-scale problems. ADP is defined more precisely as designs that include a critic network- a network whose output is an approximation of the J function, or to its derivatives, or to something very closely related to these two, the action network in an adaptive critic system is adapted so as to maximize J in the near-term future. To maximize future utility subject to constraints, you can simply train the action network to obey those constraints when maximizing J. The validity of dynamic programming itself is not affected by such constraints.

Dynamic programming is used to solve for another function, J, which serves as a secondary or strategic utility function. The key theorem is that any strategy of action that maximizes J in the short term will also maximize the sum of U over all future times. J is a function of $R(t)$, where $R(t)$ is complete state description of the plant to be controlled at time t and $u(t)$ are the vector of actions. Dynamic programming converts a problem in optimization over time into a "simple" problem in maximizing J just one step ahead in time.

$$J(R(t)) = \underset{u(t)}{Max}(U(R(t), u(t)) + \frac{\langle J(R(t+1))\rangle}{1+r} - U_0 \qquad (10.28)$$

where r and U_0 are constants that are needed only in infinite-time-horizon problems (and then only sometimes), and where the angle brackets refer to expectation value.

Adaptive critic designs may be defined as design that attempt to approximate dynamic programming in the general case. The cost of running true dynamic programming is proportional to the number of possible states in the plant or environment; that number, in turn, grows exponentially with the number of variables in the environment. Therefore, approximate methods are needed even with many small-scale problems. Adaptive critic [34] designs are defined more precisely as designs that include a Critic network as shown in Figure 10.4.

It is a network whose output is an approximation of the J function, or to its derivatives, or to something very closely related to these two. The action network in an adaptive critic system is adapted so as to maximize J in the near-term future. To maximize future utility subject to constraints, one can simply train the action network, shown in Figure 10.5, to obey those constraints when maximizing J. The validity of dynamic programming itself is not affected by such constraints.

10.3.7 Variants of Adaptive Dynamic Programming

There are several Critic designs that had been proposed based on dynamic programming:

1. Heuristic dynamic programming (HDP), which adapts a Critic network whose output is an approximation of $J(R(t))$.
2. Dual Heuristic Programming (DHP), which adapts a Critic network whose outputs represent the derivatives of $J(R(t))$.

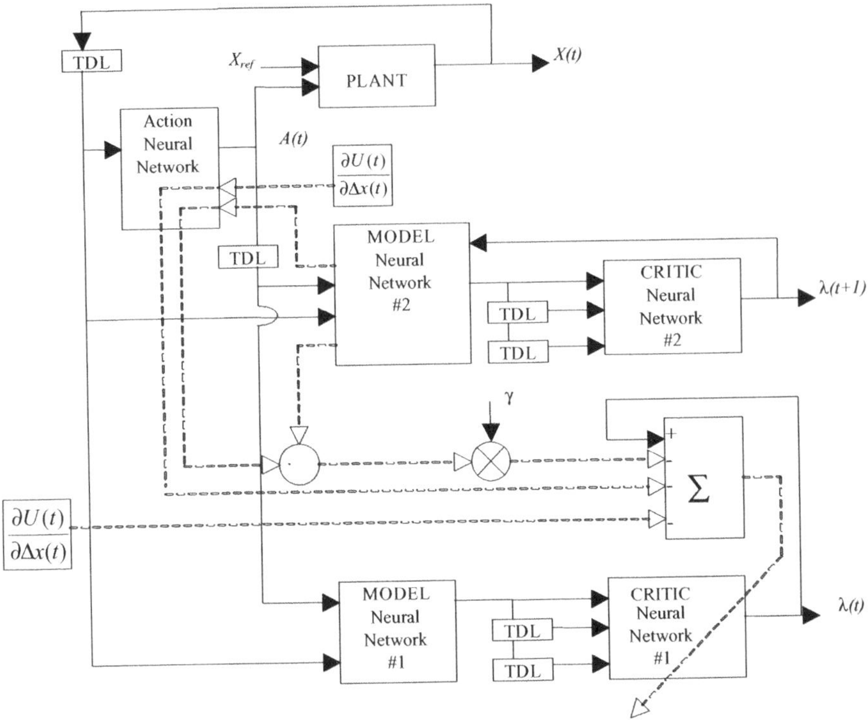

Figure 10.4. DHP critic neural network adaptation.

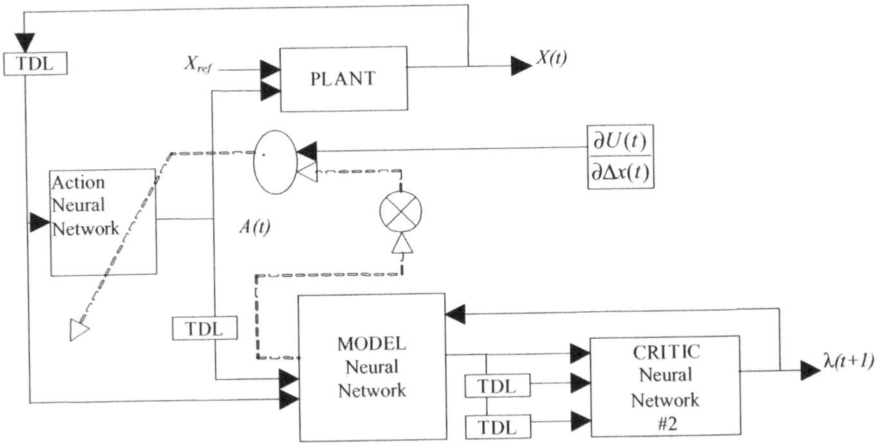

Figure 10.5. DHP Action Network adaptation.

3. Globalized Dual Heuristic Programming (GDHP), which adapts a Critic network whose output is an approximation of $J(R(t))$, but adapt it so as to minimize errors in the implied derivatives of J. GDHP tries to combine the best of HDP and DHP.

HDP intends to break down, through very slow learning, as the size of a problem grows bigger: however, DHP is more difficult to implement. The three methods listed above all yield action-independent critics, there are also ways to adapt a Critic network that inputs $R(t)$ and $u(t)$.

10.3.7.1 Neural Dynamic Programming

Neural Dynamic Programming is closely related to ADHDP [9]. One major difference is that there is no system model to predict the future system state value and consequently the cost-to-go for the next time step. Rather, by storing the previous J value together with the current J value, one can obtain the temporal difference used in training.

10.3.7.2 Heuristic Dynamic Programming (HDP)

Heuristic dynamic programming (HDP) is a procedure for adapting a network or function, $J(R(t), W)$. We have utilized an approximate the function, $J(R(t))$, which is a small perturbation of the Bellman equation:

$$J(R(t)) = \underset{u(t)}{Max}(U(R(t), u(t)) + \frac{\langle J(R(t+1))\rangle}{1+r} \tag{10.29}$$

For simplicity, we will assume problems such that we can assume $U_0 = 0$. HDP is a procedure for adapting a network or function. The steps of calculations for HDP are:

1. Obtain and store $R(t)$
2. Calculate $u(t) = A(R(t))$
3. Obtain $R(t + 1)$, either by waiting until $t + 1$ or by predicting:

$$R(t+1) = f(R(t), u(t)) \tag{10.30}$$

4. Calculate:

$$J^*(t) = U(R(t), u(t)) + J(R(t+1), W)/(1+r) \tag{10.31}$$

5. Update W in $J(R(t), W)$ based on inputs $R(t)$ and target $J^*(t)$

10.3.7.3 Dual Heuristic Programming (DHP)

DHP is based on differentiating the Bellman equation [8,36]. Before performing the differentiation, we have to decide how to handle $u(t)$. One way is simply to define the function $u(R(t))$ as that function of R, which, for every R, maximizes the right-hand side of the Bellman equation. With that definition (for the case $r = 0$), the bellman equation becomes:

$$J(R(t)) = (U(R(t), u(t)) + \langle J(R(t+1))\rangle - U_0 \tag{10.32}$$

where we must also consider how $R(t + 1)$ depends on $R(t)$ and $u(R(t))$. Differentiating, and applying the chain rule, we get:

$$\lambda_i(R(t)) = \frac{\partial J(R(t))}{\partial R_i(t)} = \frac{\partial}{\partial R_i(t)} U(R(t), u(R(t)) + \left\langle \frac{\partial J(R(t+1))}{\partial R_i(t)} \right\rangle \tag{10.33}$$

$$= \frac{\partial J(R(t),u(t))}{\partial R_i(t)} + \sum_j \frac{\partial U(R,u)}{\partial u_j} \cdot \frac{\partial u_j R(t)}{\partial R_i(t)}$$
$$+ \sum_j \left\langle \frac{\partial J(R(t+1))}{\partial R_i(t+1)} \cdot \frac{\partial R_j(t+1)}{\partial R_i(t)} \right\rangle + \sum_{j,k} \left\langle \frac{\partial J(R(t+1)}{\partial R_j(t+1)} \cdot \frac{\partial R_j(t+1)}{\partial u_k(t)} \cdot \frac{\partial u_k(t)}{\partial R_j(t)} \right\rangle \quad (10.34)$$

The salient computational steps in DHP:

1. Obtain $R(t)$, $u(t)$ and $R(t + 1)$ as was done with HDP
2. Calculate:

$$\lambda(t+1) = \lambda(R(t+1), W) \quad (10.35)$$
$$F\_u(t) = F\_Uu(R(t),u(t)) + F\_fu(R(t),u(t),\lambda(t+1)) \quad (10.36)$$
$$\lambda^*(t) = F\_f_R(R(t),u(t), \quad (10.37)$$
$$\underline{\lambda}(t+1)) + F\_U_R(R(t),u(t)) + F\_A_R(R(t),F\_u(t)) \quad (10.38)$$

3. Update W in $\lambda(R(t), W)$ based on inputs $R(t)$ and target $\lambda^*(t)$

10.3.7.4 Action Dependent Heuristic Dynamic Programming (ADHDP or "Q-learning")

If we defined a new quantity:

$$J'(R(t),u(t)) = U(R(t),u(t)) + \frac{\langle J(R(t+1))\rangle}{1+r} \quad (10.39)$$

By algebraic manipulation of the above Equations, we may derive a recurrence rule for J':

$$J'(R(t),u(t)) = U(R(t),u(t)) + \underset{u(t+1)}{Max} \frac{\langle J'(R(t+1),u(t+1))\rangle}{1+r} \quad (10.40)$$

ADHDP adapts a Critic network, $J'(R(t),u(t),W)$, which attempts to approximate J' as defined in Equation (10.40). The calculation steps in ADDHP are:

1. Obtain $R(t)$, $u(t)$ *and* $R(t + 1)$ exactly as in HDP
2. Calculate $u(t + 1) = A(R(t + 1))$
3. Calculate:

$$F\_R(t+1) = \lambda(R)(R(t+1),u(t+1),W) + F\_AR(R(t+1),\lambda(u)(R(t+1),u(t+1),W)) \quad (10.41)$$
$$\lambda_R{*}(t) = F\_f_R(R(t),u(t),F\_R(t+1)) + F\_U_R(R(t),u(t)) \quad (10.42)$$
$$\lambda_u{*}(t) = F\_f_u(R(t),u(t),F\_R(t+1)) + F\_U_u(R(t),u(t)) \quad (10.43)$$

4. Update W in the Critic based on inputs $R(t)$ and $u(t)$ and targets $\lambda^*R(t)$ *and* $\lambda^*u(t)$.

10.3.8 Comparison of ADP Variants

Tables 10.3a and 10.3b shows a comparison of three important variants of ADP based on the J-function designs and other merits and demerits in computational challenges [5].

TABLE 10.3a. Comparison of ADP J-junctions.

ADP Variant	J function Formulation
HDP	Critic network whose output is an approximation of J function: $J(R(t)) = \underset{u(t)}{Max}(U(R(t), u(t)) + \frac{\langle J(R(t+1))\rangle}{1+r}$
DHP	Adapts a Critic network whose outputs represent the derivatives of J function: $J(R(t)) = (U(R(t), u(t)) + \langle J(R(t+1))\rangle - U_0$
ADHDP	$J'(R(t), u(t)) = U(R(t), u(t)) + \underset{u(t+1)}{Max}\frac{\langle J'(R(t+1), u(t+1))\rangle}{1+r}$

TABLE 10.3b. Advantages and Disadvantages of Different ADP Variants.

ADP Variant	Advantage	Disadvantage
HDP	Easy to formulate	Problem size increases
DHP	Since DHP builds derivative terms over time directly, it reduces the probability of error introduced by backpropagation.	More difficult to implement because of derivatives of J
ADHDP	Combine HDP and DHP, and add new input to the system	Difficult to form the model.

Let us define M and P' such that:

$$P' = P + RA \tag{10.44}$$

$$M = P'^{\mathrm{T}} MP' - Q \tag{10.45}$$

Solution Summaries

a. ADHDP

It can be deduced that:

$$J' = -x(t)^{\mathrm{T}} Qx(t) + (Px(t) + Ru(t))^{\mathrm{T}} M(Px(t) + Ru(t)) \tag{10.46}$$

b. DHP

The correct value of $J(x)$ is $x(t)^{\mathrm{T}}Mx(t)$ and $\lambda = \nabla J$ such that:

$$\lambda(t+1) = 2Mx(t+1) \tag{10.47}$$

The next step is to compute the targets of $\lambda(t+1)$ as generated by DHP and compare them against the correct values.

Propagation of $\lambda(t+1)$ through the DHP model yields the first term of the expected value:

$$\langle P^{\mathrm{T}}(2Mx(t+1))\rangle = 2P^{\mathrm{T}}M(P+RA)x(t) \tag{10.48}$$

The second term is the gradient of:

$$U(x(t)) = -2Qx(t) \tag{10.49}$$

The third term found my propagating $\lambda(t + 1)$ through the model back to $u(t)$, and then through the Action network, yielding an expected value of:

$$\langle A^T(R^T(2Mx(t+1)))\rangle = 2A^T R^T M(P + RA)x(t) \tag{10.50}$$

Summing the 3 terms yields the correct final expectation.

$$\langle \lambda^*(t)\rangle = 2Mx(t) \tag{10.51}$$

The details of ADP concepts and other useful information for problem solving can be found in [35].

10.4 APPLICATIONS OF NEXT GENERATION OPTIMIZATION TO POWER SYSTEMS

10.4.1 Overview

The conventional Optimal Power Flow tools lack two basic ingredients that are essential for the smooth operation of the power system. One is foresight, which includes the capability of existing OPF to predict the future in terms of asset valuation and economic rate of return on investment in power system infrastructure subject to various system dynamics and network constraints. The other is an explicit optimization technique to handle perturbation and noise.

Classical illustrative methods of the proposed methods in Table 10.4 used in the classroom environment will be discussed.

TABLE 10.4. Chart of Power System Problem and Hybrid Optimization Techniques.

Selected Power System Challenges	Optimization Methods							
	DA	Optimal Control	Risk Assessment	IS	DP	ADP	AHP	Game Theory	Classical Methods (LP, NLP, IP, etc.)
Reliability	¦		¦				¦	¦	
Fault Analysis/3Rs	¦			¦		¦			
Unit Commitment	¦	¦		¦	¦	¦	¦	¦	
DSOPF		¦		¦		¦			¦
Control Coordination	¦			¦	¦	¦		¦	¦
Stability and DSA		¦	¦	¦		¦			
State Estimation		¦	¦					¦	

Legend: 3R's: Reconfiguration, Restoration, and Remedial Control DA: Decision Analysis AHP: Analytical Hierarchical Processes IS: Intelligent Systems DP: Dynamic Programming ADP: Adaptive Dynamic Programming

10.4.2 Framework for Implementation of DSOPF

There is a need for a generalized framework for solving the many classes of power system problems where programmers, domain experts, etc. can submit their challenge problem. The collective knowledge will by publish and posted on the web for further dissemination. Figure 10.6 below shows the general framework for application of ADP to develop a new class of OPF problem called DSOPF [5], it is divided into three modules.

Module 1: Read power system parameters and obtain distribution function for state estimation of measurement errors inherent in data, ascertain and improve

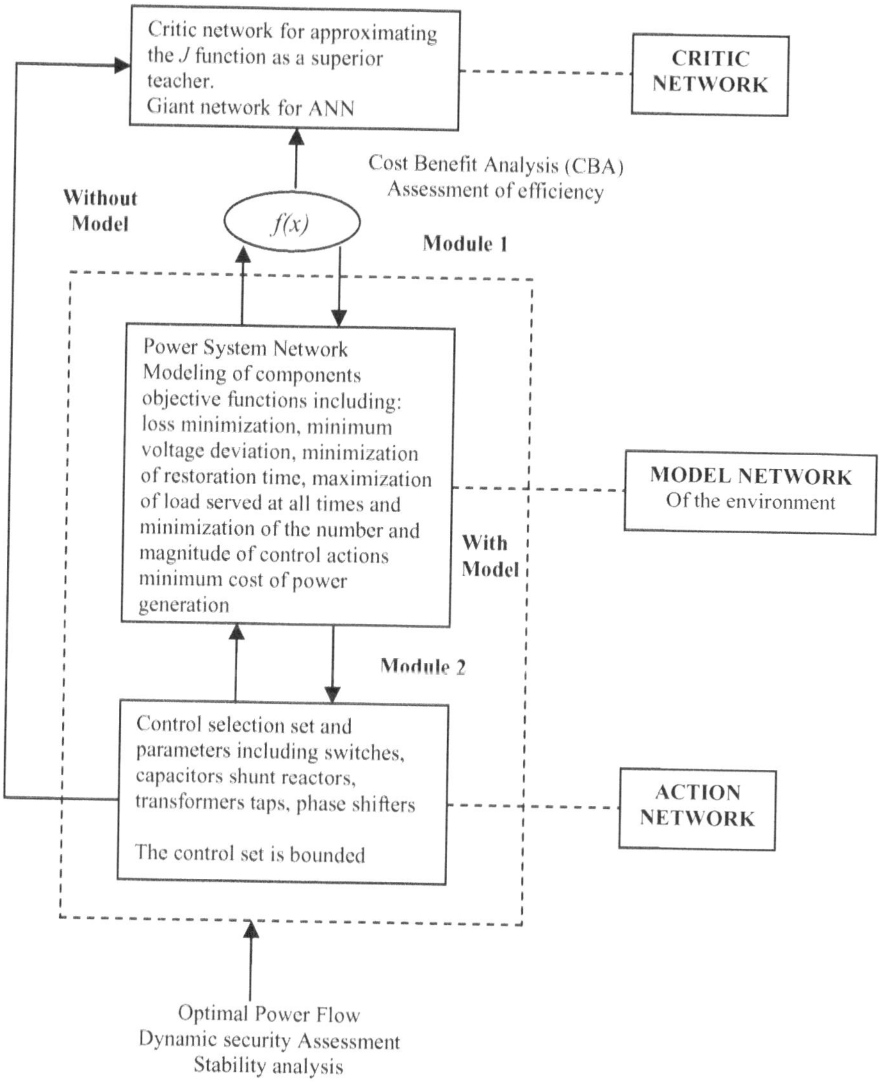

Figure 10.6. Framework of ADP applications to power systems [35].

accuracy of data. Infer relationships between the past data and future ones of unknown period using time series and dynamical systems and in all cases determine the time dependent model approximation behavior of the systems generation the data. Define the model and with the uncertainties, this step includes defining the problem objective and constraint functions for each problem.

Module 2: Determine the feasibility region of operation of the power systems and the emergency state with corresponding violations under different contingencies. Enumerate and schedule different control options over time for different contingency scenarios. Coordinate the controls and perform post optimizations of additional changes. Evaluate results and perform sensitivity analysis studies.

Module 3: For post-optimization process, evaluation and assessment of control options during contingencies are necessary. This module handles the post optimization process by through cost benefit analysis to evaluate the various controls (cost effectiveness and efficiency). In the power system parlance, a big network, which will perform this evaluation, is essential and indispensable). The critic network from ADP techniques will help realize the dual goals of cost effectiveness and efficiency of the solution via the optimization process.

10.4.3 Applications of DSOPF to Power Systems Problems

In this section of the chapter, we show two solved examples of applications of new optimization techniques to power systems research work listed in Table 10.4 and we present here some of the ongoing research work at CESaC, Howard University for illustrative purposes.

10.4.3.1 Power System Unit Commitment (UC) Problem The objective function of the unit commitment problem can be formulated as the sum of costs of all the units over time, and presented mathematically as [13,37]:

$$F = \sum_{t=1}^{T}\sum_{i=1}^{N}[u_i(t)F_i(E_i(t)) + S_i(t)] \tag{10.52}$$

The constraint models for the unit commitment optimization problem are as follows:

- System energy balance

$$0.5\sum_{i=1}^{N}[u_i(t)P_{g_i}(t) + u_i(t-1)P_{g_i}(t-1)] = P_D(t) \tag{10.53}$$

- Energy and Power Exchange

$$E_i(t) = 0.5[P_{g_i}(t) + P_{g_i}(t-1)] \tag{10.54}$$

- System spinning reserve requirements

$$\sum_{i=1}^{N} u_i(t)P_{gi}(t) \geq P_D(t) + P_R(t) \tag{10.55}$$

- Unit generation limits

$$P_{gi}^{\min} \leq P_{gi}(t) \leq P_{gi}^{\max} \tag{10.56}$$

With $t \in \{1, T\}$ and $t \in \{1, N\}$ in all cases where:

F : total operation cost on the power system
$E_i(t)$: energy output of the ith unit at hour t
$F_i(E_i(t))$: fuel cost of the ith unit at hour t
$u_i(t)$: ratio of generation output and capability
N : total number of units in the power system
T : total time under which UC is performed
$P_{gi}(t)$: Power output of the ith unit at hour t
P_{gi}^{max} : Maximum power output of the ith unit
P_{gi}^{min} : Minimum power output of the ith unit
$S_i(t)$: Start-up cost of the ith unit at hour t

In the reserve constraints, there are various classifications for reserve and these include units on Spinning Reserve and Units on Cold Reserve under the conditions of banked boiler or cold start.

Lagrange Relaxation is being used regularly to solve UC problems. It is much more beneficial for utilities with a large number of units since the degree of sub-optimality goes to zero as the number of units increases. It has also the advantage of being easily modified to model characteristics of specific utilities. It is relatively easy to add unit constraints. The main disadvantage of Lagrangian Relaxation is its inherent sub-optimality.

$$\begin{aligned} L(\lambda, \mu, v) = &\sum_{t=1}^{T}\sum_{i=1}^{N}[C_i(P_{gi}(t)) + S_i(x_i(t))] \\ &+ \lambda(t)\left(P_d(t) + P_R(t) - \sum P_{gi}\right) + \mu(t)\left(P_{gi}^{max} - P_{gi}\right) \end{aligned} \tag{10.57}$$

Where $\lambda(t)$, $\mu(t)$ are the multipliers associated with the requirement for Time t.

10.4.3.2 Solution Approach Using ADP Variant for the Unit Commitment (UC) Problem

ADP is able to optimize the system over time under conditions of noise and uncertainty. If optimal operation samples are used to train the networks of the ADP, the Neural Network can learn how to commit or adapt the generators and follow the operators' patterns. When load is changed, it can change the operation according to the load changing. Figure 10.7 shows the schematic diagram for implementations of HDP.

The input of the action network is the states of generators and the action is how to adjust the output of generators. The output J presents the cost-to-go function and the task is to minimize the J function.

In this diagram, the input is the state variable of the network, and it is the cost of generation vector. It can be presented as $X = [C(P_{gi})]$. And the output is control variables of units, and it is the adjustment of unit generation, presented as: $u = [\Delta P_g]$. The utility function is local cost, so it is a cost function about unit generation within any time interval. It can be presented as $U = f(P, t)$.

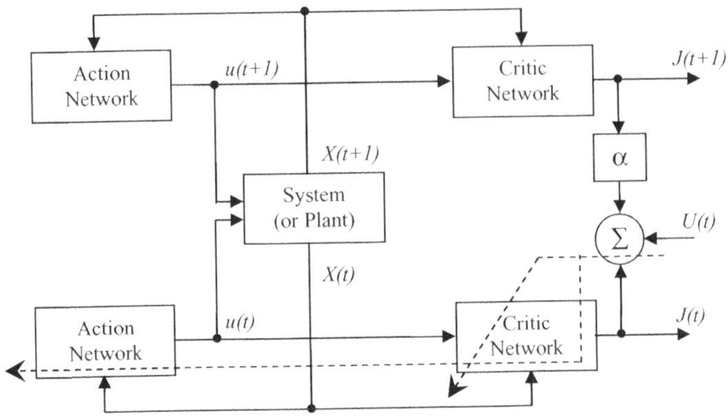

Figure 10.7. The scheme of implementation of HDP.

After transposing the power system variables using the guidelines above, the schema of implementation of HDP include the following computations:

The error of the critic network is:

$$e_C(t) = \gamma J(t) - J(t+1) - U(t) \tag{10.58}$$

and the updating weight using:

$$w_C(t+1) = w_C(t) + \Delta w_C(t) \tag{10.59}$$

and

$$\Delta w_C(t) = \eta e_C \left[-\frac{\partial e_C(t)}{\partial w_C(t)} \right] \tag{10.60}$$

where

$$\frac{\partial E_C}{\partial w_{Cij}^{(1)}} = \frac{\partial E_C}{\partial e_C} \cdot \frac{\partial e_C}{\partial y_{Ck}} \cdot \frac{\partial h_{Ck}}{\partial h'_{Ck}} \cdot \frac{\partial h'_{Cj}}{\partial w_{Cij}^{(1)}} \tag{10.61}$$

$$= \gamma e_C \cdot \left[\frac{1}{2}\left(1 - h_{Cj}^2\right) \right] \cdot w_{Cj}^{(2)} x_i \tag{10.62}$$

$$\frac{\partial E_C}{\partial w_{Cjk}^{(2)}} = \frac{\partial E_C}{\partial e_C} \cdot \frac{\partial e_C}{\partial y_{Ck}} \cdot \frac{\partial y_{Ck}}{\partial w_{Cjk}^{(2)}} = \gamma e_C y_{Ck} \tag{10.63}$$

I : Number of elements in R vector
J : Number of hidden layer node
K : Number of output layer node
M : Number of elements in u (action) vector
h'_C : Hidden layer input nodes
h_C : Hidden layer output nodes
y'_C : Output layer input nodes
y_C : Output layer output nodes
$w_C^{(1)}$: Weights between input and hidden layers
$w_C^{(2)}$: Weights between hidden and output layers
x : Input layer nodes

The error of the action network is computed as:

$$e_A(t) = J(t) - U(t) \tag{10.64}$$

and the updating weight is:

$$w_A(t+1) = w_A(t) + \Delta w_A(t) \tag{10.65}$$

and

$$\Delta w_A(t) = \eta e_A \left[-\frac{\partial e_A(t)}{\partial w_A(t)} \right] \tag{10.66}$$

where

$$\frac{\partial E_A}{\partial w_{Ajk}^{(2)}} = \frac{\partial E_A}{\partial e_A} \cdot \frac{\partial e_A}{\partial J_k} \cdot \frac{\partial J_k}{\partial y_{Ak}} \cdot \frac{\partial y_{Ak}}{\partial y'_{Ak}} \cdot \frac{\partial y'_{Ak}}{\partial w_{Ajk}^{(2)}} \tag{10.67}$$

$$= \gamma e_A h_{Aj} \cdot \left[\frac{1}{2}\left(1 - h_{Aj}^2\right) \right] \cdot \left[\sum_{j=1}^{J} w_{Cj}^{(2)} \frac{1}{2}\left(1 - h_{Cj}^2\right) w_{Cij}^{(1)} \right] \tag{10.68}$$

$$\frac{\partial E_A}{\partial w_{Aij}^{(1)}} = \frac{\partial E_A}{\partial e_A} \cdot \frac{\partial e_A}{\partial J_k} \cdot \frac{\partial J_k}{\partial y_{Ak}} \cdot \frac{\partial y_{Ak}}{\partial y'_{Ak}} \cdot \frac{\partial y'_{Ak}}{\partial w_{Aij}^{(1)}} \tag{10.69}$$

$$= \gamma e_A w_{Ajk}^{(2)} x_i \cdot \left[\frac{1}{2}\left(1 - h_{Aj}^2\right) \right] \cdot \left[\frac{1}{2}\left(1 - y_{Ak}^2\right) \right] \cdot \left[\sum_{j=1}^{J} w_{Cj}^{(2)} \frac{1}{2}\left(1 - h_{Cj}^2\right) w_{Cij}^{(1)} \right] \tag{10.70}$$

The structure of the neural network in HDP is shown in Figure 10.8.

The corresponding calculation steps are as follows:

Step 1: Use the sample data to pre-train the action network. The error is the difference between the output and the real value.

Step 2: Use the sample data to train the critic network with the pre-trained and unchanged action network. Use Equations (10.58)~(10.63) to update the weights. Then begin to apply the mature ADP network in the real work.

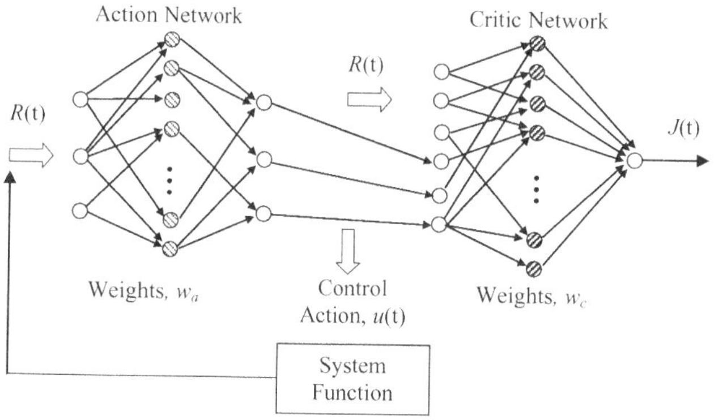

Figure 10.8. The structure of the neural network in HDP.

Step 3: Input the current state data $X(t)$ to the action network.

Step 4: Get the output $u(t)$ of the action network. Input $u(t)$ to the system function to get the state of next time $X(t + 1)$.

Step 5: Use the state of next time $X(t + 1)$ to get the action of next time $u(t + 1)$.

Step 6: Input the action and state of different time $u(t)$, $X(t)$ and $u(t + 1)$, $X(t + 1)$ to different critic network respectively and J function for different time $J(t)$, $J(t + 1)$ are obtained.

Step 7: Backpropagate and update the weights of the critic and action network using Equations (10.58)~(10.70). Then time $t = t + 1$. Go to step 3.

10.4.3.3 Results of ADP Computation for the Unit Commitment (UC) Problem Figure 10.9 shows the load duration curve used for this small five-bus test system. There are three generators in the system and the network parameters and cost function for this simple parameter in this example is given in [42].

Figure 10.10 shows the control action impact on the J function of output versus expected function, [J]. The closeness of the line graphs indicate that the ADP method generates correct results.

After training, the HDP can give the generation plan, which is very close to the optimal plan. The HDP method can deal with the dynamic process of UC, and easily to get a global optimal solution, which is difficult for classical optimization methods. Figure 10.11 shows that generation schedule of three generators system.

In Figure 10.11, X1, X2, and X3 present the output of the three generators respectively, and [X1], [X2], and [X3] present the expected (or say, optimal) output of the three generators respectively.

UC problem is a large-scale, mixed-integer, and dynamic optimization problem. The ADP method is employed for solving the unit commitment problem over time and obtains the global optimization solution with the constraints in load dynamics and topology changes.

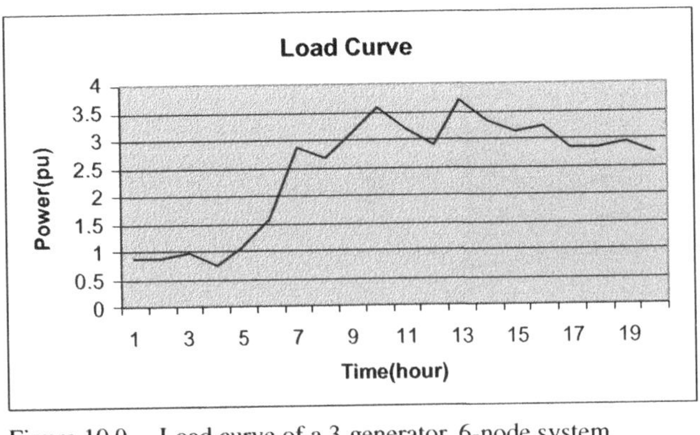

Figure 10.9. Load curve of a 3-generator, 6-node system.

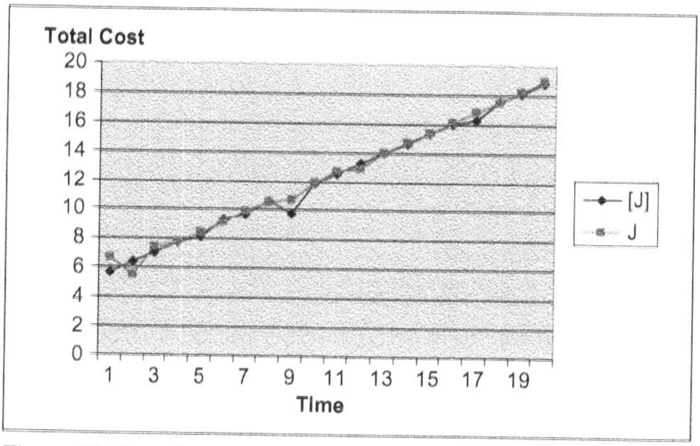

Figure 10.10. Comparison of expected [J] vs. Actual J.

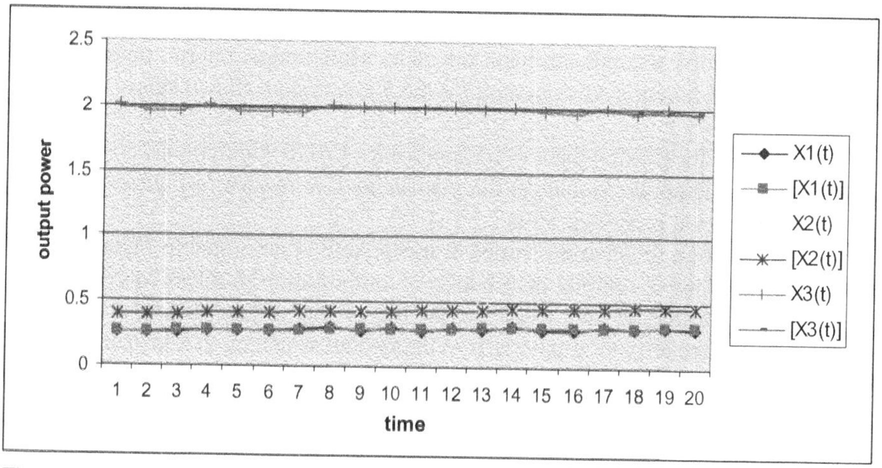

Figure 10.11. Generation schedule for the UC problem solved using ADP.

10.4.3.4 ADP for Optimal Network Reconfiguration Distribution networks are generally configured radially for effective and non-complicated protection schemes. Under normal operation conditions, distribution feeders may be reconfigured to satisfy objectives of minimum distribution line losses, optimum voltage profile and relieve the overloads in the network. Power system reconfiguration problem has the objectives:

- Minimum distribution line losses
- Optimum voltage profile
- Relieve the overloads in the network

The minimum distribution line loss optimization problem of the reconfigured distribution systems is formulated as follows:

$$\text{Minimize} \quad \sum |z_b i_b| \tag{10.71}$$

s.t.

$$[A]\boldsymbol{i} = \boldsymbol{I} \tag{10.72}$$

Where:

z_b : Impedance of the branch

$\boldsymbol{I}_b$: complex current flow in the branch b

i : m-vector of complex branch currents

A : $n \times m$ network incidence matrix, whose entries is:
= +1 if branch b starts from the node p
= −1 if the branch b starts from the node b
= zero if the branch is not connected to the node p

m : Total number of the branches

n : Total number of network nodes

$\boldsymbol{I}$: n-vector of complex nodal injection currents

The illustrative example problem solved by using integer interior point method presented in [44], here the ADP method for the 5-bus system shown below in Figure 10.12 is utilized.

It involves the development of a framework of ADP which involves (a) action network, (b) critic network, and (c) the plant model, as shown in Figure 10.13 for network distribution reconfiguration.

The algorithm to solve this problem using ADP is presented in Figure 10.14.

In order to solve optimal distribution reconfiguration problem by ADP algorithm, we need to model and specify each part of the system structure shown in Figure 10.13. There are four major parts in the system structure: action vectors, state

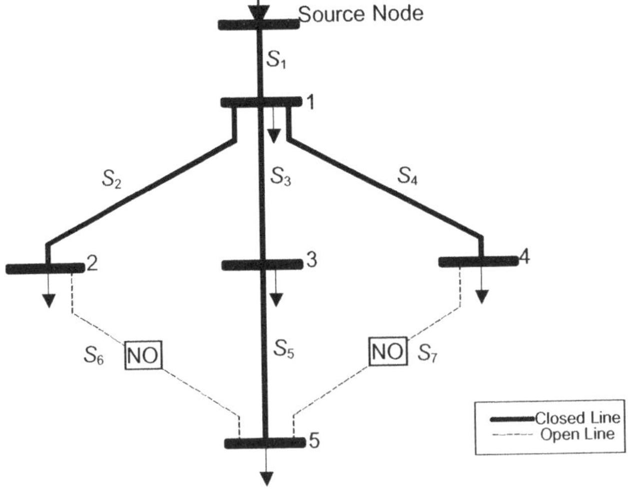

Figure 10.12. Small power system for reconfiguration problem.

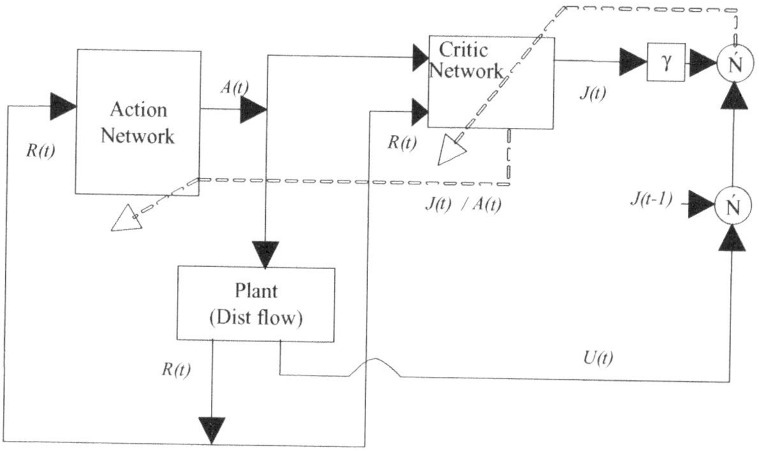

Figure 10.13. ADP structure for the network reconfiguration problem.

vectors, immediate rewards, and the plant. The system is tested with a five-bus and a 32-bus system. We discuss the different parts of the ADP implementation structure as follows:

Rewards (Utility function)

Optimal reconfiguration involves selection of the best set of branches to be opened, one from each loop, such that the resulting radial distribution systems has the desired performance. Amongst the several performance criteria considered for optimal network reconfiguration, the one selected is the minimization of real power losses. Application of the ADP to optimal reconfiguration of radial distribution systems is linked to the choice of an immediate reward U, such that the iterative value of J is minimized, while the minimization of total power losses is satisfied over the whole planning period. Thus, we compute the immediate reward as:

$$U = -Total\ Losses \tag{10.73}$$

Action vectors

If each control variable A_i is discretized in d_{u_i} levels (e.g. branches to be opened one at each loop of radial distribution systems), the total number of action-vectors affecting the load flow is:

$$A = \prod_{i=1}^{m} d_{u_i} \tag{10.74}$$

Here, m expresses the total number of control variables (e.g. total number of branches to be switched out).

The control variables comprise the sets of branches to be opened, one from each loop. From the network above, we can easily deduce from the simple system the entire set of action vectors that can maintain the radial structure of the network. The combinations are:

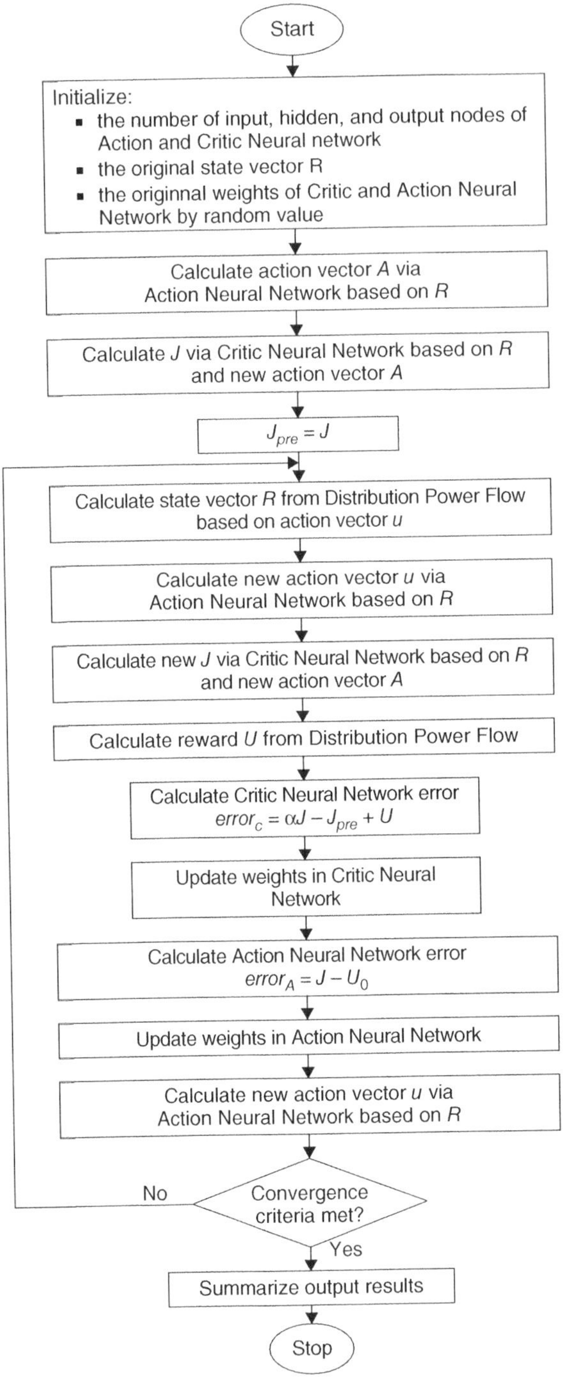

Figure 10.14. Flowchart for ADP-based Optimal Reconfiguration strategy.

A_1: {open switches 2, 3 close all other switches}
A_2: {open switches 6, 3 close all other switches}
A_3: {open switches 2, 5 close all other switches}
A_4: {open switches 6, 5 close all other switches}
A_5: {open switches 2, 4 close all other switches}
A_6: {open switches 3, 4 close all other switches}
A_7: {open switches 6, 4 close all other switches}
A_8: {open switches 5, 4 close all other switches}
A_9: {open switches 2, 7 close all other switches}
A_{10}: {open switches 3, 7 close all other switches}
A_{11}: {open switches 6, 7 close all other switches}
A_{12}: {open switches 5, 7 close all other switches}

10.4.3.5 Results of ADP Computation for the Network Reconfiguration Problem The purpose of the algorithm presented is to find the optimal switches status combination, for the five-bus case. The program was used to determine the optimal solution, which is Action Vector 15. In Figure 10.15, the minimization of the losses as action vectors is shown for the optimal switching sequence.

After the initialization, the action network generates the first action vector by random number, the action vector then input into the critic vector with state variables. With the output of critic network J and immediate cost U, the new error for action and critic network could be obtained. The weights in those two networks then can be updated based on backpropagation rules. After sufficient iterations, the

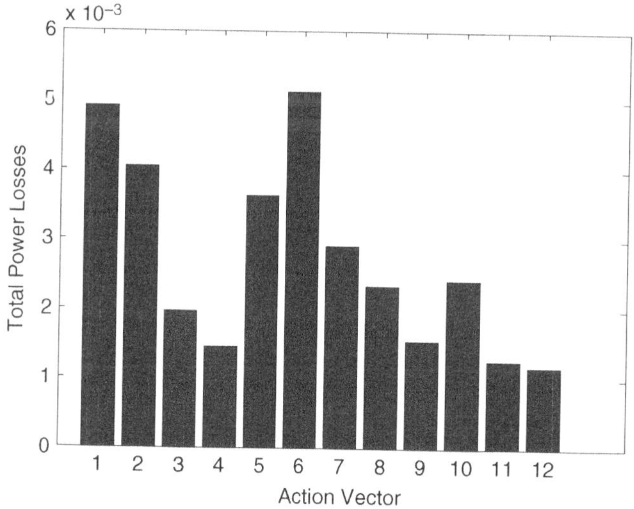

Figure 10.15. Action vector performance during system training.

system will output the result. In our case, it is the optimal action vector, which is the best switches status combination with the minimum losses.

Optimal training of the weights of ADP Action Vectors were obtained and used to minimize losses in the reconfigured network. We recommend extending this study to large-scale aerospace power system while addressing the multi-objective challenges of restoration, reconfiguration, and remedial control.

10.5 GRAND CHALLENGES IN NEXT GENERATION OPTIMIZATION AND RESEARCH NEEDS

The section of the paper presents some immediate concerns and research needs for development of the next generation optimization techniques.

- **Decision Analysis Methods and Hierarchical Programming**—There is a need for defining an acceptably and meaningful possibility of probabilistic events based on effective decision attributes. Also, the ability of decision support process such as DA and hierarchical programming to handle multi-objective power system optimization problems with fuzziness in the constraints required some attention by operation researchers. Further work to include multiple objectives under uncertainty needs to be investigated.
- **Game Theory and Risk Assessment**—Utilization of concepts from next generation of optimization techniques such as ADP to allocate costs of decision-making and risk assessment is an open research field. Also, as power market becomes more interactive with increase participation of market players, the game theoretic approaches will be favored in some types of analysis and settlement. Risk assessment [17] that incorporated public perception will be important to integrate in future tools for optimal power flow.
- **Adaptive Dynamic Programming (ADP)**—It is a challenge to define training set of data and testing of Action and Critic Networks. There is a new evolutionary programming with other optimization methods. These should be used to complement the disadvantages of using the commonly used back propagation technique used in Adaptive Dynamic Programming.
- **Dynamic Stochastic Optimal Power Flow (DSOPF)**—Incorporation of stability, dynamics, and voltage stability sensitivities as constraints in extending the capability of ADP to solve a constrained OPF with uncertainty and dynamic changes, referred to as the next generation OPF is needed. DSOPF will require the use of the framework presented in the previous section. Several of the problems listed in Table 10.4 will be tested using this new variant of OPF.
- **Testbeds and Benchmark Systems**—The development of computational tools for power system applications requires extensive testing and validation for efficiency, speed, accuracy, reliability and robustness. We will require data

and/or users to test the final product based on the uniqueness of there test system being studied under wide ranges of normal, alert, emergency, and restoration conditions.

10.6 CONCLUDING REMARKS AND BENCHMARK PROBLEMS

This course, *Next Generation Optimization for Power Systems*, utilizes research experiences and innovations in systems engineering from various communities aimed at providing examples and insights to solving the grand challenge problems of power networks. In this paper, we presented an overview of known optimization techniques, their strengths and weaknesses, and provided decision analyses and game theoretic tools for system engineering enthusiasts. We provided formal insights to selected optimization problem formulation, algorithms, and illustrative examples.

From our research, new advances are needed to update the capability of these tools to solve practical grand challenge problems of modern electric power networks. For example, next generation optimization techniques must be capable of:

- Handle stochastic and dynamics changes in practical systems.
- Handle experiences and preferences of the domain expert or user in making realistic, intelligent decisions under uncertainty.
- Reduce the complexity and/or the computational burden of problem sets in order to reach optimal solutions in the shortest timeframe possible.
- Handle various levels of hierarchy in decision-making in economics, engineering, etc.

In an ongoing research work, we hope to continue development on a MATLAB-based environment for generalizing the dynamic stochastic OPF, the variants of ADP methods, Decision Analysis tools and others for solving different test beds and bench mark in civilian and military power networks. The results of our experience in teaching this course and test cases from colleagues will form the basis for the book entitled "Next Generation Optimization for Power Systems" to be used as interdisciplinary by power engineers as well as system engineers, and the computational intelligence community.

ACKNOWLEDGMENTS

The author wish to acknowledge the funding received from the National Science Foundation (NSF) in support of this research activity. The NSF support came from the Division of Electrical and Communication Systems (ECS)—Grant Award No. (ECS-0224873) and the Division of Design, Manufacture and Industrial Innovation (DMII)—Grant Award No. (DMII-053907) for the development of this research and education project.

We also would like to thank the supporting students and staff at the Center of Energy Systems and Control (CESaC), Howard University, who support the research efforts.

BIBLIOGRAPHY

[1] "Grand Challenges to Optimal Power Flow Towards Dynamic Stochastic Optimal Power Flow", in *Learning and Approximation in Real Systems*, IEEE Press, 2004.

[2] J.A. Momoh, M.E. El-Hawary, and R. Adapa. "A Review of Selected Optimal Power Flow Literature to 1993, Part I: Nonlinear and Quadratic Programming Approaches," *IEEE Trans. on Power Systems*, Vol. 14, No. 1, pp. 96–104, Feb. 1999.

[3] J.A. Momoh, M.E. El-Hawary, and R. Adapa. "A Review of Selected Optimal Power Flow Literature to 1993 Part II: Newton Linear Programming and Interior Point Methods", *IEEE Trans. on PWRS*, April 1998.

[4] J. Si and Y. Wang. "On-Line Learning Control by Association and Reinforcement", *IEEE Trans. on Neural Networks*, Vol. 12, pp. 264–276, Mar 2001.

[5] J. Si, A.G. Barto, W.B. Powell, and D. Wunsch II. *Handbook of Learning and Approximate Dynamic Programming*. New Jersey: Wiley, 2004.

[6] P.J. Werbos. *The Roots of Backpropagation form Order Derivates to Neural Networks and Political Forecasting*. New York: Wiley, 1994.

[7] P. Werbos and A.D.P. Goals. "Opportunities, and Principles," in J. Si, A.G. Barto, W. Buckler Powell, and D. Wunsch (Eds.). *Handbook of Learning and Approximate Dynamic Programming*. John Wiley & Sons.

[8] P.J. Werbos. "Approximate Dynamic Programming for Real-Time Control and Neural Modeling," in D.A. White and D. Sofge (Eds.). *Handbook of Intelligent Control*. Van Nostrand Reinhold, pp. 493–525, 1992.

[9] P.J. Werbos. "Neurocontrol and Supervised Learning, an Overview and Evaluation," in D.A. White and D. Sofge (Eds.). *Handbook of Intelligent Control*, pp. 65–89, Van Nostrand Reinhold, 1992.

[10] M.H. Mickle and T.W. Sze. *Optimization in Systems Engineering*. Scranton: Intext, 1972.

[11] A.P. Sage and C.C. White III. *Optimum Systems Control*. 2nd Ed., New Jersey: Prentice-Hall, 1977.

[12] H.T. Chieh. *Applied Optimization Theory and Optimum Control*. Taiwan: Feng Chia University, 1990.

[13] J.A. Momoh. *Electric Power System Application of Optimization*. New York: Marcel Dekker, 2001.

[14] N.S. Rau. *Optimization Principles: Practical Application to the Operation and Markets of the Electric Power Industry*. New Jersey: Wiley, 2003.

[15] T. Au and T.E. Stelson. *Introduction to Systems Engineering, Deterministic Models*. Reading: Addison-Wesley, 1969.

[16] H. Raiffa. *Decision Analysis: Introductory Lectures on Choices under Uncertainty*. Reading: Addison-Wesley, 1968.

[17] E.J. Henley and H. Kumamoto. *Probabilistic Risk Assessment: Reliability Engineering, Design, and Analysis*. New Jersey: IEEE Press, 1992.

[18] T.L. Saaty. *Decision Making With Dependence and Feedback: The Analytic Network Process*. 1st Ed., Pittsburg: Saaty, 1996.

[19] K.E. Kinnear Jr. *Advances in Genetic Programming*. Cambridge: MIT, 1994.

[20] D.J. Smith. *Reliability Engineering*. New York: Barnes and Noble, 1972.

[21] R.A. Waller and V.T. Covello. *Low-Probability High-Consequence Risk Analysis: Issues, Methods and Case Studies*. New York: Plenum, 1984.

[22] M. Negnevitsky. *Artificial Intelligence: A guide to Intelligent Systems*. Harlow: Addison Wesley, 2002.

[23] J.O. Gray and D.G. Caldwell. *Advanced Robotics & Intelligent Machines*. London: IEE, 1996.

[24] K. Warwick, A. Ekwue, and R. Aggarwal. *Artificial Intelligence Techniques in Power Systems*. London: IEE, 1997.

[25] M.C. Cullingford, S.M. Shah, and J.H. Gittus. *Implication of Probabilistic Risk Assessment*. Essex: Elsevier Applied Science, 1987.

[26] P.P. Wakker. *Additive Representations of Preferences: A New Foundation of Decision Analysis*. Dordrecht: Kluwer Academic, 1989.

[27] L.B. Lave. *Quantitative Risk Assessment in Regulation*. Washington, DC: Brookings Institution, 1982.

[28] T.L. Saaty and K.P. Kearns. *Analytical Planning: The Organization of Systems*. Oxford: Pergamon Press, 1985.

[29] N.G. Hingorani and L. Gyugyi. *Understanding FACTS: Concepts and Technology of Flexiable AC Transmission Systems*. New York: IEEE Press, 2000.

[30] J.D. Johnson and A.B. Whinston. *Advances in Artificial Intelligence in Economics, Finance, and Management*, vol. 1. Greenwich: Jai Press, 1994.

[31] E.A.C. Crouch and R. Wilson. *Risk/Benefit Analysis*. Cambridge: Ballinger, 1982.

[32] C.C. Jaeger, O. Renn, E.A. Rosa, and T. Webler. *Risk, Uncertainty, and rational Action*. London: Earthscan, 2001.

[33] W.L. Winston. *Operations Research: Applications and Algorithms*. Boston: Duxbury, 1987.

[34] D.V. Prokhorov and D.C. Wunsch. "Adaptive Critic Designs," *IEEE Trans. Neural Network*, vol. 8, pp. 997–1007, Sep 1997.

[35] J. Si, A.G. Barto, W.B. Powell and D.C. Wunsch. *Handbook of Learning and Approximate Dynamic Programming*. IEEE Press Series on Computational Intelligence, 2004.

[36] G.K. Venayagamoorthy, R.G. Harley and D.C. Wunsch. "Dual Heuristic Programming Excitation Neuro-Control for Generators in a Multi-machine Power System", *IEEE Transaction on Industry Applications*, Vol. 39, pp. 382–394, Mar/Apr 2003.

[37] N.P. Padhy. "Unit Commitment—A Bibliographical Survey", *IEEE Trans. Power Systems*, Vol. 19, pp. 1196–1205, May 2004.

[38] J.M. Lee and J.H. Lee. "Approximate Dynamic Programming Strategies and Their Applicability for Process Control: A Review and Future Directions", *International Journal of Control, Automation, and Systems*, vol. 2, no. 3, pp. 263–278, Sep 2004.

[39] Proceedings of the Electric Power Networks Efficiency and Security (EPNES) NSF/ONR-Sponsored Research Initiative, 2004 EPNES Grantees Workshop, Mayaguez, Puerto Rico, pp. 13–14, July 2004.

[40] Proceedings of the Electric Power Networks Efficiency and Security (EPNES), NSF/ONR-Sponsored Research Initiative, 2003 EPNES Grantees Workshop, Orlando, Florida, pp. 23–24, October 2003.

[41] J.A. Momoh, S.X. Guo, E.C. Ogbuobiri, and R. Adapa. "*The Quadratic Interior Point Method for Solving Power System Optimization Problems*," IEEE/PES, 1993 Summer Meeting (93 SM 475-4 PWRS), Vancouver, B.C., Canada, July 18–22, 1993.

[42] J.A. Momoh and Y. Zhang. "Unit Commitment Using Adaptive Dynamic Programming," Proceedings of the IEEE Intelligent Systems Application to Power Systems (ISAP) Conference, Accepted for publication, November 6–10, 2005.

[43] J.A. Momoh, D. Sun, and B. Stott. "Challenges to Optimal Power Flow", *IEEE Transactions on Power Systems*, Vol. 12, No. 1, pp. 444–445, April 1997.

[44] J.A. Momoh and A.C. Caven. "Distribution System Reconfiguration Scheme Using Integer Interior Point Programming Technique", *Transmission and Distribution Conference and Exposition, 2003 IEEE PES*, Vol. 1, pp. 234–241, Sept. 2003.

INDEX

Economic Market Design and Planning for Electric Power Systems, Edited by James Momoh and Lamine Mili

Books in the IEEE Press Series on Power Engineering

Principles of Electric Machines with Power Electronic Applications, Second Edition
M.E. El-Hawary

Pulse Width Modulation for Power Converters: Principles and Practice
D. Grahame Holmes and Thomas Lipo

Analysis of Electric Machinery and Drive Systems, Second Edition
Paul C. Krause, Oleg Wasynczuk, and Scott D. Sudhoff

Risk Assessment for Power Systems: Models, Methods, and Applications
Wenyuan Li

Optimization Principles: Practical Applications to the Operations of Markets of the Electric Power Industry
Narayan S. Rau

Electric Economics: Regulation and Deregulation
Geoffrey Rothwell and Tomas Gomez

Electric Power Systems: Analysis and Control
Fabio Saccomanno

Electrical Insulation for Rotating Machines: Design, Evaluation, Aging, Testing, and Repair
Greg Stone, Edward A. Boulter, Ian Culbert, and Hussein Dhirani

Signal Processing of Power Quality Disturbances
Math H. J. Bollen and Irene Y. H. Gu

Instantaneous Power Theory and Applications to Power Conditioning
Hirofumi Akagi, Edson H. Watanabe and Mauricio Aredes

Maintaining Mission Critical Systems in a 24/7 Environment
Peter M. Curtis

Elements of Tidal-Electric Engineering
Robert H. Clark

Handbook of Large Turbo-Generator Operation and Maintenance, Second Edition
Geoff Klempner and Isidor Kerszenbaum

Introduction to Electrical Power Systems
Mohamed E. El-Hawary

Modeling and Control of Fuel Cells: Distributed Generation Applications
M. Hashem Nehrir and Caisheng Wang

Power Distribution System Reliability: Practical Methods and Applications
Ali A. Chowdhury and Don O. Koval

Economic Market Design and Planning for Electric Power Systems
James Momoh and Lamine Mili

Operation and Control of Electric Energy Processing Systems
James Momoh and Lamine Mili

Printed and bound by CPI Group (UK) Ltd, Croydon, CR0 4YY

16/09/2026

14963412-0003